Atomic Mass Values for Selected Elements

Element	Atomic Mass
Aluminum	26.98
Boron	10.81
Bromine	79.90
Carbon	12.01
Chlorine	35.45
Fluorine	18.99
Hydrogen	1.008
Iodine	126.9
Lithium	6.941
Magnesium	24.30
Nitrogen	14.01
Oxygen	15.99
Phosphorus	30.97
Potassium	39.09
Silicon	28.09
Sodium	22.99
Sulfur	32.07

CONCENTRATED ACIDS AND BASES

REAGENT	HCl	HNO_3	H_2SO_4	HCOOH	CH_3COOH	$NH_3(NH_4OH)$
Specific Gravity	1.18	1.41	1.84	1.20	1.06	0.90
% Acid or Base (by weight)	37.3	70.0	96.5	90.0	99.7	29.0
Molecular Weight	36.47	63.02	98.08	46.03	60.05	17.03
Molarity of Concentrated Acid or Base	12	16	18	23.4	17.5	15.3
Normality of Concentrated Acid or Base	12	16	36	23.4	17.5	15.3
Volume of Concentrated Reagent Required to Prepare 1 liter of 1M Solution (ml)	83	64	56	42	58	65
Volume of Concentrated Reagent Required to Prepare 1 liter of 10% Solution (ml)*	227	101	56	93	95	384
Molarity of a 10% Solution*	2.74	1.59	1.02	2.17	1.67	5.87

*Percent solutions by weight

Microscale and Macroscale Techniques in the Organic Laboratory

Microscale and Macroscale Techniques in the Organic Laboratory

Donald L. Pavia

Gary M. Lampman

George S. Kriz

Western Washington University
Bellingham, Washington

Randall G. Engel

North Seattle Community College
Seattle, Washington

THOMSON

BROOKS/COLE

Australia • Canada • Mexico • Singapore • Spain
United Kingdom • United States

Publisher: Emily Barrosse
Acquisitions Editor: Angus McDonald
Marketing Strategist: Pauline Mula
Developmental Editor: Sandi Kiselica
Project Editor: TSI Publishing Services
Production Manager: Charlene Squibb
Art Director: Jonel Sofian

Cover Credit: Geoff Tompkinson/Science Photo Library

MICROSCALE AND MACROSCALE TECHNIQUES IN THE ORGANIC LABORATORY, First edition
ISBN: 0-03-034311-9
Library of Congress Catalog Card Number: 200109364

Brooks/Cole — Thomson Learning
10 Davis Drive
Belmont CA 94002-3098
USA

For information about our products, contact us:
Thomson Learning Academic Resource Center
1-800-423-0563
http://www.wadsworth.com

For permission to use material from this text, contact us by
Web: http://www.thomsonrights.com
Fax: 1-800-730-2215
Phone: 1-800-730-2214

Printed in the United States of America
10 9 8 7 6 5 4 3 2

WARNING ABOUT SAFETY PRECAUTIONS

Some of the procedures described in this laboratory manual involve a degree of risk on the part of the instructor and student. Although performing experiments is generally safe for the college laboratory, unanticipated and potentially dangerous reactions are possible for a number of reasons, such as improper measurement or handling of chemicals, improper use of laboratory equipment, failure to follow laboratory safety procedures, and other causes. Neither the publisher nor the authors can accept any responsibility for personal injury or property damage resulting from the use of this publication.

This book is dedicated to the memory of John Vondeling

Preface

Microscale and Macroscale Techniques in the Organic Laboratory is dedicated to describing the most important techniques used in the modern organic chemistry laboratory. The book is written to accommodate experiments using both microscale and macroscale glassware. The microscale glassware described here features ground-glass joints in the most common joint size, that being ᶋ 14/10. However, most of the discussion is also appropriate for use with other types of microscale glassware. The macroscale glassware described in this textbook is appropriate for use with virtually any ground-glass macroscale glassware kit.

Most of the material presented here has already been published in one of our two textbooks, *Introduction to Organic Laboratory Techniques: A Microscale Approach, third edition* or *Introduction to Organic Laboratory Techniques: A Small-scale Approach, first edition.* Many of the chapters from these textbooks have been rewritten to present the two types of techniques, microscale and macroscale, in a well-organized, coherent manner within the same chapter. In all cases, we have chosen to present the macroscale version of a technique first, because these techniques are generally easier to describe, and our feeling is that most students, even those using microscale techniques exclusively, should have some familiarity with macroscale glassware and methods. For those instructors wishing to direct the students' reading to cover primarily either macroscale or microscale techniques, we have provided a detailed Table of Contents to make it easier to assign appropriate sections.

In addition to describing all of the basic organic laboratory techniques, we have also included chapters about most of the modern instrumental methods. Included here are nuclear magnetic resonance spectroscopy (proton and ^{13}C), infrared spectroscopy, mass spectrometry, polarimetry, refractometry, and the various types of chromatography (column, thin-layer, gas, and HPLC).

In addition to chapters on techniques, we have also included chapters on up-to-date laboratory safety and advance preparation and laboratory records. There are two chapters on the chemical literature, including a completely new chapter on how to find physical property data using handbooks and catalogues. We have also included a chapter on solubility, because solubility principles form the basis for many of the organic laboratory techniques. Finally, there is a chapter on molecular modeling and computational chemistry, because these topics are finding their way into most modern laboratory settings.

Answers to the problems at the ends of the chapters will be made available to qualified instructors *via* the World Wide Web.

ACKNOWLEDGEMENTS

We owe our sincere thanks to the many colleagues who have used our textbooks and who have offered their suggestions for changes and improvements in our discussions. Although we cannot mention everyone who has made important contributions, we must make special mention of Jim Vyvyan (Western Washington University), Frank Deering (North Seattle Community College), and Jim Patterson (University of Washington). We are particularly grateful to Professor Ron Starkey (University of Wisconsin, Green Bay), who has made many valuable suggestions that have significantly improved our textbooks. Production of this book was capably handled by TSI Publishing Services. We thank all who contributed, with special thanks to our Senior Development Editor, Sandi Kiselica, and to Tom Robinson, of TSI Publishing Services.

We must make note of the recent passing of our publisher, John Vondeling. John has been an important part of our team since we began writing our first organic laboratory textbook in 1974. Largely because of his vision and power of persuasion, we have written a total of 10 books. If we had more time and energy, John would have liked to have seen a couple more! John has made his mark on the way chemistry is taught throughout the country, because of the wide acceptance of Saunders and Harcourt chemistry textbooks at all levels. He is dearly missed, but his influence will be felt for many years to come.

If you wish to contact us with comments, questions, or suggestions, we have a special electronic mail address for this purpose. (*plke@chem.wwu.edu*), or you can reach us at:

pavia@chem.wwu.edu
lampman@chem.wwu.edu
kriz@chem.wwu.edu
tawnydog@earthlink.net

We encourage you to visit our home page at *http://atom.chem.wwu.edu/dept/staff/org/plkhome.html*. You may also wish to visit the Brooks/Cole web site at http://www.brookscole.com/chemistry.

Finally, we must thank our families and special friends, especially Neva-Jean Pavia, Marian Lampman, Carolyn Kriz, and Earl Engel, for their encouragement, support, and patience.

Donald L. Pavia
Gary M. Lampman
George S. Kriz
Randall G. Engel

March 2001

Contents Overview

Table of Contents

Introduction: Welcome to Organic Chemistry!

Organic chemistry can be fun, and we hope to prove it to you. The work in your organic chemistry laboratory course will teach you a lot. The personal satisfaction that comes with performing a sophisticated experiment skillfully and successfully will be great.

To get the most out of the laboratory course, you must try to understand both the purpose and the principles behind each experiment you do. You must also try to organize your time effectively *before* each laboratory period.

ORGANIZATION OF THE TEXTBOOK

Consider briefly how this textbook is organized. The are four introductory chapters that follow this Welcome: a chapter on laboratory safety, a chapter on advance preparation and laboratory records, a chapter describing important pieces of laboratory glassware, and a chapter that outlines how handbooks and catalogues can be used to obtain important information about the physical properties of a compound. Beyond these introductory chapters, the textbook contains 20 chapters consisting of a series of detailed instructions and explanations dealing with the techniques of organic chemistry. These 20 chapters are followed by four additional chapters that serve as an introduction to the essential features of the important spectrosopic methods used in the organic chemistry laboratory. A chapter on molecular modeling and computational chemistry is followed by a chapter that outlines how to use the chemical literature.

ADVANCE PREPARATION

It is essential to plan carefully for each laboratory period so you can keep abreast of the material you will learn in your organic chemistry laboratory course. You should not treat the experiments as a novice cook would treat *The Good Housekeeping Cookbook*. You should come to the laboratory with a plan for how to use your time and with some understanding of what you are about to do. A really good cook does not follow the recipe line by line with a finger, nor does a good mechanic fix your car with the instruction manual in one hand and a wrench in the other. In addition, you probably will not learn much if you follow the instructions blindly, without understanding them. We cannot emphasize strongly enough that you should come to the laboratory *prepared*.

If there are items or techniques that you do not understand, do not hesitate to ask questions. You will learn more, however, if you work things out on your own. Do not rely on others to do your thinking for you.

Read Chapter 2, *Advance Preparation and Laboratory Records*, first. Although your instructor will undoubtedly have a preferred format for keeping records, much of the material in Chapter 2 will help you learn to think constructively about laboratory experiments in advance. It would also save time if, as soon as possible, you read Chapters 4 through 9. These techniques are basic to all experiments that you will do in your laboratory course. The laboratory class will begin with experiments almost immediately, and a thorough familiarity with this particular material will save you much valuable laboratory time.

It is also very important to read Chapter 1, *Laboratory Safety*. It is your responsibility to know how to perform the experiments safely and how to understand and evaluate risks associated with laboratory experiments. The laboratory has many potential hazards. Knowing what to do and what not to do in the laboratory is of paramount importance.

BUDGETING TIME

As mentioned in the "Advance Preparation" section of this chapter, you should read several chapters of this book *before* your first laboratory class. You should also read the assigned experiment carefully before every class. Having read the experiment will help you schedule your time wisely. You will often be doing more than one experiment at a time. At times you will have to catch up on some unfinished details of a previous experiment. For instance, it is usually not possible to determine a yield accurately or a melting point of a product immediately after you first obtain the product. Products must be free of solvent to give an accurate weight or melting point range; they must be *dried*. Usually this drying is done by leaving the product in an open container on your desk or in your locker. Then, when you have time, you can determine these missing data using a dry sample. Through careful planning you can set aside the time required to perform these miscellaneous experimental details.

THE PURPOSE

The main purpose of an organic laboratory course is to teach you the techniques necessary to deal with organic chemicals. You will also learn the techniques needed to separate and purify organic compounds. If the appropriate experiments are included in your course, you may also learn how to identify unknown compounds. The experiments themselves are only the vehicle for learning these techniques. Your instructor may provide laboratory lectures and demonstrations explaining the techniques, but the burden is on you to master them by familiarizing yourself with these chapters.

In addition to good laboratory technique and methods for carrying out basic laboratory procedures, you will also learn from your laboratory course how to

1. Work safely
2. Take data carefully
3. Record relevant observations
4. Use your time effectively
5. Assess the efficiency of your experimental method
6. Plan for the isolation and purification of the substance you prepare
7. Solve problems and think like a chemist.

This textbook discusses the important laboratory techniques of organic chemistry. In the traditional approach to teaching this subject, called **macroscale (or standard scale)**, the quantities of chemical used are on the order of 5–100 g and glassware is designed to contain up to 500 mL of liquid. Some *macroscale* textbooks have reduced the scale of experiments so that smaller amounts of chemicals (1–10 g and less than 50 mL of liquid) are used, but the glassware and methods remain identical to those used in traditional macroscale experiments.

Another approach, a **microscale** approach, differs from the traditional laboratory approach in that the experiments use *very* small amounts of chemicals (0.050–1.000 g) and glassware is designed to contain less than 25 mL of liquid. Some microscale glassware is very different from macroscale glassware, and there are a few techniques that are unique to the microscale laboratory. Because of the widespread use of each of these methods, this textbook describes *both* macroscale and microscale techniques, side by side. It is our hope that this textbook will be useful to students regardless of the style of glassware and equipment that they will encounter in their classroom.

CHAPTER 1

Laboratory Safety

In any laboratory course, familiarity with the fundamentals of laboratory safety is critical. Any chemistry laboratory, particularly an organic chemistry laboratory, can be a dangerous place in which to work. Understanding potential hazards will serve you well in minimizing that danger. It is ultimately your responsibility, along with your laboratory instructor, to make sure that all laboratory work is carried out in a safe manner.

1.1 SAFETY GUIDELINES

It is vital that you take necessary precautions in the organic chemistry laboratory. Your laboratory instructor will advise you of specific rules for the laboratory in which you work. The following list of safety guidelines should be observed in all organic chemistry laboratories.

A. Eye Safety

Always Wear Approved Safety Glasses or Goggles. It is essential to wear eye protection whenever you are in the laboratory. Even if you are not actually carrying out an experiment, a person near you might have an accident that could endanger your eyes. Even dish washing may be hazardous. We know of cases in which a person has been cleaning glassware only to have an undetected piece of reactive material explode, throwing fragments into the person's eyes. To avoid such accidents, wear your safety glasses or goggles at all times.

Learn the Location of Eyewash Facilities. If there are eyewash fountains in your laboratory, determine which one is nearest to you before you start to work. If any chemical enters your eyes, go immediately to the eyewash fountain and flush your eyes and face with large amounts of water. If an eyewash fountain is not available, the laboratory will usually have at least one sink fitted with a piece of flexible hose. When the water is turned on, this hose can be aimed upward and the water can be directed into the face, working much like an eyewash fountain. To avoid damaging the eyes, the water flow rate should not be set too high, and the water temperature should be slightly warm.

B. Fires

Use Care with Open Flames in the Laboratory. Because an organic chemistry laboratory course deals with flammable organic solvents, the danger of fire is frequently present. Because of this danger, DO NOT SMOKE IN THE LABORATORY. Furthermore, use extreme caution when you light matches or use any open flame. Always check to see whether your neighbors on either side, across the bench, and behind you are using flammable solvents. If so, either wait or move to a safe location, such as a fume hood, to use your open flame. Many flammable organic substances are the source of dense vapors that can travel for some distance down a bench. These vapors present a fire danger, and you should be careful, because the source of those vapors may be far away from you. Do not use the bench sinks to dispose of flammable solvents. If your bench has a trough running along it, pour only *water* (no flammable solvents!) into it. The troughs and sinks are designed to carry water—not flammable materials—from the condenser hoses and aspirators.

Learn the Location of Fire Extinguishers, Fire Showers, and Fire Blankets. For your own protection in case of a fire, you should immediately determine the location of the nearest fire extinguisher, fire shower, and fire blanket. You should learn how to operate these safety devices, particularly the fire extinguisher. Your instructor can demonstrate this.

If there is a fire, the best advice is to get away from it and let the instructor or laboratory assistant take care of it. DON'T PANIC! Time spent in thought before action is never wasted. If it is a small fire in a container, it can usually be extinguished quickly by placing a wire gauze screen with a ceramic fiber center or, possibly, a watch glass over the mouth of the container. It is good practice to have a wire screen or watch glass handy whenever you are using a flame. If this method does not extinguish the fire and if help from an experienced person is not readily available, then extinguish the fire yourself with a fire extinguisher.

Should your clothing catch on fire, DO NOT RUN. Walk *purposefully* toward the fire shower station or the nearest fire blanket. Running will fan the flames and intensify them.

C. Organic Solvents: Their Hazards

Avoid Contact with Organic Solvents. It is essential to remember that most organic solvents are *flammable* and will burn if they are exposed to an open flame or a match. Remember also that on repeated or excessive exposure, some organic solvents may be toxic, carcinogenic (cancer causing), or both. For example, many chlorocarbon solvents, when accumulated in the body, result in liver deterioration similar to cirrhosis caused by excessive use of ethanol. The body does not easily rid itself of chlorocarbons nor does it *detoxify* them; they build up over time and may cause future illness. Some chlorocarbons are also suspected of being carcinogens. MINIMIZE YOUR EXPOSURE. Long-term exposure to benzene may cause a form of leukemia. Do not sniff benzene, and avoid spilling it on yourself. Many other solvents, such as chloroform and ether, are good anesthetics and will put you to sleep if you breathe too much of them. They subsequently cause nausea. Many of these solvents have a synergistic effect with ethanol, meaning that they enhance its effect. Pyridine causes temporary impotence. In other words, organic solvents are just as dangerous as corrosive chemicals, such as sulfuric acid, but manifest their hazardous nature in other, more subtle ways.

If you are pregnant, you may want to consider taking this course at a later time. Some exposure to organic fumes is inevitable, and any possible risk to an unborn baby should be avoided.

Minimize any direct exposure to solvents, and treat them with respect. The laboratory room should be well ventilated. Normal cautious handling of solvents should not result in any health problem. If you are trying to evaporate a solution in an open container, you must do the evaporation in the hood. Excess solvents should be discarded in a container specifically intended for waste solvents, rather than down the drain at the laboratory bench.

A sensible precaution is to wear gloves when working with solvents. Gloves made from polyethylene are inexpensive and provide good protection. The disadvantage of polyethylene gloves is that they are slippery. Disposable surgical gloves provide a better grip on glassware and other equipment, but they do not offer as much protection as polyethylene gloves. Nitrile gloves offer better protection (see page 9).

Do Not Breathe Solvent Vapors. In checking the odor of a substance, be careful not to inhale very much of the material. The technique for smelling flowers is *not* advisable here; you could inhale dangerous amounts of the compound. Rather, a technique for smelling minute amounts of a substance is used. Pass a stopper or spatula moistened with the substance (if it is a liquid) under your nose. Or hold the substance away from you and waft the vapors toward you with your hand. But *never* hold your nose over the container and inhale deeply!

The hazards associated with organic solvents you are likely to encounter in the organic laboratory are discussed in detail beginning on page 17. If you use proper safety precautions, your exposure to harmful organic vapors will be minimized and should present no health risk.

D. Waste Disposal

Do Not Place Any Liquid or Solid Waste in Sinks; Use Appropriate Waste Containers. Many substances are toxic, flammable, and difficult to degrade; it is neither legal nor advisable to dispose of organic solvents or other liquid or solid reagents by pouring them down the sink.

The correct disposal method for wastes is to put them in appropriately labeled waste containers. These containers should be placed in the hoods in the laboratory. The waste containers will be disposed of safely by qualified persons using approved protocols.

Specific guidelines for disposing of waste will be determined by the people in charge of your particular laboratory and by local regulations. One system for handling waste disposal is presented here. For each experiment that you are assigned, you will be instructed to dispose of all wastes in one of the following ways:

Nonhazardous solids. Nonhazardous solids such as paper and cork can be placed in an ordinary wastebasket.

Broken glassware. Broken glassware should be put into a container specifically designated for broken glassware.

Organic solids. Solid products that are not turned in or any other organic solids should be disposed of in the container designated for organic solids.

Inorganic solids. Solids such as alumina and silica gel should be put in a container specifically designated for them.

Nonhalogenated organic solvents. Organic solvents such as diethyl ether, hexane, toluene, or any solvent that does not contain a halogen atom, should be disposed of in the container designated for nonhalogenated organic solvents.

Halogenated solvents. Methylene chloride (dichloromethane), chloroform, and carbon tetrachloride are examples of common halogenated organic solvents. Dispose of all halogenated solvents in the container designated for them.

Strong inorganic acids and bases. Strong acids such as hydrochloric, sulfuric, and nitric acid will be collected in specially marked containers. Strong bases such as sodium hydroxide and potassium hydroxide will also be collected in specially designated containers.

Aqueous solutions. Aqueous solutions will be collected in a specially marked waste container. It is not necessary to separate each type of aqueous solution (unless the solution contains heavy metals); rather, unless otherwise instructed, you may combine all aqueous solutions into the same waste container. Although many types of solutions (aqueous sodium bicarbonate, aqueous sodium chloride, and so on) may seem innocuous, and it may seem that their disposal down the sink drain is not likely to cause harm, many communities are becoming increasingly restrictive about what substances they will permit to enter municipal sewage-treatment systems. In light of this trend toward greater caution, it is important to develop good laboratory habits regarding the disposal of *all* chemicals.

Heavy metals. Many heavy metal ions such as mercury and chromium are highly toxic and should be disposed of in specifically designated waste containers.

E. Use of Flames

Even though organic solvents are frequently flammable (for example, hexane, diethyl ether, methanol, acetone, and petroleum ether), there are certain laboratory procedures for which a flame must be used. Most often these procedures involve an aqueous solution. In fact, as a general rule, use a flame to heat only aqueous solutions. Heating methods that do not use a flame are discussed in detail in Chapter 6, starting on page 79. Most organic solvents boil below 100°C, and an aluminum block, heating mantle, sand bath, or water bath may be used to heat these solvents safely. Common organic solvents are listed in Chapter 10, Table 10.3, page 134. Solvents marked in that table with boldface type will burn. Diethyl ether, pentane, and hexane are especially dangerous, because in combination with the correct amount of air, they may explode.

Some commonsense rules apply to using a flame in the presence of flammable solvents. Again, we stress that you should check to see whether anyone in your vicinity is using flammable solvents before you ignite any open flame. If someone is using a flammable solvent, move to a safer location before you light your flame. Your laboratory should have an area set aside for using a burner to prepare micropipets or other pieces of glassware.

The drainage troughs or sinks should never be used to dispose of flammable organic solvents. They will vaporize if they are low boiling and may encounter a flame further down the bench on their way to the sink.

F. Inadvertently Mixed Chemicals

To avoid unnecessary hazards of fire and explosion, *never pour any reagent back into a stock bottle.* There is always the chance that you may accidentally pour back some foreign substance that will react explosively with the chemical in the stock bottle. Of course, by pouring reagents back into the stock bottles you may introduce impurities that could spoil the experiment for the person using the stock reagent after you. Pouring things back into bottles is not only a dangerous practice, it is also inconsiderate. This also means that you should not take more chemicals than you need.

G. Unauthorized Experiments

Never undertake any unauthorized experiments. The risk of an accident is high, particularly if the experiment has not been completely checked to reduce hazards. Never work alone in the laboratory. The laboratory instructor or supervisor must always be present.

H. Food in the Laboratory

Because all chemicals are potentially toxic, avoid accidentally ingesting any toxic substance; therefore, never eat or drink any food while in the laboratory. There is always the possibility that whatever you are eating or drinking may become contaminated with a potentially hazardous material.

I. Clothing

Always wear closed shoes in the laboratory; open-toed shoes or sandals offer inadequate protection against spilled chemicals or broken glass. Do not wear your best clothing in the laboratory, because some chemicals can make holes in or permanent stains on your clothing. To protect yourself and your clothing, it is advisable to wear a full-length laboratory apron or coat.

When working with chemicals that are very toxic, wear some type of gloves. Disposable gloves are inexpensive, offer good protection, provide acceptable "feel," and can be bought in many departmental stockrooms and college bookstores. Disposable latex surgical or polyethylene gloves are the least expensive type of glove; they are satisfactory when working with inorganic reagents and solutions. Better protection is afforded by disposable "nitrile" gloves. This type of glove provides good protection against organic chemicals and solvents. Heavier nitrile gloves are also available.

Finally, hair that is shoulder length or longer should be tied back. This precaution is especially important if you are working with a burner.

J. First Aid: Cuts, Minor Burns, and Acid or Base Burns

If any chemical enters your eyes, immediately irrigate the eyes with copious quantities of water. Tempered (slightly warm) water, if available, is preferable. Be sure that the eyelids are kept open. Continue flushing the eyes in this way for 15 minutes.

In case of a cut, wash the wound well with water, unless you are specifically instructed to do otherwise. If necessary, apply pressure to the wound to stop the flow of blood.

Minor burns caused by flames or contact with hot objects may be soothed by immediately immersing the burned area in cold water or cracked ice until you no longer feel a burning sensation. Applying salves to burns is discouraged. Severe burns must be examined and treated by a physician. For chemical acid or base burns, rinse the burned area with copious quantities of water for at least 15 minutes.

If you accidentally ingest a chemical, call the local poison control center for instructions. Do not drink anything until you have been told to do so. It is important that the examining physician be informed of the exact nature of the substance ingested.

1.2 RIGHT-TO-KNOW LAWS

The federal government and most state governments now require that employers provide their employees with complete information about hazards in the workplace. These regulations are often referred to as *Right-to-Know Laws.* At the federal level, the Occupational Safety and Health Administration (OSHA) is charged with enforcing these regulations.

In 1990, the federal government extended the Hazard Communication Act, which established the Right-to-Know Law, to include a provision that requires the establishment of a Chemical Hygiene Plan at all academic laboratories. Every college and university chemistry department should have a Chemical Hygiene Plan. Having this plan means that all the safety regulations and laboratory safety procedures should be written in a manual. The plan also provides for the training of all employees in laboratory safety. Your laboratory instructor and assistants should have this training.

One of the components of Right-to-Know Laws is that employees and students have access to information about the hazards of any chemicals with which they are working. Your instructor will alert you to dangers to which you need to pay particular attention. However, you may want to seek additional information. Two excellent sources of information are labels on the bottles that come from a chemical manufacturer and Material Safety Data Sheets (MSDSs). The MSDSs are also provided by the manufacturer and must be kept available for all chemicals used at educational institutions.

A. Material Safety Data Sheets

Reading an MSDS for a chemical can be a daunting experience, even for an experienced chemist. MSDSs contain a wealth of information, some of which must be decoded to understand. The MSDS for methanol is shown on pages 11–15. Only the information that might be of interest to you is described in the paragraphs that follow.

Section 1. The first part of Section 1 identifies the substance by name, formula, and various numbers and codes. Most organic compounds have more than one name. In this case, the systematic [or International Union of Pure and Applied Chemistry (IUPAC)] name is methanol, and the other names are common names or are from an older system of nomenclature. The Chemical Abstract Service Number (CAS No.) is often used to identify a substance, and it may be used to access extensive information about a substance found in many computer databases or in the library.

Section 3. The Baker SAF-T-DATA System is found on all MSDSs and bottle labels for chemicals supplied by J. T. Baker, Inc. For each category listed, the number indicates the degree of hazard. The lowest number is 0 (very low hazard) and the highest number is 4 (extreme hazard). The **Health** category refers to the danger involved when a substance is inhaled, ingested, or absorbed. **Flammability** indicates the tendency of a substance to burn. **Reactivity** refers to how reactive a substance is with air, water, or other substances. The last category, **Contact,** refers to how hazardous a substance is when it comes in contact with external parts of the body. Note that this rating scale is applicable only to Baker MSDSs and labels; other rating scales with different meanings are also in common use.

Section 4. This section provides helpful information for emergency and first aid procedures.

Section 6. This part of the MSDS deals with procedures for handling spills and disposal. The information could be very helpful, particularly if a large amount of the chemical was spilled. More information about disposal is also given in Section 13.

MSDS Number: M2015   **Effective Date: 12/8/96**

 **Material Safety Data Sheet**

24 Hour Emergency Telephone: 908-859-2151
CHEMTREC: 1-800-424-9300

National Response in Canada
CANUTEC: 613-996-6666

From: **Mallinckrodt Baker, Inc.**
        222 Red School Lane
        Phillipsburg, NJ 08865

Outside U.S. and Canada
Chemtrec: 202-483-7616

NOTE: CHEMTREC, CANUTEC and National
Response Center emergency numbers to be
used only in the event of chemical
emergencies involving a spill, leak, fire,
exposure or accident involving chemicals.

All non-emergency questions should be directed to Customer Service (1-800-582-2537) for assistance.

METHYL ALCOHOL

1. Product Identification

Synonyms: Wood alcohol; methanol; carbinol
CAS No: 67-56-1
Molecular Weight: 32.04
Chemical Formula: CH_3OH
Product Codes: **J.T. Baker:**

5217, 5370, 5794, 5807, 5811, 5842, 5869, 9049, 9063, 9066, 9067, 9069, 9070, 9071,
9073, 9075, 9076, 9077, 9091, 9093, 9096, 9097, 9098, 9263, 9893
Mallinckrodt:

3004, 3006, 3016, 3017, 3018, 3024, 3041, 3701, 4295, 5160, 8814, H080, H488, H603,
V079, V571

2. Composition/Information on Ingredients

Ingredient	CAS No.	Percent	Hazardous
Methyl Alcohol	67-56-1	100%	Yes

3. Hazards Identification

Emergency Overview

POISON! DANGER! VAPOR HARMFUL. MAY BE FATAL OR CAUSE BLINDNESS IF SWALLOWED.
HARMFUL IF INHALED OR ABSORBED THROUGH SKIN. CANNOT BE MADE NONPOISONOUS.
FLAMMABLE LIQUID AND VAPOR. CAUSES IRRITATION TO SKIN, EYES AND RESPIRATORY TRACT.
AFFECTS THE LIVER.

J.T. Baker SAF-T-DATA(tm) Ratings
(Provided here for your convenience)

Health:	Flammability:	Reactivity:	Contact:
3 - Severe (Poison)	4 - Extreme (Flammable)	1 - Slight	1 - Slight
Lab Protection Equip:	GOGGLES & SHIELD; LAB COAT & APRON; VENT HOOD; PROPER GLOVES; CLASS B EXTINGUISHER		
Storage Color Code:	Red (Flammable)		

Potential Health Effects

Inhalation:

A slight irritant to the mucous membranes. Toxic effects exerted upon nervous system, particularly the optic nerve. Once absorbed into the body, it is very slowly eliminated. Symptoms of overexposure may include headache, drowsiness, nausea, vomiting, blurred vision, blindness, coma, and death. A person may get better but then worse again up to 30 hours later.

Ingestion:

Toxic. Symptoms parallel inhalation. Can intoxicate and cause blindness. Usual fatal dose: 100–125 milliliters.

Skin Contact:

Methyl alcohol is a defatting agent and may cause skin to become dry and cracked. Skin absorption can occur; symptoms may parallel inhalation exposure.

Eye Contact:

Irritant. Continued exposure may cause eye lesions.

Chronic Exposure:

Marked impairment of vision and enlargement of the liver have been reported. Repeated or prolonged exposure may cause skin irritation.

Aggravation of Pre-existing Conditions:

Persons with pre-existing skin disorders or eye problems or impaired liver or kidney function may be more susceptible to the effects of the substance.

4.   First Aid Measures

Inhalation:

Remove to fresh air. If not breathing, give artificial respiration. If breathing is difficult, give oxygen. Call a physician.

Ingestion:

Induce vomiting immediately as directed by medical personnel. Never give anything by mouth to an unconscious person.

Skin Contact:

Remove any contaminated clothing. Wash skin with soap or mild detergent and water for at least 15 minutes. Get medical attention if irritation develops or persists.

Eye Contact:

Immediately flush eyes with plenty of water for at least 15 minutes, lifting lower and upper eyelids occasionally. Get medical attention immediately.

5.   Fire Fighting Measures

Fire:

Flash point: 12°C (54°F) CC
Autoignition temperature: 464°C (867°F)
Flammable limits in air % by volume:
lel: 7.3; uel: 36
Flammable.

Explosion:

Above flash point, vapor-air mixtures are explosive within flammable limits noted above. Moderate explosion hazard and dangerous fire hazard when exposed to heat, sparks or flames. Sensitive to static discharge.

Fire Extinguishing Media:

Water spray, dry chemical, alcohol foam, or carbon dioxide.

Special Information:

In the event of a fire, wear full protective clothing and NIOSH-approved self-contained breathing apparatus with full facepiece operated in the pressure demand or other positive pressure mode. Use water spray to blanket fire, cool fire exposed containers, and to flush non-ignited spills or vapors away from fire. Vapors can flow along surfaces to distant ignition source and flash back.

6. Accidental Release Measures

Ventilate area of leak or spill. Remove all sources of ignition. Wear appropriate personal protective equipment as specified in Section 8. Isolate hazard area. Keep unnecessary and unprotected personnel from entering. Contain and recover liquid when possible. Use non-sparking tools and equipment. Collect liquid in an appropriate container or absorb with an inert material (e. g., vermiculite, dry sand, earth), and place in a chemical waste container. Do not use combustible materials, such as saw dust. Do not flush to sewer! J. T. Baker SOLUSORB® solvent adsorbent is recommended for spills of this product.

7. Handling and Storage

Protect against physical damage. Store in a cool, dry, well-ventilated location, away from any area where the fire hazard may be acute. Outside or detached storage is preferred. Separate from incompatibles. Containers should be bonded and grounded for transfers to avoid static sparks. Storage and use areas should be No Smoking areas. Use non-sparking type tools and equipment, including explosion proof ventilation. Containers of this material may be hazardous when empty since they retain product residues (vapors, liquid); observe all warnings and precautions listed for the product.

8. Exposure Controls/Personal Protection

Airborne Exposure Limits:

For Methyl Alcohol:
- OSHA Permissible Exposure Limit (PEL):
    200 ppm (TWA)
- ACGIH Threshold Limit Value (TLV):
    200 ppm (TWA), 250 ppm (STEL) skin

Ventilation System:

A system of local and/or general exhaust is recommended to keep employee exposures below the Airborne Exposure Limits. Local exhaust ventilation is generally preferred because it can control the emissions of the contaminant at its source, preventing dispersion of it into the general work area. Please refer to the ACGIH document, "Industrial Ventilation, A Manual of Recommended Practices," most recent edition, for details.

Personal Respirator (NIOSH Approved)

If the exposure limit is exceeded, wear a supplied air, full-facepiece respirator, airlined hood, or full-facepiece self-contained breathing apparatus.

Skin Protection:

Rubber or neoprene gloves and additional protection including impervious boots, apron, or coveralls, as needed in areas of unusual exposure.

Eye Protection:

Use chemical safety goggles. Maintain eye wash fountain and quick-drench facilities in work area.

9. Physical and Chemical Properties

Appearance:	**Boiling Point:**
Clear, colorless liquid.	64.5°C (147°F)
Odor:	**Melting Point:**
Characteristic odor.	−98°C (−144°F)
Solubility:	**Vapor Density (Air=1):**
Miscible in water.	1.1
Specific Gravity:	**Vapor Pressure (mm Hg):**
0.8	97 @ 20°C (68°F)
pH:	**Evaporation Rate (BuAc=1):**
No information found.	5.9
% Volatiles by volume @ 21°C (70°F):	
100	

10. Stability and Reactivity

Stability:
Stable under ordinary conditions of use and storage.

Hazardous Decomposition Products:
May form carbon dioxide, carbon monoxide, and formaldehyde when heated to decomposition.

Hazardous Polymerization:
Will not occur.

Incompatibilities:
Strong oxidizing agents such as nitrates, perchlorates or sulfuric acid. Will attack some forms of plastics, rubber, and coatings. May react with metallic aluminum and generate hydrogen gas.

Conditions to Avoid:
Heat, flames, ignition sources and incompatibles.

11. Toxicological Information

Methyl Alcohol (Methanol) Oral rat LD50: 5628 mg/kg; inhalation rat LC50: 64000 ppm/4H; skin rabbit LD50: 15800 mg/kg; Irritation data-standard Draize test: skin, rabbit: 20 mg/24 hr. Moderate; eye, rabbit: 100 mg/24 hr. Moderate; Investigated as a mutagen, reproductive effector.

Cancer Lists			
	---NTP Carcinogen---		
Ingredient	Known	Anticipated	IARC Category
Methyl Alcohol (67-56-1)	No	No	None

12. Ecological Information

Environmental Fate:
When released into the soil, this material is expected to readily biodegrade. When released into the soil, this material is expected to leach into groundwater. When released into the soil, this material is expected to quickly evaporate. When released into the water, this material is expected to have a half-life between 1 and 10 days. When released into water, this material is expected to readily biodegrade. When released into the air, this material is expected to exist in the aerosol phase with a short half-life. When released into the air, this material is expected to be readily degraded by reaction with photochemically produced hydroxyl radicals. When released into air, this material is expected to have a half-life between 10 and 30 days. When released into the air, this material is expected to be readily removed from the atmosphere by wet deposition.

Environmental Toxicity:
This material is expected to be slightly toxic to aquatic life.

13. Disposal Considerations

Whatever cannot be saved for recovery or recycling should be handled as hazardous waste and sent to a RCRA approved incinerator or disposed in a RCRA approved waste facility. Processing, use or contamination of this product may change the waste management options. State and local disposal regulations may differ from federal disposal regulations.

Dispose of container and unused contents in accordance with federal, state and local requirements.

14. Transport Information

Domestic (Land, D.O.T.)

Proper Shipping Name:	METHANOL		
Hazard Class:	3		
UN/NA:	UN1230	**Packing Group:**	II

Information reported for product/size: 350LB
International (Water, I.M.O.)
    Proper Shipping Name: METHANOL
    Hazard Class: 3.2, 6.1
    UN/NA: UN1230   **Packing Group:**   II
    Information reported for product/size: 350LB

15. Regulatory Information

Chemical Inventory Status

						---Canada---		
Ingredient	TSCA	EC	Japan	Australia	Korea	DSL	NDSL	Phil.
Methyl Alcohol (67-56-1)	Yes	Yes	Yes	Yes	Yes	Yes	No	Yes

Federal, State & International Regulations

	---SARA 302---		------SARA 313-------			-RCRA-	-TSCA-
Ingredient	RQ	TPQ	List	Chemical Catg.	CERCLA	261.33	8(d)
Methyl Alcohol (67-56-1)	No	No	Yes	No	5000	U154	No

Chemical Weapons Convention: No   **TSCA 12(b):** No   **CDTA:** No

SARA 311/312: Acute: Yes   Chronic: Yes   Fire: Yes   Pressure: No   Reactivity: No   (Pure / Liquid)

Australian Hazchem Code: 2PE   **Australian Poison Schedule:** S6

WHMIS: This MSDS has been prepared according to the hazard criteria of the Controlled Products Regulations (CPR) and the MSDS contains all of the information required by the CPR.

16. Other Information

NFPA Ratings:
Health: 1 Flammability: 3 Reactivity: 0

Label Hazard Warning:
POISON! DANGER! VAPOR HARMFUL. MAY BE FATAL OR CAUSE BLINDNESS IF SWALLOWED. HARMFUL IF INHALED OR ABSORBED THROUGH SKIN. CANNOT BE MADE NONPOISONOUS. FLAMMABLE LIQUID AND VAPOR. CAUSES IRRITATION TO SKIN, EYES AND RESPIRATORY TRACT. AFFECTS THE LIVER.

Label Precautions:
Keep away from heat, sparks and flame.
Keep container closed.
Use only with adequate ventilation.
Wash thoroughly after handling.
Avoid breathing vapor.
Avoid contact with eyes, skin and clothing.

Label First Aid:
If swallowed, induce vomiting immediately as directed by medical personnel. Never give anything by mouth to an unconscious person. In case of contact, immediately flush eyes or skin with plenty of water for at least 15 minutes while removing contaminated clothing and shoes. Wash clothing before reuse. If inhaled, remove to fresh air. If not breathing give artificial respiration. If breathing is difficult, give oxygen. In all cases get medical attention immediately.

Product Use:
Laboratory Reagent.

Revision Information:
New 16 section MSDS format, all sections have been revised.

Disclaimer:
Mallinckrodt Baker, Inc. provides the information contained herein in good faith but makes no representation as to its comprehensiveness or accuracy. This document is intended only as a guide to the appropriate precautionary handling of the material by a properly trained person using this product. Individuals receiving the information must exercise their independent judgment in determining its appropriateness for a particular purpose. MALLINCKRODT BAKER, INC. MAKES NO REPRESENTATIONS OR WARRANTIES, EITHER EXPRESS OR IMPLIED, INCLUDING WITHOUT LIMITATION ANY WARRANTIES OR MERCHANTABILITY, FITNESS FOR A PARTICULAR PURPOSE WITH RESPECT TO THE INFORMATION SET FORTH HEREIN OR THE PRODUCT TO WHICH THE INFORMATION REFERS. ACCORDINGLY, MALLINCKRODT BAKER, INC. WILL NOT BE RESPONSIBLE FOR DAMAGES RESULTING FROM USE OF OR RELIANCE UPON THIS INFORMATION.

Prepared By: Strategic Services Division
      Phone Number: (314) 539-1600 (U.S.A.)

Section 8.   Much valuable information is found in Section 8. To help you understand this material, some of the more important terms used in this section are defined:

Threshold Limit Value (TLV).   The American Conference of Governmental Industrial Hygienists (ACGIH) developed the TLV. This is the maximum concentration of a substance in air that a person should be exposed to on a regular basis. It is usually expressed in ppm or mg/m^3. Note that this value assumes that a person is exposed to the substance 40 hours per week, on a long-term basis. This value may not be particularly applicable in the case of a student performing an experiment in a single laboratory period.

Permissible Exposure Limit (PEL).   This has the same meaning as TLV; however, PELs were developed by OSHA. Note that for methanol the TLV and PEL are both 200 ppm.

Section 10.   The information contained in Section 10 refers to the stability of the compound and the hazards associated with mixing of chemicals. It is important to consider this information before carrying out an experiment not previously done.

Section 11.   More information about the toxicity is given in this section. Another important term must first be defined:

Lethal Dose, 50% Mortality (LD$_{50}$).   This is the dose of a substance that will kill 50% of the animals administered a single dose. Different means of administration are used, such as oral, intraperitoneal (injected into the lining of the abdominal cavity), subcutaneous (injected under the skin), and applied to the surface of the skin. The LD$_{50}$ is usually expressed in milligrams (mg) of substance per kilogram (kg) of animal weight. The *lower* the value of LD$_{50}$ the more toxic the substance. It is assumed that the toxicity in humans will be similar.

Without considerably more knowledge about chemical toxicity, the information in Sections 8 and 11 is most useful for comparing the toxicity of one substance with another. For example, the TLV for methanol is 200 ppm, whereas the TLV for benzene is 10 ppm. Clearly, performing an experiment involving benzene would require much more stringent precautions than an experiment involving methanol. One of the LD$_{50}$ values for methanol is 5628 mg/kg. The comparable LD$_{50}$ value of aniline is 250 mg/kg. Clearly, aniline is much more toxic, and because it is easily absorbed through the skin it presents a significant hazard.

It should also be mentioned that both TLV and PEL ratings assume that the worker comes in contact with a substance on a repeated and long-term basis. Thus, even if a chemical has a relatively low TVL or PEL, it does not mean that using it for one experiment will present a danger to you. Furthermore, by performing experiments using small amounts of chemicals and with proper safety precautions, your exposure to organic chemicals in this course will be minimal.

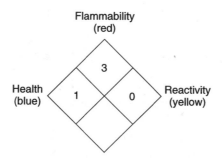

Section 16.   Section 16 contains the National Fire Protection Association (NFPA) rating. This is similar to the Baker SAF-T-DATA (discussed in Section 1), except that the number represents the hazards when a fire is present. The order here is **Health, Flammability,** and **Reactivity.** Often this is presented in graphic form on a label (see figure). The small diamonds are often color coded: blue for health, red for flammability, and yellow for reactivity. The bottom diamond (white) is sometimes used to display graphic symbols denoting unusual reactivity, hazards, or special precautions to be taken.

B. Bottle Labels

Reading the label on a bottle can be a very helpful way of learning about the hazards of a chemical. The amount of information varies greatly, depending on which company supplied the chemical.

Apply some common sense when you read MSDSs and bottle labels. Using these chemicals does not mean you will experience the consequences that can potentially result from exposure to each chemical. For example, an MSDS for sodium chloride states, "Exposure to this product may have serious adverse health effects." Despite the apparent severity of this cautionary statement, it would not be reasonable to expect people to stop using sodium chloride in a chemistry experiment or to stop sprinkling a small amount of it (as table salt) on eggs to enhance their flavor. In many cases, the consequences described in MSDSs from exposure to chemicals are somewhat overstated, particularly for students using these chemicals to perform a laboratory experiment.

1.3   COMMON SOLVENTS

Most organic chemistry experiments involve an organic solvent at some step in the procedure. A list of common organic solvents follows, with a discussion of toxicity, possible carcinogenic properties, and precautions that you should use when handling these solvents. A tabulation of the compounds currently suspected of being carcinogens appears at the end of this chapter.

Acetic Acid.   Glacial acetic acid is corrosive enough to cause serious acid burns on the skin. Its vapors can irritate the eyes and nasal passages. Care should be exercised not to breathe the vapors and not to allow them to escape into the laboratory.

Acetone.   Relative to other organic solvents, acetone is not very toxic. It is flammable, however. Do not use acetone near open flames.

Benzene.   Benzene can damage bone marrow; it causes various blood disorders, and its effects may lead to leukemia. Benzene is considered a serious carcinogenic hazard. It is absorbed rapidly through the skin and also poisons the liver and kidneys. In addition, benzene is flammable. Because of its toxicity and its carcinogenic properties, benzene should not be used in the laboratory; you should use some less dangerous solvent instead. Toluene is considered a safer alternative solvent in procedures that specify benzene.

Carbon Tetrachloride.   Carbon tetrachloride can cause serious liver and kidney damage as well as skin irritation and other problems. It is absorbed rapidly through the skin. In high concentrations, it can cause death, as a result of respiratory failure. Moreover, carbon tetrachloride is suspected of being a carcinogenic material. Although this solvent has the advantage of being nonflammable (in the past, it was used on occasion as a fire extinguisher), it should not be used routinely in the laboratory since it causes health problems. If no reasonable substitute exists, however, it must be used in small quantities, as in preparing samples for infrared (IR) and nuclear magnetic resonance (NMR) spectroscopy. In such cases, you must use it in a hood.

Chloroform.   Chloroform is similar to carbon tetrachloride in its toxicity. It has been used as an anesthetic. However, chloroform is currently on the list of suspected carcinogens. Because of this, do not use chloroform routinely as a solvent in the laboratory. If it is occasionally necessary to use chloroform as a solvent for special samples, then you must use it in a hood. Methylene chloride is usually found to be a safer substitute in procedures that specify chloroform as a solvent. Deuterochloroform, $CDCl_3$, is a common solvent for NMR spectroscopy. Caution dictates that you should treat it with the same respect as chloroform.

1,2-Dimethoxyethane (ethylene glycol dimethyl ether or monoglyme).   Because it is miscible with water, 1,2-dimethoxyethane is a useful alternative to solvents such as dioxane and tetrahydrofuran, which may be more hazardous. 1,2-Dimethoxyethane is flammable and should not handled near an open flame. On long exposure to light and oxygen, explosive peroxides may form. 1,2-Dimethoxyethane is a possible reproductive toxin.

Dioxane.   Dioxane has been used widely because it is a convenient, water-miscible solvent. It is now suspected, however, of being carcinogenic. It is also toxic, affecting the central nervous system, liver, kidneys, skin, lungs, and mucous membranes. Dioxane is also flammable and tends to form explosive peroxides when it is exposed to light and air. Because of its carcinogenic properties, it is no longer used in the laboratory unless absolutely necessary. Either 1,2-dimethoxyethane or tetrahydrofuran is a suitable, water-miscible alternative solvent.

Ethanol.   Ethanol has well-known properties as an intoxicant. In the laboratory, the principal danger arises from fires, because ethanol is a flammable solvent. When using ethanol, take care to work where there are no open flames.

Diethyl Ether (ether).   The principal hazard associated with diethyl either is fire or explosion. Ether is probably the most flammable solvent found in the laboratory. Because ether vapors are much more dense than air, they may travel along a laboratory bench for a considerable distance from their source before being ignited. Before using ether, it is very important to be sure that no one is working with matches or any open flame. Ether is not a particularly toxic solvent, although in high enough concentrations it can cause drowsiness and perhaps nausea. It has been used as a general anesthetic. Ether can form highly explosive peroxides when exposed to air. Consequently, you should never distill it to dryness.

Hexane.   Hexane may be irritating to the respiratory tract. It can also act as an intoxicant and a depressant of the central nervous system. It can cause skin irritation because it is an excellent solvent for skin oils. The most serious hazard, however, comes from its flammability. The precautions recommended for using diethyl ether in the presence of open flames apply equally to hexane.

Ligroin.   *See* Hexane.

Methanol.   Much of the material outlining the hazards of ethanol applies to methanol. Methanol is more toxic than ethanol; ingestion can cause blindness and even death. Because methanol is more volatile, the danger of fires is more acute.

Methylene Chloride (dichloromethane).   Methylene chloride is not flammable. Unlike other members of the class of chlorocarbons, it is not currently considered a serious carcinogenic hazard. Recently, however, it has been the subject of much serious investigation, and there have been proposals to regulate it in industrial situations in which workers have high levels of exposure on a day-to-day basis. Methylene chloride is less toxic than chloroform and carbon tetrachloride. It can cause liver damage when ingested, however, and its vapors may cause drowsiness or nausea.

Pentane.   *See* Hexane.

Petroleum Ether.   *See* Hexane.

Pyridine.   Some fire hazard is associated with pyridine. However, the most serious hazard arises from its toxicity. Pyridine may depress the central nervous system; irritate the skin and respiratory tract; damage the liver, kidneys, and gastrointestinal system; and even cause temporary sterility. You should treat pyridine as a highly toxic solvent and handle it only in the fume hood.

Tetrahydrofuran.   Tetrahydrofuran may cause irritation of the skin, eyes, and respiratory tract. It should never be distilled to dryness, because it tends to form potentially explosive peroxides on exposure to air. Tetrahydrofuran does present a fire hazard.

Toluene.   Unlike benzene, toluene is not considered a carcinogen. However, it is at least as toxic as benzene. It can act as an anesthetic and damage the central nervous system. If benzene is present as an impurity in toluene, expect the usual hazards associated with benzene. Toluene is also a flammable solvent, and the usual precautions about working near open flames should be applied.

You should not use certain solvents in the laboratory because of their carcinogenic properties. Benzene, carbon tetrachloride, chloroform, and dioxane are among these solvents. For certain applications, however, notably as solvents for infrared or NMR spectroscopy, there may be no suitable alternative. When it is necessary to use one of these solvents, safety precautions are recommended, or refer to the discussion in Chapters 26–28.

Because relatively large amounts of solvents may be used in a large organic laboratory class, your laboratory supervisor must take care to store these substances safely. Only the amount of solvent needed for a particular experiment should be kept in the laboratory. The preferred location for bottles of solvents being used during a class period is in a hood. When the solvents are not being used, they should be stored in a fireproof storage cabinet for solvents. If possible, this cabinet should be ventilated into the fume hood system.

1.4   CARCINOGENIC SUBSTANCES

A **carcinogen** is a substance that causes cancer in living tissue. In determining whether a substance is carcinogenic, the normal procedure is to expose laboratory animals to high dosages over a long period. It is not clear whether short-term exposure to these chemicals carries a comparable risk, but it is prudent to use these substances with special precautions.

Many regulatory agencies have compiled lists of carcinogenic substances or substances suspected of being carcinogenic. Because these lists are inconsistent, compiling a definitive list of carcinogenic substances is difficult. The following table includes common substances that are found in many of these lists.

Acetamide	4-Methyl-2-oxetanone (ß-butyrolactone)
Acrylonitrile	1-Naphthylamine
Asbestos	2-Naphthylamine
Benzene	*N*-Nitroso compounds
Benzidine	2-Oxetanone (ß-propiolactone)

Carbon tetrachloride	Phenacetin
Chloroform	Phenylhydrazine and its salts
Chromic oxide	Polychlorinated biphenyl (PCB)
Coumarin	Progesterone
Diazomethane	Styrene oxide
1,2-Dibromoethane	Tannins
Dimethyl sulfate	Testosterone
p-Dioxane	Thioacetamide
Ethylene oxide	Thiourea
Formaldehyde	o-Toluidine
Hydrazine and its salts	Trichloroethylene
Lead (II) acetate	Vinyl chloride

REFERENCES

Aldrich Catalog and Handbook of Fine Chemicals. Milwaukee, WI: Aldrich Chemical Co., current edition.

Armour, M. A. *Hazardous Laboratory Chemicals: Disposal Guide.* Boca Raton, FL: CRC Press, 1991.

Fire Protection Guide on Hazardous Materials, 10th ed. Quincy, MA: National Fire Protection Association, 1991.

Flinn Chemical Catalog Reference Manual. Batavia, IL: Flinn Scientific, Inc., current edition.

Gosselin, R. E., Smith, R. P., and Hodge, H. C. *Clinical Toxicology of Commercial Products,* 5th ed. Baltimore: Williams & Wilkins, 1984.

Lenga, R. E., ed. *The Sigma-Aldrich Library of Chemical Safety Data.* Milwaukee, WI: Sigma-Aldrich Corp., 1985.

Lewis, R. J. *Carcinogenically Active Chemicals: A Reference Guide.* New York: Van Nostrand Reinhold, 1990.

Merck Index, 13th ed. Rahway, NJ: Merck and Co., 2001.

Prudent Practices for Disposal of Chemicals from Laboratories. Washington, DC: Committee on Hazardous Substances in the Laboratory, National Research Council, National Academy Press, 1983.

Prudent Practices for Handling Hazardous Chemicals in Laboratories. Washington, DC: Committee on Hazardous Substances in the Laboratory, National Research Council, National Academy Press, 1981.

Renfrew, M. M., ed. *Safety in the Chemical Laboratory.* Easton, PA: Division of Chemical Education, American Chemical Society, 1967–1991.

Safety in Academic Chemistry Laboratories, 4th ed. Washington, DC: Committee on Chemical Safety, American Chemical Society, 1985.

Sax, N. I., and Lewis, R. J. *Dangerous Properties of Industrial Materials,* 7th ed. New York: Van Nostrand Reinhold, 1988.

Sax, N. I., and Lewis, R. J., eds. *Rapid Guide to Hazardous Chemicals in the Work Place,* 2nd ed. New York: Van Nostrand Reinhold, 1990.

Useful Safety-Related Internet Addresses

Enviro-Net site: http://www.enviro-net.com/technical/msds (This site mirrors the University of Utah site in this list, and it has a user-friendly interface.)

Fisher Scientific, Inc., site: http://www.fisher1.com (This site provides information about each chemical in the Fisher Scientific catalogue; it includes MSDS sheets for each chemical.)

Northwest Fisheries site: http://research.nwfsc.noaa.gov/msds.html

Oregon State University gopher site: gopher://gaia.ucs.orst.edu:70/11/osuis/osudo/ehs (Using this site may be slightly cumbersome, but it provides lots of information.)

Sigma-Aldrich site: http://www.sigma.sial.com (this site provides MSDS information for substances that are found in the Sigma or Aldrich catalogues).

Text-Trieve Internet Services site: http://halcyon.com/ttrieve/msdshome.html (This site provides useful links to other Internet sites that specialize in chemical safety information.)

University of Utah MSDS site: http://atlas.chem.utah.edu:70/11/MSDS

CHAPTER 2

Advance Preparation and Laboratory Records

In the Introduction to this book, we mentioned the importance of advance preparation for laboratory work. Presented in this chapter are some suggestions about what specific information you should try to obtain in your advance studying. Because much of this information must be obtained while preparing your laboratory notebook, the two subjects, advance study and notebook preparation, are developed simultaneously.

An important part of any laboratory experience is learning to maintain very complete records of every experiment undertaken and every item of data obtained. Far too often, careless recording of data and observations has resulted in mistakes, frustration, and lost time due to needless repetition of experiments. If reports are required, you will find that proper collection and recording of data can make your report writing much easier.

Because organic reactions are seldom quantitative, special problems result. Frequently, reagents must be used in large excess to increase the amount of product. Some reagents are expensive, and, therefore, care must be used in measuring the amounts of these substances. Very often, many more reactions take place than you desire. These extra reactions, or **side reactions**, may form products other than the desired product. These are called **side products**. For all these reasons, you must plan your experimental procedure carefully before undertaking the actual experiment.

2.1 THE NOTEBOOK

For recording data and observations during experiments, use a *bound notebook*. The notebook should have consecutively numbered pages. If it does not, number the pages immediately. A spiral-bound notebook or any other notebook from which the pages can be removed easily is not acceptable, because the possibility of losing the pages is great.

All data and observations must be recorded in the notebook. Paper towels, napkins, toilet tissue, or scratch paper tend to become lost or destroyed. It is bad laboratory practice to record information on such random and perishable pieces of paper. All entries must be recorded in *permanent ink*. It can be frustrating to have important information disappear from the notebook because it was recorded in washable ink or pencil and could not survive a flood caused by the student at the next position on the bench. Because you will be using your notebook in the laboratory, the book will probably become soiled or stained by chemicals, filled with scratched-out entries, or even slightly burned. That is expected and is a normal part of laboratory work.

Your instructor may check your notebook at any time, so you should always have it up to date. If your instructor requires reports, you can prepare them quickly from the material recorded in the laboratory notebook.

2.2 NOTEBOOK FORMAT

A. Advance Preparation

Individual instructors vary greatly in the type of notebook format they prefer; such variation stems from differences in philosophies and experience. You must obtain specific directions from your own instructor for preparing a notebook. Certain features, however, are common to most notebook formats. The following discussion indicates what might be included in a typical notebook.

It will be very helpful and you can save much time in the laboratory if for each experiment you know the main reactions, the potential side reactions, the mechanism, and the stoichiometry, and you understand fully the procedure and the theory underlying it before you come to the laboratory. Understanding the procedure by which the desired product is to be separated from undesired materials is also very important. If you examine each of these topics before coming to class, you will be prepared to do the experiment efficiently. You will have your equipment and reagents already prepared when they are to be used. Your reference material will be at hand when you need it. Finally, with your time efficiently organized, you will be able to take advantage of long reaction or reflux periods to perform other tasks, such as doing shorter experiments or finishing previous ones.

For experiments in which a compound is synthesized from other reagents, that is, **preparative experiments**, it is essential to know the main reaction. To perform stoichiometric calculations, the equation for the main reaction should be balanced. Therefore, before you begin the experiment, your notebook should contain the balanced equation for the pertinent reaction. Using the preparation of isopentyl acetate, or banana oil, as an example, you should write the following:

$$
\underset{\textbf{Acetic acid}}{CH_3-\overset{\displaystyle O}{\overset{\displaystyle \|}{C}}-OH} \;+\; \underset{\textbf{Isopentyl alcohol}}{CH_3-\overset{\displaystyle CH_3}{\overset{\displaystyle |}{CH}}-CH_2-CH_2-OH} \;\;\xrightarrow{\;H^+\;}
$$

$$
\underset{\textbf{Isopentyl acetate}}{CH_3-\overset{\displaystyle O}{\overset{\displaystyle \|}{C}}-O-CH_2-CH_2-\overset{\displaystyle CH_3}{\overset{\displaystyle |}{CH}}-CH_3} \;+\; H_2O
$$

Also enter in the notebook the possible side reactions that divert reagents into contaminants (side products), before beginning the experiment. You will have to separate these side products from the major product during purification.

You should list physical constants such as melting points, boiling points, densities, and molecular weights in the notebook when this information is needed to perform an experiment or to do calculations. These data are located in sources such as the *Handbook of Chemistry and Physics*, the *Merck Index*, Lange's *Handbook of Chemistry*, or *Aldrich Handbook of Fine Chemicals*. Write physical constants required for an experiment in your notebook before you come to class.

Advance preparation may also include examining some subjects, information not necessarily recorded in the notebook, that should prove useful in understanding the experiment. Included among these subjects are an understanding of the mechanism of the reaction, an examination of other methods by which the same compound might be prepared, and a detailed study of the experimental procedure. Many students find that an outline of the procedure, prepared *before* they come to class, helps them use their time more efficiently once they begin the experiment. Such an outline could very well be prepared on some loose sheet of paper, rather than in the notebook itself.

Once the reaction has been completed, the desired product does not magically appear as purified material; it must be isolated from a frequently complex mixture of side products, unreacted starting materials, solvents, and catalysts. You should try to outline a **separation scheme** in your notebook for isolating the product from its contaminants. At each stage you should try to understand the reason for the particular instruction given in the experimental procedure. This not only will familiarize you with the basic separation and purification techniques used in organic chemistry but it also will help you understand when to use these techniques. Such an outline might take the form of a flowchart. For example, see the separation scheme for isopentyl acetate (Fig. 2.1). Careful attention to understanding the separation may, besides familiarizing you with the procedure by which the desired product is separated from impurities in your particular experiments, also prepare you for original research, in which no experimental procedure exists.

In designing a separation scheme, note that the scheme outlines those steps undertaken once the reaction period has been concluded. For this reason, the represented scheme does not include steps such as the addition of the reactants (isopentyl alcohol and acetic acid) and the catalyst (sulfuric acid) or the heating of the reaction mixture.

For experiments in which a compound is isolated from a particular source and is not prepared from other reagents, some information described in this section will not be applicable. Such experiments are called **isolation experiments.** A typical isolation experiment involves isolating a pure compound from a natural source. Examples include isolating caffeine from tea or isolating cinnamaldehyde from cinnamon. Although isolation experiments require somewhat different advance preparation, this advance study may include looking up physical constants for the compound isolated and outlining the isolation procedure. A detailed examination of the separation scheme is very important here because it is the heart of such an experiment.

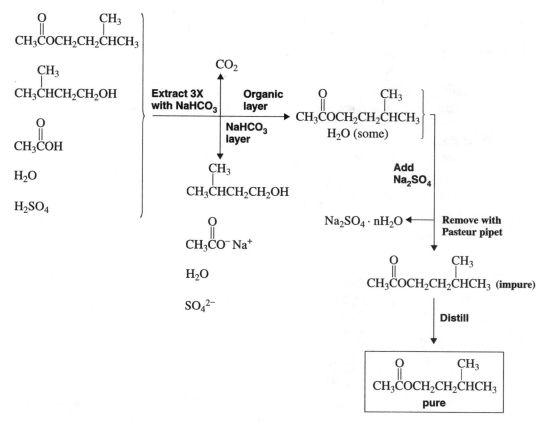

Figure 2.1 Separation scheme for isopentyl acetate.

B. Laboratory Records

When you begin the actual experiment, keep your notebook nearby so you will be able to record those operations you perform. When working in the laboratory, the notebook serves as a place in which to record a rough transcript of your experimental method. Data from actual weighings, volume measurements, and determinations of physical constants are also noted. This section of your notebook should *not* be prepared in advance. The purpose is not to write a recipe, but rather to record what you *did* and what you *observed*. These observations will help you write reports without resorting to memory. They will also help you or other workers repeat the experiment in as nearly as possible the same way. The sample notebook pages found in Figures 2.2 and 2.3 illustrate the type of data and observations that should be written in your notebook.

When your product has been prepared and purified, or isolated if it is an isolation experiment, record pertinent data such as the melting point or boiling point of the substance, its density, its index of refraction, and the conditions under which spectra were determined.

THE PREPARATION OF ISOPENTYL ACETATE (BANANA OIL)

MAIN REACTION

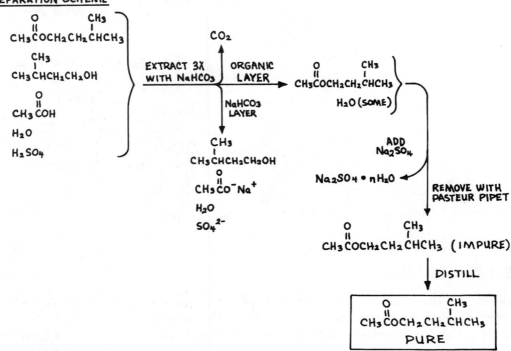

TABLE OF PHYSICAL CONSTANTS

	MW	BP	DENSITY
ISOPENTYL ALCOHOL	88.2	132°C	0.813 g/mL
ACETIC ACID	60.1	118	1.06
ISOPENTYL ACETATE	130.2	142	0.876

SEPARATION SCHEME

DATA AND OBSERVATIONS

0.70 mL OF ISOPENTYL ACETATE WAS ADDED TO A PRE WEIGHED 5-mL CONICAL VIAL:

VIAL + ALCOHOL   25.524 g
VIAL   24.955 g
0.569 g ISOPENTYL ALCOHOL

ACETIC ACID (1.4 mL) AND THREE DROPS OF CONCENTRATED H_2SO_4 (USING A PASTEUR PIPET) WERE ALSO ADDED TO THE CONICAL VIAL ALONG WITH A SMALL BOILING STONE. A WATER-COOLED CONDENSER TOPPED WITH A DRYING TUBE CONTAINING A LOOSE PLUG OF GLASS WOOL WAS ATTACHED TO THE VIAL. THE REACTION MIXTURE WAS REFLUXED IN AN ALUMINUM BLOCK (ABOUT 155°) FOR 75 MIN. AND THEN COOLED TO ROOM TEMPERATURE. THE COLOR OF THE REACTION MIXTURE WAS BROWNISH-YELLOW.

Figure 2.2 A sample notebook, page 1.

THE BOILING STONE WAS REMOVED AND THE REACTION MIXTURE WAS EXTRACTED THREE TIMES WITH 1.0 mL OF 5% $NaHCO_3$. THE BOTTOM AQUEOUS LAYER WAS REMOVED AND DISCARDED AFTER EACH EXTRACTION. DURING THE FIRST TWO EXTRACTIONS, MUCH CO_2 GAS WAS GIVEN OFF. THE ORGANIC LAYER WAS A LIGHT YELLOW COLOR. IT WAS TRANSFERRED TO A DRY CONICAL VIAL, AND TWO FULL MICROSPATULAS OF ANHYDROUS Na_2SO_4 WERE ADDED TO DRY THE CRUDE PRODUCT. IT WAS ALLOWED TO SET WITH OCCASIONAL STIRRING FOR 10 MINS.

THE DRY PRODUCT WAS TRANSFERRED TO A 3-mL CONICAL VIAL AND A BOILING STONE WAS ADDED. A DISTILLATION APPARATUS USING A HICKMAN STILL, A WATER-COOLED CONDENSER, AND A DRYING TUBE PACKED WITH $CaCl_2$ WAS ASSEMBLED. THE SAMPLE WAS HEATED IN AN ALUMINUM BLOCK AT ABOUT 180°C. THE LIQUID BEGAN BOILING AFTER ABOUT FIVE MINS, BUT NO DISTILLATE APPEARED IN THE HICKMAN STILL UNTIL ABOUT 20 MINS. LATER. ONCE THE PRODUCT BEGAN COLLECTING IN THE HICKMAN STILL, THE DISTILLATION REQUIRED ONLY ABOUT TWO MINS. TO COMPLETE. ABOUT 1-2 DROPS REMAINED IN THE DISTILLING VIAL. THE ISOPENTYL ACETATE WAS TRANSFERRED TO A PRE-WEIGHED 3-mL CONICAL VIAL.

$$
\begin{array}{ll}
\text{VIAL + PRODUCT} & 20.428g \\
\text{VIAL} & 20.074g \\
\hline
& 0.354g \quad \text{ISOPENTYL ACETATE}
\end{array}
$$

THE PRODUCT WAS COLORLESS AND CLEAR. BP (MICRO TECHNIQUE): 140°C. THE IR SPECTRUM WAS OBTAINED.

CALCULATIONS

DETERMINE LIMITING REAGENT:

ISOPENTYL ALCOHOL $0.569g \left(\dfrac{1 \text{ MOL ISOPENTYL ALCOHOL}}{88.2g} \right) = 6.45 \times 10^{-3} \text{ MOL}$

ACETIC ACID $1.40 \text{ mL} \left(\dfrac{1.06g}{\text{mL}} \right) \left(\dfrac{1 \text{ MOL ACETIC ACID}}{60.1g} \right) = 2.47 \times 10^{-2} \text{ MOL}$

SINCE THEY REACT IN A 1:1 RATIO, ISOPENTYL ALCOHOL IS THE LIMITING REAGENT.

THEORETICAL YIELD =

$6.45 \times 10^{-3} \text{ MOL ISOPENTYL ALCOHOL} \left(\dfrac{1 \text{ MOL ISOPENTYL ACETATE}}{1 \text{ MOL ISOPENTYL ALCOHOL}} \right) \left(\dfrac{130.2g \text{ ISOPENTYL ACETATE}}{1 \text{ MOL ISOPENTYL ACETATE}} \right)$

$= 0.840g$ ISOPENTYL ACETATE

PERCENTAGE YIELD = $\dfrac{0.354g}{0.840g} \times 100 = 42.1\%$

Figure 2.3 A sample notebook, page 2.

C. Calculations

A chemical equation for the overall conversion of the starting materials to products is written on the assumption of simple ideal stoichiometry. Actually, this assumption is seldom realized. Side reactions or competing reactions will also occur, giving other products. For some synthetic reactions, an equilibrium state will be reached in which an appreciable amount of starting material is still present and can be recovered. Some of the reactant may also remain if it is present in excess or if the reaction was incomplete. A reaction involving an expensive reagent illustrates another reason for needing to know how far a particular type of reaction converts reactants to products. In such a case, it is preferable to use the most efficient method for this conversion. Thus, information about the efficiency of conversion for various reactions is of interest to the person contemplating the use of these reactions.

The quantitative expression for the efficiency of a reaction is found by calculating the **yield** for the reaction. The **theoretical yield** is the number of grams of the product expected from the reaction on the basis of ideal stoichiometry, with side reactions, reversibility, and losses ignored. To calculate the theoretical yield, it is first necessary to determine the **limiting reagent.** The limiting reagent is the reagent that is not present in excess and on which the overall yield of product depends. The method for determining the limiting reagent in the isopentyl acetate experiment is illustrated in the sample notebook pages shown in Figures 2.2 and 2.3. You should consult your general chemistry textbook for more complicated examples. The theoretical yield is then calculated from the expression

Theoretical yield = (moles of limiting reagent)(ratio)(molecular weight of product)

The ratio here is the stoichiometric ratio of product to limiting reagent. In preparing isopentyl acetate, that ratio is 1:1. One mole of isopentyl alcohol, under ideal circumstances, should yield 1 mole of isopentyl acetate.

The **actual yield** is simply the number of grams of desired product obtained. The **percentage yield** describes the efficiency of the reaction and is determined by

$$\text{Percentage yield} = \frac{\text{Actual yield}}{\text{Theoretical yield}} \times 100$$

Calculation of the theoretical yield and percentage yield can be illustrated using hypothetical data for the isopentyl acetate preparation:

$$\text{Theoretical yield} = (6.94 \times 10^{-2}\ \text{mol isopentyl alcohol}) \left(\frac{1\ \text{mol isopentyl acetate}}{1\ \text{mol isopentyl alcohol}} \right)$$

$$\times \left(\frac{130.2\ \text{g isopentyl acetate}}{1\ \text{mol isopentyl acetate}} \right)$$

$$= 9.03\ \text{g isopentyl acetate}$$

$$\text{Actual yield} = 3.81\ \text{g isopentyl acetate}$$

$$\text{Percentage yield} = \frac{3.81\ \text{g}}{9.03\ \text{g}} \times 100 = 42.2\%$$

For experiments that have the principal objective of isolating a substance such as a natural product, rather than preparing and purifying some reaction product, the **weight percentage recovery** and not the percentage yield is calculated. This value is determined by

$$\text{Weight percentage recovery} = \frac{\text{Weight of substance isolated}}{\text{Weight of original material}} \times 100$$

Thus, for instance, if 0.014 g of caffeine was obtained from 2.3 g of tea, the weight percentage recovery of caffeine would be

$$\text{Weight percentage recovery} = \frac{0.014 \text{ g caffeine}}{2.3 \text{ g tea}} \times 100 = 0.61\%$$

2.3 LABORATORY REPORTS

Various formats for reporting the results of the laboratory experiments may be used. You may write the report directly in your notebook in a format similar to the sample notebook pages included in this section. Alternatively, your instructor may require a more formal report, to be written separately from your notebook. When you do original research, these reports should include a detailed description of all the experimental steps undertaken. Frequently, the style used in scientific periodicals such as *Journal of the American Chemical Society* is applied to writing laboratory reports. Your instructor is likely to have his or her own requirements for laboratory reports and should describe the requirements to you.

2.4 SUBMISSION OF SAMPLES

In all preparative experiments, and in some isolation experiments, you will be required to submit to your instructor the sample of the substance you prepared or isolated. How this sample is labeled is very important. Again, learning a correct method of labeling bottles and vials can save time in the laboratory, because fewer mistakes will be made. More importantly, learning to label properly can decrease the danger inherent in having samples of material that cannot be identified correctly at a later date.

Solid materials should be stored and submitted in containers that permit the substance to be removed easily. For this reason, narrow-mouthed bottles or vials are not used for solid substances. Liquids should be stored in containers that will not let them escape through leakage. Be careful not to store volatile liquids in containers that have plastic caps, unless the cap is lined with an inert material such as Teflon. Otherwise, the vapors from the liquid are likely to contact the plastic and dissolve some of it, thus contaminating the substance being stored.

On the label, print the name of the substance, its melting or boiling point, the actual and percentage yields, and your name. An illustration of a properly prepared label follows:

> **Isopentyl Acetate**
> **BP 140°C**
> **Yield 3.81 g (42.2%)**
> **Joe Schmedlock**

CHAPTER 3
Laboratory Glassware

Because your glassware is expensive and you are responsible for it, you will want to give it proper care and respect. If you read this section carefully and follow the procedures presented here, you may be able to avoid some unnecessary expense. You may also save time, because cleaning problems and replacing broken glassware are time consuming.

If you are unfamiliar with the equipment found in an organic chemistry laboratory or are uncertain about how such equipment should be treated, this section provides some useful information, such as cleaning glassware and caring for glassware when using corrosive or caustic reagents. At the end of this section are illustrations that show and name most of the equipment you are likely to find in your drawer or locker.

3.1 CLEANING GLASSWARE

Glassware can be cleaned easily if you clean it immediately after use. It is good practice to do your "dishwashing" right away. With time, organic tarry materials left in a container begin to attack the surface of the glass. The longer you wait to clean glassware, the more extensively this interaction will have progressed. If you wait, cleaning is more difficult, because water will no longer wet the surface of the glass as effectively. If you cannot wash your glassware immediately after use, soak the dirty pieces of glassware in soapy water. A half-gallon plastic container is convenient for soaking and washing glassware. Using a plastic container also helps prevent the loss of small pieces of equipment.

Various soaps and detergents are available for washing glassware. They should be tried first when washing dirty glassware. Organic solvents can also be used, because the residue remaining in dirty glassware is likely to be soluble. After the solvent has been used, the glass item probably will have to be washed with soap and water to remove the residual solvent. When you use solvents to clean glassware, use caution, because the solvents are hazardous (see Chapter 1). Use fairly small amounts of a solvent for cleaning purposes. Usually less than 5 mL (or 1–2 mL for microscale glassware) will be sufficient. Acetone is commonly used, but it is expensive. Your **wash acetone** can be used effectively several times before it is "spent." Once your acetone is spent, dispose of it as your instructor directs. If acetone does not work, other organic solvents such as methylene chloride or toluene can be used.

▨▨▨ **CAUTION** Acetone is very flammable. Do not use it around flames.

For troublesome stains and residues that adhere to the glass despite your best efforts, use a mixture of sulfuric acid and nitric acid. Cautiously add about 20 drops of concentrated sulfuric acid and 5 drops of concentrated nitric acid to the flask or vial.

▨▨▨ **CAUTION** You must wear safety glasses when you are using a cleaning solution made from sulfuric acid and nitric acid. Do not allow the solution to come into contact with your skin or clothing. It will cause severe burns on your skin and create holes in your clothing. The acids may also react with the residue in the container.

Swirl the acid mixture in the container for a few minutes. If necessary, place the glassware in a warm water bath and heat it cautiously to accelerate the cleaning process. Continue heating the glassware until any sign of a reaction ceases. When the cleaning procedure is completed, decant the mixture into an appropriate waste container.

▨▨▨ **CAUTION** Do not pour the acid solution into a waste container that is intended for organic wastes.

Rinse the piece of glassware thoroughly with water and then wash it with soap and water. For most common organic chemistry applications, any stains that survive this treatment are not likely to cause difficulty in subsequent laboratory procedures.

If the glassware is contaminated with stopcock grease, rinse the glassware with a small amount (1–2 mL) of methylene chloride. Discard the rinse solution into an appropriate waste container. Once the grease is removed, wash the glassware with soap or detergent and water.

3.2 DRYING GLASSWARE

The easiest way to dry glassware is to let it stand overnight. Store vials, flasks, and beakers upside down on a piece of paper towel to permit the water to drain from them. Drying ovens can be used to dry glassware if they are available and if they are not being used for other purposes. Rapid drying can be achieved by rinsing the glassware with acetone and air drying it or placing it in an oven. First, thoroughly drain the glassware of water. Then rinse it with one or two *small* portions (1–2 mL) of acetone. Do not use any more acetone than is suggested here. Return the used acetone to an acetone waste container for recycling. After you rinse the glassware with acetone, dry it by placing it in a drying oven for a few minutes or allow it to air dry at room temperature. The acetone can also be removed by aspirator suction. In some laboratories, it may be possible to dry the glassware by blowing a *gentle* stream of dry air into the container. (Your laboratory instructor will indicate if you should do this.) Before drying the glassware with air, make sure that the air line is not filled with oil. Otherwise, the oil will be blown into the container, and you will have to clean it again. It is not necessary to blast the acetone out of the glassware with a wide-open stream of air; a gentle stream of air is just as effective and will not startle other people in the room.

Do not dry your glassware with a paper towel unless the towel is lint free. Most paper will leave lint on the glass that can interfere with subsequent procedures. Sometimes it is not necessary to dry a piece of equipment thoroughly. For example, if you are going to place water or an aqueous solution in a container, it does not need to be completely dry.

3.3 GROUND-GLASS JOINTS

It is likely that the glassware in your organic kit has **standard-taper ground-glass joints.** For example, the Claisen head in Figure 3.1A consists of an inner (male) ground-glass joint at the bottom and two outer (female) joints at the top. Each end is ground to a precise size, which is designated by the symbol ⊤ followed by two numbers. A common joint size in many macroscale organic glassware kits is ⊤ 19/22; in microscale glassware kits the most common joint size is ⊤ 14/10 (Figure 3.1B). The

A   B

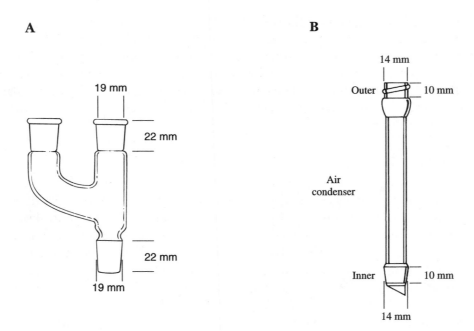

Figure 3.1 Illustration of inner and outer joints, showing dimensions. **(A)** A Claisen head with ⊤19/22 joints. **(B)** An air condenser with ⊤14/10 joints.

first number indicates the diameter (in millimeters) of the joint at its widest point, and the second number refers to its length (see Figure 3.1). One advantage of standard-taper joints is that the pieces fit together snugly and form a good seal. In addition, standard-taper joints allow all glassware components with the same joint size to be connected, thus permitting the assembly of a wide variety of apparatuses. One disadvantage of glassware with ground-glass joints, however, is that it is expensive.

3.4 CONNECTING GROUND-GLASS JOINTS

Macroscale.   It is a simple matter to connect pieces of macroscale glassware using standard-taper ground-glass joints. Figure 3.2B illustrates the connection of a condenser to a round-bottom flask. At times, however, it may be difficult to secure the connection so that it does not come apart unexpectedly. Figure 3.2A shows a plastic clip that serves to secure the connection. Methods to secure ground-glass connections with macroscale apparatus, including the use of plastic clips, are covered in Chapter 7.

Microscale.   Some pieces of microscale glassware with ground-glass joints also have threads cast into the outside surface of the outer joints (see the top of the air condenser in Figure 3.1B). The threaded joint allows a plastic screw cap with a hole in the top to be used to fasten two pieces of glassware together securely. The plastic cap is slipped over the inner joint of the upper piece of glassware, followed by a rubber O-ring (see Chapter 7, Figure 7.5, page 82). The O-ring should be pushed down so that it fits snugly on top of the ground-glass joint. The inner ground-glass joint is then fitted into the outer joint of the bottom piece of glassware. The screw cap is tightened without excessive force to attach the entire apparatus firmly together. The O-ring provides an additional seal that makes this joint airtight. With this connecting

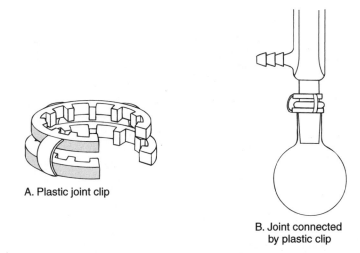

A. Plastic joint clip

B. Joint connected
by plastic clip

Figure 3.2 Connection of ground-glass joints [the use of a plastic clip **(A)** is also shown **(B)**].

system, it is unnecessary to use any type of grease to seal the joint. The O-ring *must be used* to obtain a good seal and to decrease the chances of breaking the glassware when you tighten the plastic cap.

It is important to make sure no solid or liquid is on the joint surfaces. Either of these will decrease the efficiency of the seal, and the joints may leak. With microscale glassware, the presence of solid particles could cause the ground-glass joints to break when the plastic cap is tightened. Also, if the apparatus is to be heated, material caught between the joint surfaces will increase the tendency for the joints to stick. If the joint surfaces are coated with liquid or adhering solid, you should wipe the surfaces with a cloth or a lint-free paper towel before assembling.

3.5 CAPPING FLASKS, CONICAL VIALS, AND OPENINGS

Macroscale.   The sidearms in two-necked or three-necked round-bottom flasks can be capped using the ⵟ 19/22 ground-glass stoppers that are part of a normal macroscale organic kit. Figure 3.3 shows such a stopper being used to cap the sidearm of a three-necked flask.

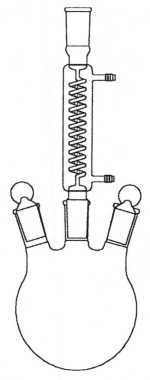

Figure 3.3 Capping a sidearm with a ⵟ19/22 stopper.

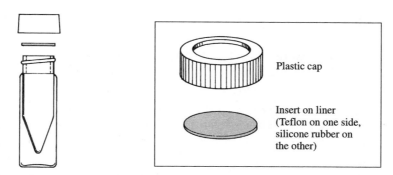

Figure 3.4 A conical reaction vial (the inset shows an expanded view of the cap with its Teflon insert).

Microscale.   The plastic screw caps used to join two pieces of glassware together can also be used to cap conical vials (see Figure 3.4) or other openings. A Teflon insert, or liner, fits inside the cap to cover the hole when the cap is used to seal a vial. Only one side of the liner is coated with Teflon. This side should always face the inside of the vial. (Note that the O-ring is not used when the cap is used to seal a vial.) To seal a vial, it is necessary to tighten the cap firmly, but not too tightly. It is possible to crack the vial if you apply too much force. Some Teflon liners have a soft backing material (silicone rubber) that allows the liner to compress slightly when the cap is screwed down. It is easier to cap a vial securely with these liners without breaking the vial than with liners that have a harder backing material.

The Teflon stoppers that are part of the typical microscale glassware kit can also be used to stopper openings in conical vials or flasks. The use of the Teflon stoppers is similar to the use of ground-glass stoppers with macroscale glassware.

3.6 SEPARATING GROUND-GLASS JOINTS

When ground-glass joints become "frozen" or stuck together, you are faced with the often vexing problem of separating them. The techniques for separating ground-glass joints, or for removing stoppers that are stuck in the openings of flasks and vials, are the same for both macroscale and microscale glassware.

The most important thing you can do to prevent ground-glass joints from becoming frozen is to disassemble the glassware as soon as possible after a procedure is completed. Even when this precaution is followed, ground-glass joints may become stuck tightly together. The same is true of glass stoppers in bottles or conical vials. Since certain items of microscale glassware may be small and very fragile, it is relatively easy to break a piece of glassware when trying to pull two pieces apart. If the pieces do not separate easily, you must be careful when you try to pull them apart. The best way is to hold the two pieces, with both hands touching, as close as possible to the joint. With a firm grasp, try to loosen the joint with a slight twisting motion (do not twist very hard). If this does not work, try to pull your hands apart without pushing sideways on the glassware.

If it is not possible to pull the pieces apart, the following methods may help. A frozen joint can sometimes be loosened if you tap it *gently* with the wooden handle of a spatula. Then, try to pull it apart as already described. If this procedure fails, you may try heating the joint in hot water or a steam bath. If heating fails, the instructor may be able to advise you. As a last resort, you may try heating the joint in a flame. You should not try this unless the apparatus is hopelessly stuck, because heating by flame often causes the joint to expand rapidly and crack or break. If you use a flame, make sure the joint is clean and dry. Heat the outer part of the joint slowly, in the yellow portion of a low flame, until it expands and separates from the inner section. Heat the joint very slowly and carefully, or it may break.

3.7 ETCHING GLASSWARE

Glassware that has been used for reactions involving strong bases such as sodium hydroxide or sodium alkoxides must be cleaned thoroughly *immediately* after use. If these caustic materials are allowed to remain in contact with the glass, they will etch the glass permanently. The etching makes later cleaning more difficult, because dirt particles may become trapped within the microscopic surface irregularities of the etched glass. Furthermore, the glass is weakened, so the lifetime of the glassware is shortened. If caustic materials are allowed to come into contact with ground-glass joints without being removed promptly, the joints will become fused or "frozen." It is extremely difficult to separate fused joints without breaking them.

3.8 ATTACHING RUBBER TUBING TO EQUIPMENT

When you attach rubber tubing to the glass apparatus or when you insert glass tubing into rubber stoppers, first lubricate the rubber tubing or the rubber stopper with either water or glycerin. Without such lubrication, it can be difficult to attach rubber tubing to the sidearms of items of glassware such as condensers and filter flasks. Furthermore, glass tubing may break when it is inserted into rubber stoppers. Water is a good lubricant for most purposes. Do not use water as a lubricant when it might contaminate the reaction. Glycerin is a better lubricant than water and should be used when there is considerable friction between the glass and rubber. If glycerin is the lubricant, be careful not to use too much.

3.9 DESCRIPTION OF EQUIPMENT

Figures 3.5–3.8 include examples of glassware and equipment that are commonly used in the organic laboratory. Your glassware and equipment may vary slightly from the pieces shown on pages 37–41. Note that separate illustrations showing macroscale glassware and microscale glassware have been provided.

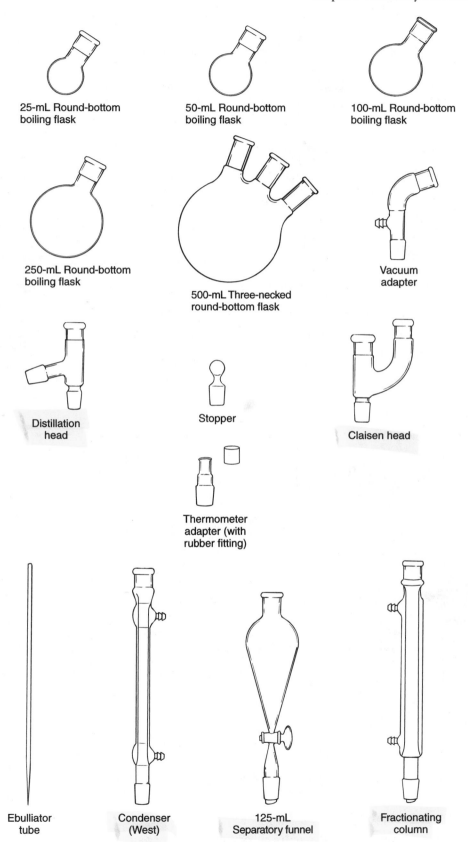

25-mL Round-bottom
boiling flask

50-mL Round-bottom
boiling flask

100-mL Round-bottom
boiling flask

250-mL Round-bottom
boiling flask

500-mL Three-necked
round-bottom flask

Vacuum
adapter

Distillation
head

Stopper

Claisen head

Thermometer
adapter (with
rubber fitting)

Ebulliator
tube

Condenser
(West)

125-mL
Separatory funnel

Fractionating
column

Figure 3.5 Components of the macroscale organic laboratory kit.

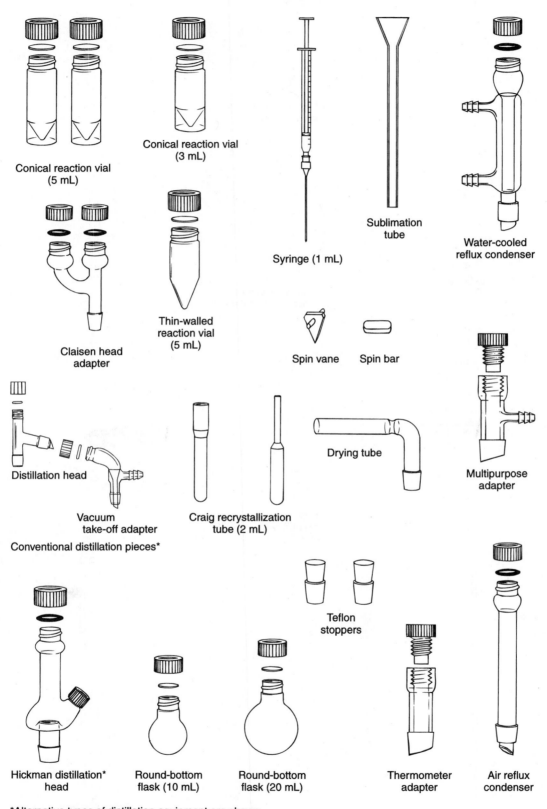

Conical reaction vial
(5 mL)

Conical reaction vial
(3 mL)

Syringe (1 mL)

Sublimation
tube

Water-cooled
reflux condenser

Claisen head
adapter

Thin-walled
reaction vial
(5 mL)

Spin vane   Spin bar

Distillation head

Vacuum
take-off adapter

Craig recrystallization
tube (2 mL)

Drying tube

Multipurpose
adapter

Conventional distillation pieces*

Teflon
stoppers

Hickman distillation*
head

Round-bottom
flask (10 mL)

Round-bottom
flask (20 mL)

Thermometer
adapter

Air reflux
condenser

*Alternative types of distillation equipment are shown.

Figure 3.6 Components of the microscale organic laboratory kit.

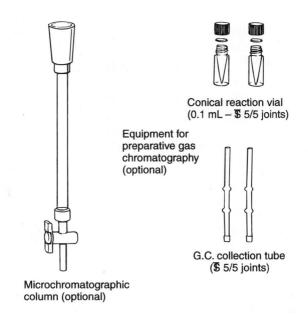

Conical reaction vial
(0.1 mL – 𝕊 5/5 joints)

Equipment for
preparative gas
chromatography
(optional)

G.C. collection tube
(𝕊 5/5 joints)

Microchromatographic
column (optional)

Figure 3.7 Optional pieces of microscale glassware. *Note:* The optional pieces of equipment shown are not part of the standard microscale kit. They must be purchased separately.

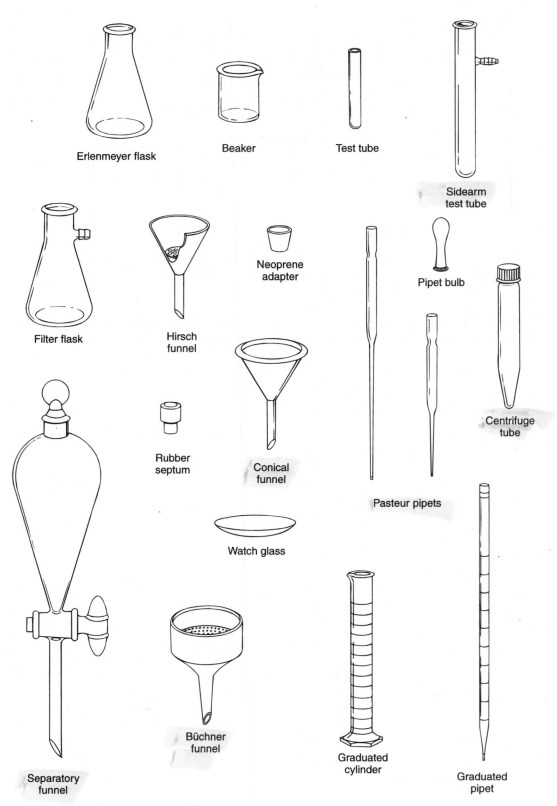

Figure 3.8 Equipment commonly used in the organic chemistry laboratory.

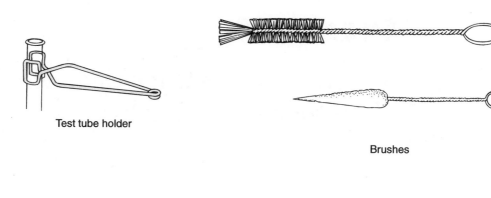

Test tube holder

Brushes

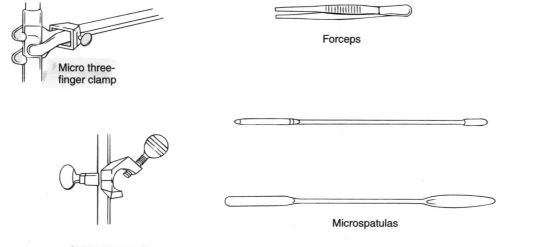

Micro three-finger clamp

Forceps

Clamp holder

Microspatulas

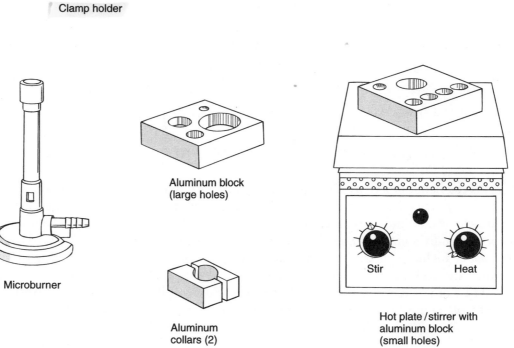

Microburner

Aluminum block
(large holes)

Aluminum
collars (2)

Stir   Heat

Hot plate / stirrer with
aluminum block
(small holes)

CHAPTER 4

How to Find Data for Compounds: Handbooks and Catalogs

The best way to find information quickly on organic compounds is to consult a handbook. In this chapter, we will discuss the use of the *CRC Handbook of Chemistry and Physics, Lange's Handbook of Chemistry*, the *Merck Index*, and the *Aldrich Handbook of Fine Chemicals*. Complete citations to these handbooks are provided in Chapter 30. Depending on the type of handbook consulted, the following information may be found:

Name and common synonyms

Formula

Molecular weight

Boiling point for a liquid or melting point for a solid

Beilstein reference

Solubility data

Density

Refractive index

Flash point

Chemical Abstract Service (CAS) registry number

Toxicity data

Uses and synthesis

4.1 *CRC HANDBOOK OF CHEMISTRY AND PHYSICS*

This is the handbook that is most often consulted for data on organic compounds. Although a new edition of the handbook is published each year, the changes that are made are often minor. An older copy of the handbook will often suffice for most purposes. In addition to the extensive tables of properties of organic compounds, the *CRC Handbook* includes sections on nomenclature and ring structures, an index of synonyms, and an index of molecular formulas.

The nomenclature used in this book most closely follows the Chemical Abstracts system of naming organic compounds. This system differs, but only slightly, from

Table 4.1 Examples of Names of Compounds in the *CRC Handbook*

Name of Organic Compound	Location in *CRC Handbook*
1-Chloropentane	Pentane, 1-chloro-
1,4-Dichlorobenzene	Benzene, 1,4-dichloro-
4-Chlorotoluene	Benzene, 1-chloro-4-methyl-
Ethanoic acid	Acetic acid
tert-Butyl acetate (ethanoate)	Acetic acid, 1,1-dimethylethyl ester
Ethyl propanoate	Propanoic acid, ethyl ester
Isopentyl alcohol	1-Butanol, 3-methyl-
Isopentyl acetate (banana oil)	1-Butanol, 3-methyl-, acetate
Salicylic acid	Benzoic acid, 2-hydroxy-
Acetylsalicyclic acid (aspirin)	Benzoic acid, 2-acetyloxy-

standard IUPAC nomenclature. Table 4.1 lists some examples of how some commonly encountered compounds are named in this handbook. The first thing you will notice is that this handbook is not like a dictionary. Instead, you must first identify the *parent* name of the compound of interest. The parent names are found in alphabetical order. Once the parent name is identified and found, then you look for the particular substituent or substituents that may be attached to this parent.

For most compounds, it is easy to find what you are looking for as long as you know the parent name. Alcohols are, as expected, named by IUPAC nomenclature. Notice in Table 4.1 that the branched-chain alcohol, isopentyl alcohol, is listed as 1-butanol, 3-methyl.

Esters, amides, and acid halides are usually named as derivatives of the parent carboxylic acid. Thus, in Table 4.1, you find ethyl propanoate listed under the parent carboxylic acid, propanoic acid. If you have trouble finding a particular ester under the parent carboxylic acid, try looking under the alcohol part of the name. For example, isopentyl acetate was not listed under acetic acid, as expected, but instead is found under the alcohol part of the name (see Table 4.1). Fortunately, this handbook has a Synonym Index that nicely locates isopentyl acetate for you in the main part of the handbook.

Once you locate the compound by its name, you will find the following useful information:

CRC number — This is an identification number for the compound. You can use this number to find the molecular structure located elsewhere in the handbook. This is especially useful when the compound has a complicated structure.

Name and synonym — The Chemical Abstracts name and possible synonyms.

Mol. form. — Molecular formula for the compound.

Mol. wt. — Molecular weight.

CAS RN	Chemical Abstract Service Registry Number. This number is very useful for locating additional information on the compound in the primary chemical literature (see Chapter 30, Section 30.11).
mp/°C	Melting point of the compound in degrees Celsius.
bp/°C	Boiling point of the compound in degrees Celsius. A number without a superscript indicates that the recorded boiling point was obtained at 760 mm Hg pressure (atmospheric pressure). A number with a superscript indicates that the boiling point was obtained at reduced pressure. For example, an entry of 234; 122[16] would indicate that the compound boils at 234°C at 760 mm Hg and 122°C at 16 mm Hg pressure.
Den/g cm^{-3}	Density of a liquid. A superscript indicates the temperature in degrees Celsius at which the density was obtained.
n_D	Refractive index determined at a wavelength of 589 nm, the yellow line in a sodium lamp (D line). A superscript indicates the temperature at which the refractive index was obtained (see Chapter 24).

Solubility	Solubility classification	Solvent abbreviations
	1 = insoluble	ace = acetone
	2 = slightly soluble	bz = benzene
	3 = soluble	chl = chloroform
	4 = very soluble	EtOH = ethanol
	5 = miscible	eth = ether
	6 = decomposes	hx = hexane

Beil. ref.	Beilstein reference. An entry of 4-02-00-00157 would indicate that the compound is found in the 4th supplement in Volume 2, with no subvolume, on page 157 (see Chapter 30, Section 30.10 for details on the use of Beilstein).
Merck No.	*Merck Index* number in the 11th edition of the handbook. These numbers change each time a new edition of the *Merck Index* is issued.

Examples of sample handbook entries for isopentyl alcohol (1-butanol, 3-methyl) and isopentyl acetate (1-butanol, 3-methyl, acetate) are shown in Table 4.2.

4.2 *LANGE'S HANDBOOK OF CHEMISTRY*

This handbook tends not to be as available as the *CRC Handbook,* but it has some interesting differences and advantages. *Lange's Handbook* has synonyms listed at the bottom of each page, along with structures of more complicated molecules. The most

Table 4.2 Properties of isopentyl alcohol and isopentyl acetate as listed in the *CRC Handbook*

No.	Name Synonym	Mol. Form. Mol. Wt.	CAS RN mp/°C	Merck No. bp/°C	Beil. Ref. den/g cm⁻³	Solubility n_D
3627	1-Butanol, 3-methyl	$C_5H_{12}O$	123-51-3	5081	4-01-00-01677	ace 4; eth 4; EtOH 4
	Isopentyl alcohol	88.15	–117.2	131.1	0.8104[20]	1.4053[20]
3631	1-Butanol, 3-methyl, acetate	$C_7H_{14}O_2$	123-92-2	4993	4-02-00-00157	H_2O 2; EtOH 5; eth 5; ace 3
	Isopentyl acetate	130.19	–78.5	142.5	0.876[15]	1.4000[20]

noticeable difference is in how compounds are named. For many compounds, the system lists names as they would appear in a dictionary. Table 4.3 lists examples of how some commonly encountered compounds are named in this handbook. Most often, you do not need to identify the *parent* name. Unfortunately, *Lange's Handbook* often uses common names that are becoming obsolete. For example, propionate is used rather than propanoate. Nevertheless, this handbook often names compounds as a practicing organic chemist would tend to name them. Notice how easy it is to find the entries for isopentyl acetate and acetylsalicylic acid (aspirin) in this handbook.

Once you locate the compound by its name, you will find the following useful information:

Lange's number   This is an identification number for the compound.

Name   See examples in Table 4.3.

Formula   Structures are drawn out. If they are complicated, then the structures are shown at the bottom of the page.

Formula weight   Molecular weight of the compound.

Table 4.3 Examples of Names of Compounds in *Lange's Handbook*

Name of Organic Compound	Location in *Lange's Handbook*
1-Chloropentane	1-Chloropentane
1,4-Dichlorobenzene	1,4-Dichlorobenzene
4-Chlorotoluene	4-Chlorotoluene
Ethanoic acid	Acetic acid
tert-Butyl acetate (ethanoate)	*tert*-Butyl acetate
Ethyl propanoate	Ethyl propionate
Isopentyl alcohol	3-Methyl-1-butanol
Isopentyl acetate (banana oil)	Isopentyl acetate
Salicylic acid	2-Hydroxybenzoic acid
Acetylsalicyclic acid (aspirin)	Acetylsalicylic acid

Beilstein reference	An entry of 2, 132 would indicate that the compound is found in Volume 2 of the Main Work on page 132. An entry of 3^2, 188 would indicate that the compound is found in Volume 3 of the second supplement on page 188 (see Chapter 30, Section 30.10 for details on the use of Beilstein).
Density	Density is usually expressed in units of g/mL or g/cm^3. A superscript indicates the temperature at which the density was measured. If the density is also subscripted, usually 4°, it indicates that the density was measured at a certain temperature relative to water at its maximum density, 4°C. Most of the time you can simply ignore the subscripts and superscripts.
Refractive index	A superscript indicates the temperature at which the refractive index was determined (see Chapter 24).
Melting point	Melting point of the compound in degrees Celsius. When a "d" or "dec" appears with the melting point, it indicates that the compound decomposes at the melting point. When decomposition occurs, you will often observe a change in color of the solid.
Boiling point	Boiling point of the compound in degrees Celsius. A number without a superscript indicates that the recorded boiling point was obtained at 760 mm Hg pressure (atmospheric pressure). A number with a superscript indicates that the boiling point was obtained at reduced pressure. For example, an entry of $102^{11 \text{ mm}}$ would indicate that the compound boils at 102°C at 11 mm Hg pressure.
Flash point	This number is the temperature in degrees Celsius at which the compound will ignite when heated in air and a spark is introduced into the vapor. There are a number of different methods that are used to measure this value, so this number varies considerably. It gives a crude indication of flammability. You may need this information when heating a substance with a hot plate. Hot plates can be a serious source of trouble because of the sparking action that can occur with switches and thermostats used in hot plates.
Solubility in 100 parts solvent	Parts by weight of a compound that can be dissolved in 100 parts by weight of solvent at room temperature. In some cases, the values given are expressed as the weight in grams that can be dissolved in 100 mL of solvent. This handbook is not consistent in describing solubility. Sometimes gram amounts are provided, but in other cases the description will be more vague, using terms such as soluble, insoluble, or slightly soluble.

Solvent abbreviations
acet = acetone
bz = benzene
chl = chloroform
aq = water
alc = ethanol
eth = ether
HOAc = acetic acid

Solubility characteristics
i = insoluble
s = soluble
sls = slightly soluble
vs = very soluble
misc = miscible

Examples of sample handbook entries for isopentyl alcohol (3-methyl-1-butanol) and isopentyl acetate are shown in Table 4.4.

4.3 THE MERCK INDEX

The Merck Index is a very useful book because it has additional information not found in the other two handbooks. This handbook, however, tends to emphasize medicinally related compounds, such as drugs and biological compounds, although it also lists many other common organic compounds. It is not revised each year; new editions are published in 5 or 6 year cycles. It does not contain all of the compounds listed in *Lange's Handbook* or the *CRC Handbook*. However, for the compounds listed, it provides a wealth of useful information. The handbook will provide you with some or all of the following data for each entry.

Merck number; this changes each time a new edition is issued

Name, including synonyms and stereochemical designation

Molecular formula and structure

Molecular weight

Percentages of each of the elements in the compound

Uses

Source and synthesis, including references to the primary literature

Optical rotation for chiral molecules

Density, boiling point, and melting point

Solubility characteristics, including crystalline form

Pharmacology information

Toxicity data

One of the problems with looking up a compound in this handbook is trying to decide the name under which the compound will be listed. For example, isopentyl alcohol can also be named as 3-methyl-1-butanol or isoamyl alcohol. In the 12th edition of the handbook it is listed under the name isopentyl alcohol (#5212) on page 886. Finding isopentyl acetate is even a more challenging task. It is located in the handbook under the name isoamyl acetate (#5125) on page 876. Often, it is easier to look up the name in the name index or to find it in the formula index.

Table 4.4 Properties of 3-Methyl-1-butanol and Isopentyl Acetate as Listed in *Lange's Handbook*

No.	Name	Formula	Formula Weight	Beilstein Reference	Density	Refractive Index	Melting Point	Boiling Point	Flash Point	Solubility in 100 Parts Solvent
m155	3-methyl-1-butanol	$(CH_3)_2CHCH_2CH_2OH$	88.15	1, 392	0.8129^{15}_4	1.4085^{15}	−117.2	132.0	45	2 aq; misc alc, bz, chl, eth, HOAc
i80	Isopentyl acetate	$CH_3COOCH_2CH_2CH(CH_3)_2$	130.19	2, 132	0.876^{15}_4	1.4007^{20}	−78.5	142.0	80	0.25 aq; misc alc, eth

The handbook has some useful appendices that include the CAS registry numbers, a biological activity index, a formula index, and a name index that also includes synonyms. When looking up a compound in one of the indexes, you need to remember that the numbers provided are compound numbers, rather than page numbers. There is also a very useful section on organic name reactions that includes references to the primary literature.

4.4 ALDRICH HANDBOOK OF FINE CHEMICALS

The *Aldrich Handbook* is actually a catalog of chemicals sold by the Aldrich Chemical Company. They include in their catalog a large body of useful data on each compound that they sell. Because the catalog is reissued each year at no cost to the user, you should be able to find an old copy when the new one is issued. As you are mainly interested in the data on a particular compound and not the price, an old volume is perfectly fine. Isopentyl alcohol is listed as 3-methyl-1-butanol and isopentyl acetate is listed as isoamyl acetate in the Aldrich catalog. The following includes some of the properties and information they list for individual compounds.

Aldrich Catalog number

Name: Aldrich uses a mixture of common and IUPAC names. It takes a bit of time to master their names. Fortunately, they do a good job of cross-referencing compounds. They have a very good molecular formula index.

CAS registry number

Structure

Synonym

Formula weight

Boiling point/melting point

Index of refraction

Density

Beilstein reference

Merck reference

Infrared spectrum reference to the Aldrich Library of FT-IR spectra

NMR spectrum reference to the Aldrich Library of ^{13}C and ^{1}H FT-NMR spectra

Literature references to the primary literature on the uses of the compound

Toxicity

Safety data and precautions

Flash point

Prices of chemicals

4.5 STRATEGY FOR FINDING INFORMATION: SUMMARY

Most students and professors find the *Merck Index* and *Lange's Handbook* easier and more "intuitive" to use than the *CRC Handbook*. You can go directly to a compound without rearranging the name according to the parent or base name followed by its substituents. Another great source of information is the *Aldrich Handbook*, which contains those compounds that are easily available from a commercial source. Many compounds are found in the *Aldrich Handbook* that you may never find in any of the other handbooks. The Sigma-Aldrich web site allows you to search by name, synonym, formula, catalog number, and CAS registry number: *http://www.sigma-aldrich.com/ saws.nsf/AldProducts?OpenFrameset*

PROBLEMS

1. Using the *Merck Index*, find and draw structures for the following compounds:
   (a) atropine
   (b) quinine
   (c) saccharin
   (d) benzo[*a*]pyrene (benzpyrene)
   (e) itaconic acid
   (f) adrenosterone
   (g) chrysanthemic acid (chrysanthemumic acid)
   (h) cholesterol
   (i) vitamin C (ascorbic acid)
2. Find the melting points for the following compounds in the *CRC Handbook*, *Lange's Handbook*, or the *Aldrich Handbook:*
   (a) biphenyl
   (b) 4-bromobenzoic acid
   (c) 3-nitrophenol
3. Find the boiling point for each compound in the references listed in Problem 2:
   (a) octanoic acid at reduced pressure
   (b) 4-chloroacetophenone at atmosphere and reduced pressure
   (c) 2-methyl-2-heptanol
4. Find the index of refraction n_D and density for the liquids listed in Problem 3.
5. Using the *Aldrich Handbook*, report the specific rotations for the enantiomers of camphor.
6. Read the section on carbon tetrachloride in the *Merck Index* and list some of the health hazards for this compound.

CHAPTER 5

Measurement of Volume and Weight

Performing successful organic chemistry experiments requires the ability to measure solids and liquids accurately. This ability involves both selecting the proper measuring device and using this device correctly.

Liquids to be used for an experiment will usually be found in small containers in a hood. For *macroscale* experiments, a graduated cylinder, a dispensing pump, or a graduated pipet will be used for measuring the volume of a liquid. For *microscale* experiments, an automatic pipet, dispensing pump or calibrated Pasteur pipet will be used for measuring the volume of a liquid. You will usually transfer the required volume of liquid to a round-bottom flask or an Erlenmeyer flask in the case of macroscale experiments, or to a conical vial or small round-bottom flask for microscale experiments.

In cases in which the liquid is a limiting reagent, you should preweigh (**tare**) the container before dispensing the liquid into it. When the container is reweighed, you obtain the actual weight for the volume of liquid you have dispensed.

When transferring the liquid to a round-bottom flask, place the flask in a beaker and tare both the flask and the beaker. The beaker keeps the round-bottom flask in an upright position and prevents spills from occurring. The same advice should be followed if a conical vial is being used.

In cases in which the liquid is not the limiting reagent, you may calculate the weight of the liquid from the volume you have delivered and the density of the liquid. You may calculate the weight from the following relationship:

$$\text{Weight (g)} = \text{Density (g/mL)} \times \text{Volume (mL)}$$

When using a graduated cylinder to measure small volumes (< 3 mL) of a limiting reagent, it is important to transfer the liquid *quantitatively* from the cylinder to the reaction vessel. This means that all of the liquid should be transferred. When the reagent is poured from the graduated cylinder, a small amount of liquid will remain in the cylinder. This remaining liquid can be transferred by pouring a small amount of the solvent being used in the reaction into the cylinder and then pouring this solution into the reaction vessel. If no solvent is used in the reaction, most of the remaining liquid in the cylinder can be transferred by using a Pasteur pipet.

Using a small amount of solvent to transfer a liquid quantitatively can also be applied in other situations. For example, if your product is dissolved in a solvent and the procedure instructs you to transfer the reaction mixture from a round-bottom flask to a separatory funnel, after pouring most of the liquid into the funnel, a small amount of solvent could be used to transfer the rest of the product quantitatively.

Solids are usually found near the balance. For *macroscale* experiments, it is usually sufficient to weigh solids on a balance that reads at least to the nearest decigram (0.01 g). For *microscale* experiments, solids must be weighed on a balance that reads to the nearest milligram (0.001 g) or tenth of a milligram (0.0001 g). To weigh a solid, place your conical vial or round-bottom flask in a small beaker and take these with you to the balance. Place a smooth piece of paper that has been folded once on the balance pan. The folded paper will enable you to pour the solid into the conical vial or flask without spilling. Use a spatula to aid the transfer of the solid to the paper. Never weigh directly into a conical vial or flask, and never pour, dump, or shake a material from a bottle. While still at the balance, carefully transfer the solid from the paper to your vial or flask. The vial or flask should be in a beaker while transferring the solid. The beaker traps any material that fails to make it into the container. It also supports the vial or flask so that it does not fall over. It is not necessary to obtain the exact amount specified in the experimental procedure, and trying to be exact requires too much time at the balance. For example, if you obtained 0.140 g of a solid, rather than the 0.136 g specified in a procedure, you could use it, but the actual amount weighed should be recorded in your notebook. Use the actual amount you weighed to calculate the theoretical yield, if this solid is the limiting agent.

Careless dispensing of liquids and solids is a hazard in any laboratory. When reagents are spilled, you may be subjected to an unnecessary health or fire hazard. In addition, you may waste expensive chemicals, destroy balance pans and clothing, and damage the environment. Always clean up any spills immediately.

5.1 GRADUATED CYLINDERS

Graduated cylinders are most often used to measure liquids for macroscale experiments (see Fig. 5.1). The most common sizes are 10 mL, 25 mL, 50 mL, and 100 mL, but it is possible that not all of these will be available in your laboratory. Volumes from about 2 to 100 mL can be measured with reasonably good accuracy provided that the correct cylinder is used. You should use the *smallest* cylinder available that can hold all of the liquid that is being measured. For example, if a procedure calls for 4.5 mL of a reagent, use a 10-mL graduated cylinder. Using a large cylinder in this case will result in a less accurate measurement. Furthermore, using any cylinder to measure less than 10% of the total capacity of that cylinder will likely result in an inaccurate measurement. Always remember that whenever a graduated cylinder is used to measure the volume of a limiting reagent, you must weigh the liquid to determine the amount used accurately. You should use a graduated pipet, a dispensing pump, or an automatic pipet for accurate transfer of liquids with a volume of less than 2 mL.

If the storage container is reasonably small (< 1.0 L) and has a narrow neck, you may pour most of the liquid into the graduated cylinder and use a Pasteur pipet to ad-

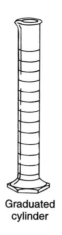

Graduated
cylinder

Figure 5.1 Graduated cylinder.

just to the final line. If the storage container is large (> 1.0 L) or has a wide mouth, two strategies are possible. First, you may use a pipet to transfer the liquid to the graduated cylinder. Alternatively, you may pour some of the liquid into a beaker first and then pour this liquid into a graduated cylinder. Use a Pasteur pipet to adjust to the final line. Remember that you should not take more than you need. Excess material should never be returned to the storage bottle. Unless you can convince someone else to take it, it must be poured into the appropriate waste container. You should be frugal in your estimation of amounts needed.

NOTE: Never return used reagents to the stock bottle.

5.2 DISPENSING PUMPS

Dispensing pumps are simple to operate, chemically inert, and quite accurate. Since the plunger assembly is made of Teflon, the dispensing pump may be used with most corrosive liquids and organic solvents. Dispensing pumps come in a variety of sizes, ranging from 1 mL to 300 mL. When used correctly, dispensing pumps can be used to deliver accurate volumes ranging from 0.1 mL to the maximum capacity of the pump. The pump is attached to a bottle containing the liquid being dispensed. The liquid is drawn up from this reservoir into the pump assembly through a piece of inert plastic tubing.

Dispensing pumps are somewhat difficult to adjust to the proper volume. Normally, the instructor or assistant will carefully adjust the unit to deliver the proper amount of liquid. As shown in Figure 5.2, the plunger is pulled up as far as it will travel to draw in the liquid from the glass reservoir. To expel the liquid from the spout into a container, you slowly guide the plunger down. With low-viscosity liquids, the weight of the plunger will expel the liquid. With more viscous liquids, however, you may need to push the plunger gently to deliver the liquid into a container. Remove the last drop of liquid on the end of the spout by touching the tip on the interior wall of

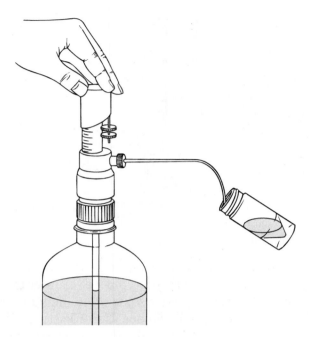

Figure 5.2 Use of a dispensing pump.

the container. When the liquid being transferred is a limiting reagent or when you need to know the weight precisely, you should weigh the liquid to determine the amount accurately.

As you pull up the plunger, look to see if the liquid is being drawn up into the pump unit. Some volatile liquids may not be drawn up in the expected manner, and you will observe an air bubble. Air bubbles commonly occur when the pump has not been used for a while. The air bubble can be removed from the pump by dispensing, and discarding, several volumes of liquid to "reprime" the dispensing pump. Also check to see if the spout is filled completely with liquid. An accurate volume will not be dispensed unless the spout is filled with liquid before you lift up the plunger.

5.3 GRADUATED PIPETS

A widely used measuring device is the graduated serological pipet. These *glass* pipets are available commercially in a number of sizes. "Disposable" graduated pipets may be used many times and discarded only when the graduations become too faint to be seen. A good assortment of these pipets consists of the following:

1.00-mL pipets calibrated in 0.01-mL divisions (1 in 1/100 mL)

2.00-mL pipets calibrated in 0.01-mL divisions (2 in 1/100 mL)

5.0-mL pipets calibrated in 0.1-mL divisions (5 in 1/10 mL)

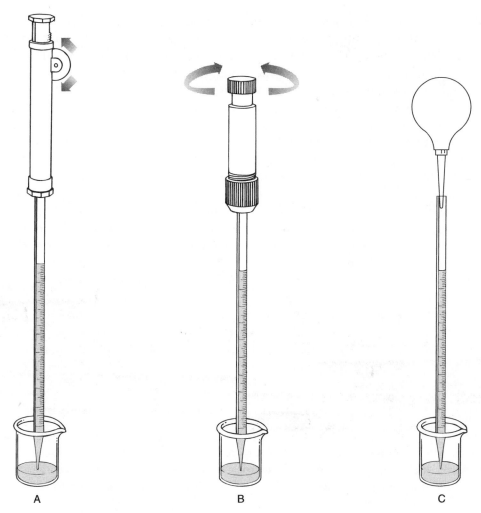

A   B   C

Figure 5.3 Pipet pumps (**A, B**) and a pipet bulb (**C**).

Never draw liquids into the pipets using mouth suction. A pipet pump or a pipet bulb, not a rubber dropper bulb, must be used to fill pipets. Two types of pipet pumps and a pipet bulb are shown in Figure 5.3. A pipet fits snugly into the pipet pump, and the pump can be controlled to deliver precise volumes of liquids. Control of the pipet pump is accomplished by rotating a knob on the pump. Suction created when the knob is turned draws the liquid into the pipet. Liquid is expelled from the pipet by turning the knob in the opposite direction. The pump works satisfactorily with organic, as well as aqueous, liquids.

The style of pipet pump shown in Figure 5.3A is available in four sizes. The top of the pipet must be inserted securely into the pump and held there with one hand to obtain an adequate seal. The other hand is used to load and release the liquid. The pipet pump shown in Figure 5.3B may also be used with graduated pipets. With this style of pipet, the top of the pipet is held securely by a rubber O-ring, and it is easily handled with one hand. You should be certain that the pipet is held securely by the O-ring before using it. Disposable pipets may not fit tightly in the O-ring because they often have smaller diameters than nondisposable pipets.

An alternative, and less expensive, approach is to use a rubber pipet bulb, shown in Figure 5.3C. Use of the pipet bulb is made more convenient by inserting a plastic automatic pipet tip into a rubber pipet bulb.[1] The tapered end of the pipet tip fits snugly into the end of a pipet. Drawing the liquid into the pipet is made easy, and it is also convenient to remove the pipet bulb and place a finger over the pipet opening to control the flow of liquid.

The calibrations printed on graduated pipets are reasonably accurate, but you should practice using the pipets in order to achieve this accuracy. When accurate quantities of liquids are required, the best technique is to weigh the reagent that has been delivered from the pipet.

The following description, along with Figure 5.4, illustrates how to use a graduated pipet. Insert the end of the pipet firmly into the pipet pump. Rotate the knob of the pipet pump in the correct direction (counterclockwise or up) to fill the pipet. Fill the pipet to a point just above the uppermost mark and then reverse the direction of rotation of the knob to allow the liquid to drain from the pipet until the meniscus is adjusted to the 0.00-mL mark. Move the pipet to the receiving vessel. Rotate the knob of the pipet pump (clockwise or down) to force the liquid from the pipet. Allow the liquid to drain from the pipet until the meniscus arrives at the mark corresponding to the volume that you wish to dispense. Be sure to touch the tip of the pipet to the inside of the container before withdrawing the pipet. Remove the pipet and drain the remaining liquid into a waste receiver. Avoid transferring the entire contents of the pipet when measuring volumes with a pipet. Remember that to achieve the greatest possible accuracy with this method, you should deliver volumes as a *difference* between two marked calibrations.

Pipets may be obtained in a number of styles, but only three types will be described here (Fig. 5.5). One type of graduated pipet is calibrated "to deliver" (TD) its total capacity when the last drop is blown out. This style of pipet, shown in Figure 5.5A, is probably the most common type of graduated pipet in use in the laboratory; it is designated by two rings at the top. Of course, it is not necessary to transfer the entire volume to a container. To deliver a more accurate volume, you should transfer an amount less than the total capacity of the pipet using the graduations on the pipet as a guide.

Another type of graduated pipet is shown in Figure 5.5B. This pipet is calibrated to deliver its total capacity when the meniscus is located on the last graduation mark near the bottom of the pipet. For example, the pipet shown in the Figure 5.5B delivers 10.0 mL of liquid when it has been drained to the point where the meniscus is located on

---

[1]This technique was described in Deckey, G. "A Versatile and Inexpensive Pipet Bulb." *Journal of Chemical Education*, 57 (July 1980): 526.

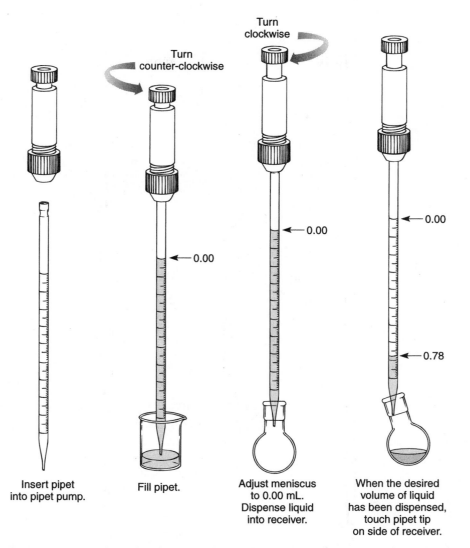

Figure 5.4 Use of a graduated pipet. (The figure shows, as an illustration, the technique required to deliver a volume of 0.78 mL from a 1.00-mL pipet.)

the 10.0-mL mark. With this type of pipet, you must not drain the entire pipet or blow it out. In contrast, notice that the pipet discussed in Figure 5.5A has its last graduation at 0.90 mL. The last 0.10-mL volume is blown out to give the 1.00-mL volume.

A nongraduated volumetric pipet is shown in Figure 5.5C. It is easily identified by the large bulb in the center of the pipet. This pipet is calibrated so that it will retain its last drop after the tip is touched on the side of the container. It must not be blown out. These pipets often have a single colored band at the top that identifies it as a "touch-off" pipet. The color of the band is keyed to its total volume. This type of pipet is commonly used in analytical chemistry.

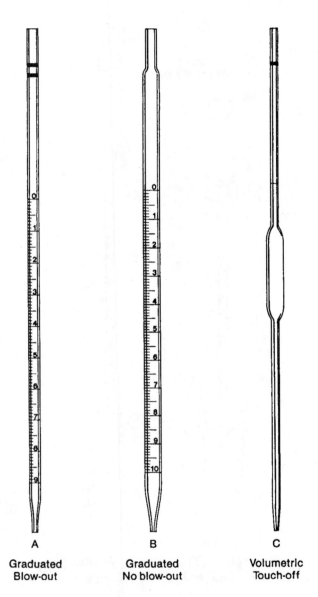

A	B	C
Graduated	Graduated	Volumetric
Blow-out	No blow-out	Touch-off

Figure 5.5 Pipets.

5.4 PASTEUR PIPETS

The Pasteur pipet is shown in Figure 5.6A with a 2-mL rubber bulb attached. There are two sizes of Pasteur pipets: a short one (5¾ inch), which is shown in the figure, and a long one (9 inch). It is important that the pipet bulb fit securely. You should not use a medicine dropper bulb because of its small capacity. A Pasteur pipet is an indispensable piece of equipment for the routine transfer of liquids. It is also used

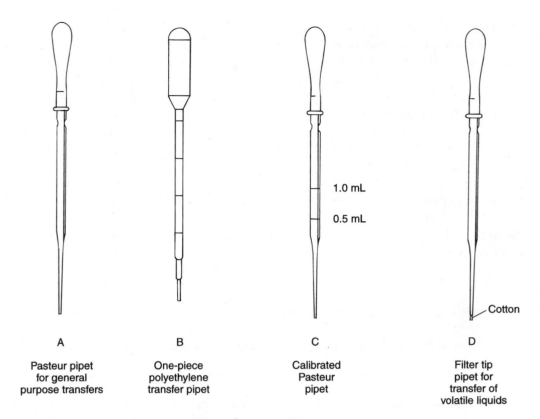

Figure 5.6 Pasteur (**A, C, D**) and transfer pipets (**B**).

for separations (Chapter 12). Pasteur pipets may be packed with cotton for use in gravity filtration (Chapter 8) or packed with an adsorbent for small-scale column chromatography (Chapter 19). Although they are considered disposable, you should be able to clean them for reuse as long as the tip remains unchipped.

A Pasteur pipet may be supplied by your instructor for dropwise addition of a particular reagent to a reaction mixture. For example, concentrated sulfuric acid is often dispensed in this way. When sulfuric acid is transferred, you should take care to avoid getting the acid into the rubber or latex dropper bulb.

The rubber dropper bulb may be avoided entirely by using one-piece transfer pipets made entirely of polyethylene (Fig. 5.6B). These plastic pipets are available in 1- or 2-mL sizes. They come from the manufacturers with approximate calibration marks stamped on them. These pipets can be used with all aqueous solutions and most organic liquids. They cannot be used with a few organic solvents or with concentrated acids.

Pasteur pipets may be calibrated for use in operations in which the volume does not need to be known precisely. Examples include measurement of solvents needed for extraction and for washing a solid obtained following crystallization. A calibrated Pasteur pipet is shown in Figure 5.6C. It is suggested that you calibrate several

5¾-inch pipets using the following procedure. On a balance, weigh 0.5 g (0.5 mL) of water into a small test tube. Select a short Pasteur pipet and attach a rubber bulb. Squeeze the rubber bulb before inserting the tip of the pipet into the water. Try to control how much you depress the bulb so that when the pipet is placed into the water and the bulb is completely released, only the desired amount of liquid is drawn into the pipet. When the water has been drawn up, place a mark with an indelible marking pen at the position of the meniscus. A more durable mark can be made by scoring the pipet with a file. Repeat this procedure with 1.0 g of water, and make a 1-mL mark on the same pipet.

Your instructor may provide you with a calibrated Pasteur pipet and bulb for transferring liquids where an accurate volume is not required. The pipet may be used to transfer a volume of 1.5 mL or less. You may find that the instructor has taped a test tube to the side of the storage bottle. The pipet is stored in the test tube with that particular reagent.

> **NOTE:** You should not assume that a certain number of drops equals a 1-mL volume. The common rule that 20 drops equal 1 mL, often used for a buret, does not hold true for a Pasteur pipet!

A Pasteur pipet may be packed with cotton to create a filter-tip pipet as shown in Figure 5.6D. This pipet is prepared by the instructions given in Chapter 8, Section 8.6, page 112. Pipets of this type are very useful in transferring volatile solvents during extractions and in filtering small amounts of solid impurities from solutions. A filter-tip pipet is very useful for removing small particles from a solution of a sample prepared for nuclear magnetic resonance (NMR) analysis.

5.5 SYRINGES

Syringes may be used to add a pure liquid or a solution to a reaction mixture. They are especially useful when anhydrous conditions must be maintained. The needle is inserted through a septum, and the liquid is added to the reaction mixture. Caution should be used with some disposable syringes as they often use solvent-soluble rubber gaskets on the plungers. A syringe should be cleaned carefully after each use by drawing acetone or another volatile solvent into it and expelling the solvent with the plunger. Repeat this procedure several times to clean the syringe thoroughly. Remove the plunger and draw air through the barrel with an aspirator to dry the syringe.

Syringes are usually supplied with volume graduations inscribed on the barrel. Large-volume syringes are not accurate enough to be used for measuring liquids in small-scale experiments. A small microliter syringe, such as that used in gas chromatography, delivers a very precise volume.

5.6 AUTOMATIC PIPETS

Automatic pipets are commonly used in microscale organic laboratories and in biochemistry laboratories. Several types of adjustable automatic pipets are shown in Figure 5.7. The automatic pipet is very accurate with aqueous solutions, but it is not as accurate with organic liquids. They are available in different sizes and can deliver accurate volumes ranging from 0.10 mL to 1.0 mL. These pipets are very expensive and must be shared by the entire laboratory. Automatic pipets should never be used with corrosive liquids, such as sulfuric acid or hydrochloric acid. *Always use the pipet with a plastic tip.*

Automatic pipets may vary in design, according to the manufacturer. The following description, however, should apply to most models. The automatic pipet consists of a handle that contains a spring-loaded plunger and a micrometer dial. The dial controls the travel of the plunger and is the means used to select the amount of liquid that the pipet is intended to dispense. Automatic pipets are designed to deliver liquids within a particular range of volumes. For example, a pipet may be designed to cover the range from 10 to 100 μL (0.010 to 0.100 mL) or from 100 to 1000 μL (0.100 to 1.000 mL).

Automatic pipets must never be dipped directly into the liquid sample without a plastic tip. The pipet is designed so that the liquid is drawn only into the tip. The liquids are never allowed to come in contact with the internal parts of the pipet. The plunger

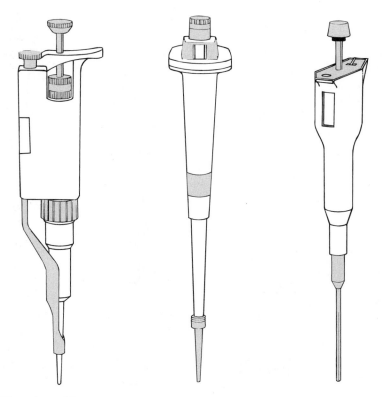

Figure 5.7 The adjustable automatic pipet.

has two *detent,* or "stop," positions used to control the filling and dispensing steps. Most automatic pipets have a stiffer spring that controls the movement of the plunger from the first to the second detent position. You will find a greater resistance as you press the plunger past the first detent.

To use the automatic pipet, follow the steps as outlined here. These steps are also illustrated in Figure 5.8. Keep the pipet in an upright position during use and return the unit to its holder, which keeps the automatic pipet in an upright position between uses. It is best to have available a dedicated pipet for each reagent being dispensed. If this is done, then the plastic tip can be left on the unit and reused. The instructor should preset the volume.

1. Select the desired volume by adjusting the micrometer control on the pipet handle.
2. Place a plastic tip on the pipet. Be certain that that tip is attached securely.
3. Push the plunger down to the first detent position. Do not press the plunger to the second position. If you press the plunger to the second detent, an incorrect volume of liquid will be delivered.
4. Dip the tip of the pipet into the liquid sample. Do not immerse the entire length of the plastic tip in the liquid. It is best to dip the tip only to a depth of about 1 cm.

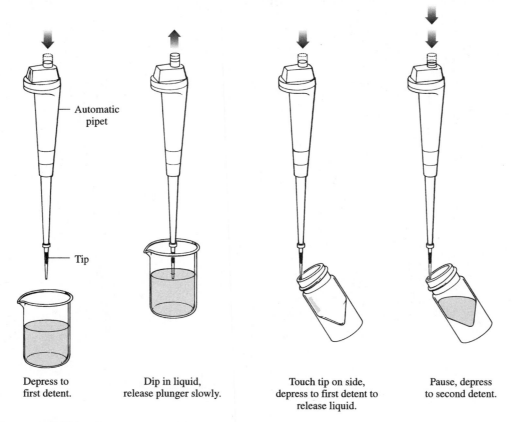

Depress to first detent.

Dip in liquid, release plunger slowly.

Touch tip on side, depress to first detent to release liquid.

Pause, depress to second detent.

Figure 5.8 Use of an automatic pipet.

5. Release the plunger *slowly*. Do not allow the plunger to snap back, or liquid may splash up into the plunger mechanism and ruin the pipet. Furthermore, rapid release of the plunger may cause air bubbles to be drawn into the pipet. At this point, the pipet has been filled.
6. Move the pipet to the receiving vessel. Touch the tip of the pipet to an interior wall of the container.
7. Slowly push the plunger down to the first detent. This action dispenses the liquid into the container.
8. Pause 1–2 seconds and then depress the plunger to its second detent position to expel the last drop of liquid. The action of the plunger may be stiffer in this range than it was up to the first detent.
9. Withdraw the pipet from the receiver. If the pipet is to be used with a different liquid, remove the pipet tip and discard it.

5.7 MEASURING VOLUMES WITH CONICAL VIALS, BEAKERS, AND ERLENMEYER FLASKS

Conical vials, beakers, and Erlenmeyer flasks all have graduations inscribed on them. Beakers and flasks can be used to give only a crude approximation of the volume. They are much less precise than graduated cylinders for measuring volume. In some cases, a conical vial may be used to estimate volumes. For example, the graduations are sufficiently accurate for measuring a solvent needed to wash a solid obtained on a Hirsch funnel after a crystallization. You should use an automatic pipet, dispensing pump, or graduated transfer pipet for accurate measurement of liquids in microscale experiments.

5.8 BALANCES

Solids and some liquids will need to be weighed on a balance that reads to at least the nearest milligram (0.001 g) for microscale experiments, or at least to the nearest decigram (0.01 g) for macroscale experiments. A top-loading balance (see Fig. 5.9) works well if the balance pan is covered with a plastic draft shield. The shield has a flap that opens to allow access to the balance pan. An analytical balance (see Fig. 5.10) may also be used. This type of balance will weigh to the nearest tenth of a milligram (0.0001 g) when provided with a glass draft shield.

Modern electronic balances have a tare device that automatically subtracts the weight of a container or a piece of paper from the combined weight to give the weight of the sample. With solids, it is easy to place a piece of paper on the balance pan, press the tare device so that the paper appears to have zero weight, and then add your solid until the balance gives the weight you desire. You can then transfer the weighed solid to a container. You should always use a spatula to transfer a solid and never pour material from a bottle. In addition, solids must be weighed on paper and not directly on the balance pan. Remember to clean any spills.

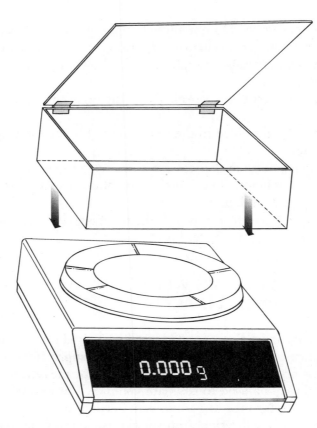

Figure 5.9 A top-loading balance with plastic draft shield.

With liquids, you should weigh the flask to determine the tare weight, transfer the liquid with a graduated cylinder, dispensing pump, or graduated pipet into the flask, and then reweigh it. With liquids, it is usually necessary to weigh only the limiting reagent. The other liquids may be transferred using a graduated cylinder, dispensing pump, or graduated pipet. Their weights can be calculated by knowing the volumes and densities of the liquids.

PROBLEMS

1. What measuring device would you use to measure the volume under each of the conditions described below? In some cases, there may be more than one answer to the question.
   (a) 25 mL of a solvent needed for a crystallization
   (b) 2.4 mL of a liquid needed for a reaction
   (c) 0.64 mL of a liquid needed for a reaction
   (d) 5 mL of a solvent needed for an extraction

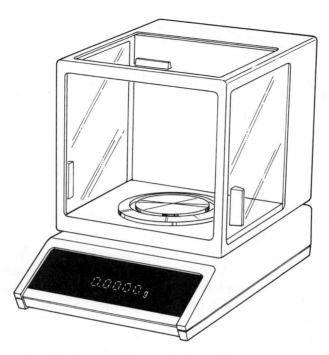

Figure 5.10 An analytical balance with glass draft shield.

2. Assume that the liquid used in Question 1b is a limiting reagent for a reaction. What should you do after measuring the volume?
3. Calculate the weight of a 2.5-mL sample of each of the following liquids:
   (a) Diethyl ether (ether)
   (b) Methylene chloride (dichloromethane)
   (c) Acetone
4. A laboratory procedure calls for 5.46 g of acetic anhydride. Calculate the volume of this reagent needed in the reaction.
5. Criticize the following techniques:
   (a) A 100-mL graduated cylinder is used to measure accurately a volume of 2.8 mL.
   (b) A one-piece polyethylene transfer pipet (Fig. 5.6B) is used to transfer precisely 0.75 mL of a liquid that is being used as the limiting reactant.
   (c) A calibrated Pasteur pipet (Fig. 5.6C) is used to transfer 25 mL of a solvent.
   (d) The volume markings on a 100-mL beaker are used to transfer accurately 5 mL of a liquid.
   (e) An automatic pipet is used to transfer 10 mL of a liquid.
   (f) A graduated cylinder is used to transfer 0.126 mL of a liquid.
   (g) A liquid is drawn up into an automatic pipet with the plunger pushed down to the second detent position.
   (h) For a small-scale reaction, the weight of a liquid limiting reactant is calculated from its density and volume.

CHAPTER 6
Heating and Cooling Methods

Most organic reaction mixtures need to be heated in order to complete the reaction. In general chemistry, you used a Bunsen burner for heating because nonflammable aqueous solutions were used. In an organic chemistry laboratory, however, the student must heat nonaqueous solutions that may contain *highly flammable* solvents. You *should not heat organic mixtures with a Bunsen burner* unless you are directed by your laboratory instructor. Open flames present a potential fire hazard. Whenever possible you should use one of the alternative heating methods, as described in the following sections.

6.1 HEATING MANTLES

A useful source of heat for most macroscale experiments is the heating mantle, illustrated in Figure 6.1. The heating mantle shown here consists of a ceramic heating shell with electric heating coils embedded within the shell. The temperature of a heating mantle is regulated with the heat controller. Although it is difficult to monitor the actual temperature of the heating mantle, the controller is calibrated so that it is fairly easy to duplicate approximate heating levels after one has gained some experience

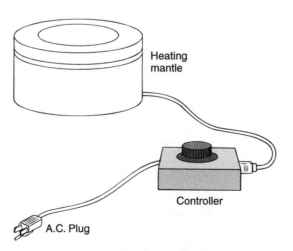

Heating mantle

Controller

A.C. Plug

Figure 6.1 A heating mantle.

with this apparatus. Reactions or distillations requiring relatively high temperatures can be easily performed with a heating mantle. For temperatures in the range of 50–80°C, you should use a water bath (Section 6.3) or a steam bath (Section 6.8).

In the center of the heating mantle shown in Figure 6.1 is a well that can accommodate round-bottom flasks of several different sizes. Some heating mantles, however, are designed to fit only specific sizes of round-bottom flasks. Some heating mantles are also made to be used with a magnetic stirrer so that the reaction mixture can be heated and stirred at the same time. Figure 6.2 shows a reaction mixture being heated with a heating mantle.

Heating mantles are very easy to use and safe to operate. The metal housing is grounded to prevent electrical shock if liquid is spilled into the well; however, flammable liquids may ignite if spilled into the well of a hot heating mantle.

⬛⬛⬛ **CAUTION**   You should be very careful to avoid spilling liquids into the well of the heating mantle. The surface of the ceramic shell may be very hot and could cause the liquid to ignite.

Raising and lowering the apparatus is a much more rapid method of changing the temperature within the flask than changing the temperature with the controller. For

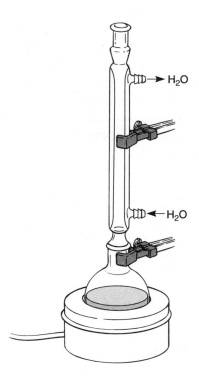

Figure 6.2 Heating with a heating mantle.

this reason, the entire apparatus should be clamped above the heating mantle so that it can be raised quickly if overheating occurs. Some laboratories may provide a lab jack or blocks of wood that can be placed under the heating mantle. In this case, the heating mantle itself is lowered and the apparatus remains clamped in the same position.

There are two situations in which it is relatively easy to overheat the reaction mixture. The first situation occurs when a larger heating mantle is used to heat a relatively small flask. You should be very careful when doing this. Many laboratories provide heating mantles of different sizes to prevent this from happening. The second situation occurs when the reaction mixture is first brought to a boil. To bring the mixture to a boil as rapidly as possible, the heat controller is often turned up higher than it will need to be set in order to keep the mixture boiling. When the mixture begins boiling very rapidly, turn the controller to a lower setting and raise the apparatus until the mixture boils less rapidly. As the temperature of the heating mantle cools down, lower the apparatus until the flask is resting on the bottom of the well.

6.2   HOT PLATES

Hot plates are a very convenient source of heat; however, it is difficult to monitor the actual temperature, and changes in temperature occur somewhat slowly. Care must be taken with flammable solvents to ensure against fires caused by "flashing" when solvent vapors come into contact with the hot plate surface. Never evaporate large quantities of a solvent by this method; the fire hazard is too great.

Some hot plates *heat constantly* at a given setting. They have no thermostat and you will have to control the temperature manually, either by removing the container being heated, or by adjusting the temperature up or down until a balance point is found. Some hot plates have a thermostat to control the temperature. A good thermostat will maintain a very even temperature. With many hot plates, however, the temperature may vary greatly (>10–20°C), depending upon whether the heater is in its "on" cycle or its "off" cycle. These hot plates will have a cycling (or oscillating) temperature, as shown in Figure 6.3. They too will have to be adjusted continually to maintain even heat.

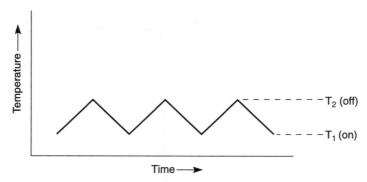

Figure 6.3 Temperature response for a hot plate with a thermostat.

Some hot plates also have built-in magnetic stirring motors that enable the reaction mixture to be stirred and heated at the same time. Their use is described in Section 6.5.

6.3 WATER BATH WITH HOT PLATE/STIRRER

A hot water bath is a very effective heat source when a temperature below 80°C is required. A beaker (250 mL or 400 mL) is partially filled with water and heated on a hot plate. A thermometer is clamped into position in the water bath. You may need to cover the water bath with aluminum foil to prevent evaporation, especially at higher temperatures. The water bath is illustrated in Figure 6.4. A mixture can be stirred with a magnetic stir bar (Chapter 7, Section 7.3, page 85). A hot water bath has some advantage over a heating mantle in that the temperature in the bath is uniform. In addition, it is sometimes easier to establish a lower temperature with a water bath than with other heating devices.

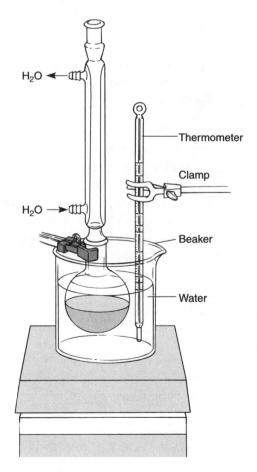

Figure 6.4 A water bath with a hot plate/stirrer.

Finally, the temperature of the reaction mixture will be closer to the temperature of the water, which allows for more precise control of the reaction conditions.

6.4 OIL BATH WITH HOT PLATE/STIRRER

In some laboratories oil baths may be available. An oil bath can be used when carrying out a distillation or heating a reaction mixture that needs a temperature above 100°C. An oil bath can be heated most conveniently with a hot plate, and a *heavy-wall* beaker[1] provides a suitable container for the oil. A thermometer is clamped into position in the oil bath. In some laboratories the oil may be heated electrically by an immersion coil. Because oil baths have a high heat capacity and heat slowly, it is advisable to heat the oil bath partially before the actual time at which it is to be used.

An oil bath with ordinary mineral oil cannot be used above 200–220°C. Above this temperature the oil bath may "flash," or suddenly burst into flame. A hot oil fire is not extinguished easily. If the oil starts smoking, it may be near its flash temperature; discontinue heating. Old oil, which is dark, is more likely to flash than new oil. Also, hot oil causes bad burns. Water should be kept away from a hot oil bath, since water in the oil will cause it to splatter. Never use an oil bath when it is obvious that there is water in the oil. If water is present, replace the oil before using the heating bath. An oil bath has only a finite lifetime. New oil is clear and colorless but, after extended use, becomes dark brown and gummy from oxidation.

Besides ordinary mineral oil, a variety of other types of oils can be used in an oil bath. Silicone oil does not begin to decompose at as low a temperature as does mineral oil. When silicone oil is heated high enough to decompose, however, its vapors are far more hazardous than mineral oil vapors. The polyethylene glycols may be used in oil baths. They are water soluble, which makes cleaning up after using an oil bath much easier than with mineral oil. One may select any one of a variety of polymer sizes of polyethylene glycol, depending on the temperature range required. The polymers of large molecular weight are often solid at room temperature. Wax may also be used for higher temperatures, but this material also becomes solid at room temperature. Some workers prefer to use a material that solidifies when not in use since it minimizes both storage and spillage problems.

6.5 ALUMINUM BLOCK WITH HOT PLATE/STIRRER

Although aluminum blocks are most commonly used in microscale organic chemistry laboratories, they can also be used with the smaller round-bottom flasks used in macroscale experiments.[2] The aluminum block shown in Figure 6.5A can be used to hold 25-, 50-, or 100-mL round-bottom flasks, as well as a thermometer. Heating will

---

[1] It is very dangerous to use a thin-wall beaker for an oil bath. Breakage due to heating can occur, spilling hot oil everywhere!

[2] The use of solid aluminum heating devices was developed by Siegfried Lodwig at Centralia College, Centralia, WA: Lodwig, S. N., *Journal of Chemical Education*, *66* (1989): 77.

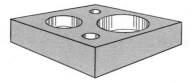 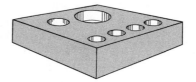

A. Large holes for 25-, 50- or
100-mL round-bottom flasks.

B. Small holes for Craig tube and 3-mL and
5-mL conical vials, and small test tubes.

Figure 6.5 Aluminum heating blocks.

occur more rapidly if the flask fits all the way into the hole; however, heating is also effective if the flask only partially fits into the hole. The aluminum block with smaller holes, as shown in Figure 6.5B, is designed for microscale glassware. It will hold a conical vial, a Craig tube or small test tubes, and a thermometer.

There are several advantages to heating with an aluminum block. The metal heats very quickly, high temperatures can be obtained, and you can cool the aluminum rapidly by removing it with crucible tongs and immersing it in cold water. Aluminum blocks are also inexpensive or can be fabricated readily in a machine shop.

Figure 6.6 shows a reaction mixture being heated with an aluminum block on a hot plate/stirrer unit. The thermometer in the figure is used to determine the temperature of the aluminum block. *Do not use a mercury thermometer:* use a thermometer containing a liquid other than mercury or use a metal dial thermometer that can be inserted into a smaller diameter hole drilled into the side of the block.[3] Make sure that the thermometer fits loosely in the hole, or it may break. Secure the thermometer with a clamp.

To avoid the possibility of breaking a glass thermometer, your hot plate may have a hole drilled into the metal plate so that a metal dial thermometer can be inserted into the unit (Fig. 6.7A). These metal thermometers, such as the one shown in Figure 6.7B, can be obtained in a number of temperature ranges. For example, a 0–250°C thermometer with 2-degree divisions can be obtained at a reasonable price. Also shown in Figure 6.7 (inset) is an aluminum block with a small hole drilled into it so that a metal thermometer can be inserted. An alternative to the metal thermometer is a digital electronic temperature-measuring device that can be inserted into the aluminum block or hot plate. It is strongly recommended that mercury thermometers be avoided when measuring the surface temperature of the hot plate or aluminum block. If a mercury thermometer is broken on a hot surface, you will introduce toxic mercury vapors into the laboratory. Nonmercury thermometers filled with high-boiling colored liquids are available as alternatives.

As already mentioned, aluminum blocks are often used in the microscale organic chemistry laboratory. The use of an aluminum block to heat a microscale reflux apparatus

[3]Garner, C. M. "A Mercury-Free Alternative for Temperature Measurement in Aluminum Blocks." *Journal of Chemical Education, 68* (1991): A244.

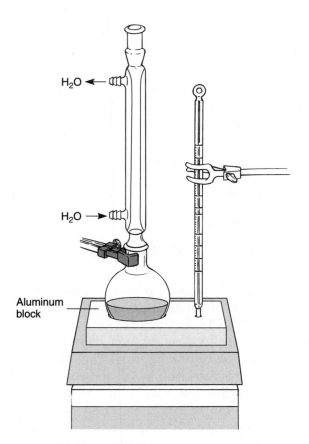

Figure 6.6 Heating with an aluminum block.

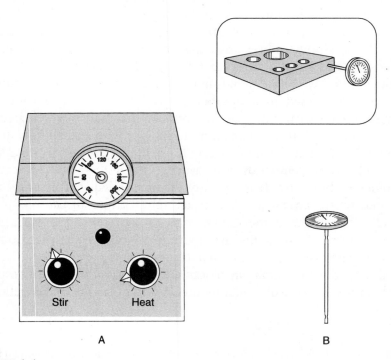

Figure 6.7 Dial thermometers.

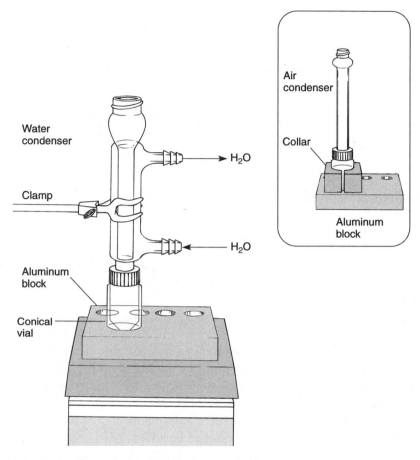

Figure 6.8 Heating with an aluminum block (microscale).

is shown in Figure 6.8. The reaction vessel in the figure is a conical vial, which is used in many microscale experiments. Also shown in Figure 6.8 is a split aluminum collar that may be used when very high temperatures are required. The collar is split to facilitate easy placement around a 5-mL conical vial. The collar helps to distribute heat further up the wall of the vial.

You should first calibrate the aluminum block so that you have an approximate idea where to set the control on the hot plate to achieve a desired temperature. Place the aluminum block on the hot plate and insert a thermometer into the small hole in the block. Select five equally spaced temperature settings, including the lowest and highest settings, on the heating control of the hot plate. Set the dial to the first of these settings and monitor the temperature recorded on the thermometer. When the thermometer reading arrives at a constant value,[4] record this final temperature, along with the dial setting. Repeat this procedure with the remaining four settings. Using these data, prepare a calibration curve for future reference.

---

[4]See, however, Section 6.2, page 68.

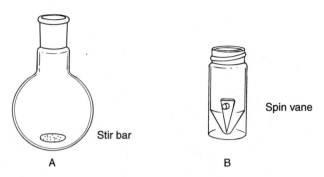

Figure 6.9 Methods of stirring in a round-bottom flask or conical vial.

It is a good idea to use the same hot plate each time, as it is very likely that two hot plates of the same type may give different temperatures with identical settings. Record the identification number printed on the unit that you are using in your notebook to ensure that you always use the same hot plate.

For many experiments, you can determine what the approximate setting on the hot plate should be from the boiling point of the liquid being heated. Because the temperature inside the flask is lower than the aluminum block temperature, you should add at least 20°C to the boiling point of the liquid and set the aluminum block at this higher temperature. In fact, you may need to raise the temperature even higher than this value in order to bring the liquid to a boil.

Many organic mixtures need to be stirred as well as heated to achieve satisfactory results. To stir a mixture, place a magnetic stir bar (Chapter 7, Fig. 7.8A, page 85) in a round-bottom flask containing the reaction mixture as shown in Figure 6.9A. If the mixture is to be heated as well as stirred, attach a water condenser as shown in Figure 6.6. With the combination stirrer/hot plate unit, it is possible to stir and heat a mixture simultaneously. With conical vials, a magnetic spin vane must be used to stir mixtures (Fig. 7.8B, page 85). This is shown in Figure 6.9B. More uniform stirring will be obtained if the flask or vial is placed in the aluminum block so that it is centered on the hot plate. Mixing may also be achieved by boiling the mixture. A boiling stone (Section 7.4, page 86) must be added when a mixture is boiled without magnetic stirring.

6.6 SAND BATH WITH HOT PLATE/STIRRER

The sand bath is used in some microscale laboratories to heat organic mixtures. It can also be used as a heat source in some macroscale experiments. Sand provides a clean way of distributing heat to a reaction mixture. To prepare a sand bath for microscale use, place about a 1-cm depth of sand in a crystallizing dish and then set the dish on a hot plate/stirrer unit. The apparatus is shown in Figure 6.10. Clamp the thermometer into position in the sand bath. You should calibrate the sand bath in a manner similar to that used with the aluminum block (see previous section). Because sand heats more slowly than an aluminum block, you will need to begin heating the sand bath well before using it.

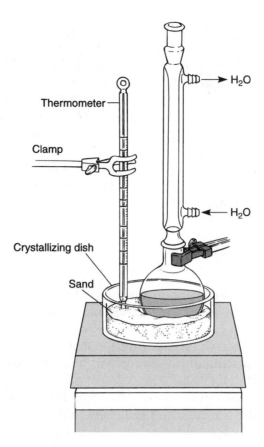

Figure 6.10 Heating with a sand bath.

Do not heat the sand bath much above 200°C or you may break the dish. If you need to heat at very high temperatures, you should use a heating mantle or an aluminum block rather than a sand bath. With sand baths, it may be necessary to cover the dish with aluminum foil to achieve a temperature near 200°C. Because of the relatively poor heat conductivity of sand, a temperature gradient is established within the sand bath. It is warmer near the bottom of the sand bath and cooler near the top for a given setting on the hot plate. To make use of this gradient, you may find it convenient to bury the flask or vial in the sand to heat a mixture more rapidly. Once the mixture is boiling, you can then slow the rate of heating by raising the flask or vial. These adjustments may be made easily and do not require a change in the setting on the hot plate.

6.7 FLAMES

The simplest technique for heating mixtures is to use a Bunsen burner. Because of the high danger of fires, however, the use of the Bunsen burner should be strictly limited to those cases for which the danger of fire is low or for which no reasonable alternative source of heat is available. A flame should generally be used only to heat aqueous solutions or solutions with very high boiling points. You should always check with your instructor about using a burner. If you use a burner at your bench, great care should be taken to ensure that others in the vicinity are not using flammable solvents.

In heating a flask with a Bunsen burner, you will find that using a wire gauze can produce more even heating over a broader area. The wire gauze, when placed under the object being heated, spreads the flame to keep the flask from being heated in one small area only.

Bunsen burners may be used to prepare capillary micropipets for thin-layer chromatography or to prepare other pieces of glassware requiring an open flame. For these purposes, burners should be used in designated areas in the laboratory and not at your laboratory bench.

6.8 STEAM BATHS

The steam cone or steam bath is a good source of heat when temperatures around 100°C are needed. Steam baths are used to heat reaction mixtures and solvents needed for crystallization. A steam cone and a portable steam bath are shown in Figure 6.11. These methods of heating have the disadvantage that water vapor may be introduced, through condensation of steam, into the mixture being heated. A slow flow of steam may minimize this difficulty.

Because water condenses in the steam line when it is not in use, it is necessary to purge the line of water before the steam will begin to flow. This purging should be accomplished before the flask is placed on the steam bath. The steam flow should be started with a high rate to purge the line; then, the flow should be reduced to the desired rate. When using a portable steam bath, be certain that condensate (water) is drained into a sink. Once the steam bath or cone is heated, a slow steam flow will

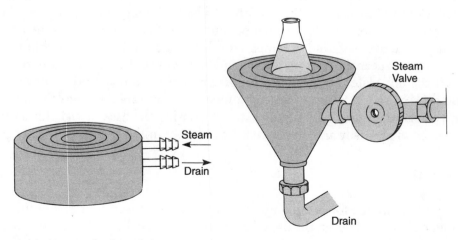

Figure 6.11 A steam bath and a steam cone.

maintain the temperature of the mixture being heated. There is no advantage to having a Vesuvius on your desk! An excessive steam flow may cause problems with condensation in the flask. This condensation problem can often be avoided by selecting the correct place at which to locate the flask on top of the steam bath.

The top of the steam bath consists of several flat concentric rings. The amount of heat delivered to the flask being heated can be controlled by selecting the correct sizes of these rings. Heating is most efficient when the largest opening that will still support the flask is used. Heating large flasks on a steam bath while using the smallest opening leads to slow heating and wastes laboratory time.

6.9 COLD BATHS

At times, you may need to cool an Erlenmeyer flask or round-bottom flask below room temperature. A cold bath is used for this purpose. The most common cold bath is an **ice bath,** which is a highly convenient source of 0°C temperatures. An ice bath requires water along with ice to work well. If an ice bath is made up of only ice, it is not a very efficient cooler since the large pieces of ice do not make good contact with the flask. Enough water should be present with ice so that the flask is surrounded by water but not so much that the temperature is no longer maintained at 0°C. In addition, if too much water is present, the buoyancy of a flask resting in the ice bath may cause it to tip over. There should be enough ice in the bath to allow the flask to rest firmly.

For temperatures somewhat below 0°C, you may add some solid sodium chloride to the ice-salt-water bath. The ionic salt lowers the freezing point of the ice, so that temperatures in the range of 0 to –10°C can be reached. The lowest temperatures are reached with ice-water mixtures that contain relatively little water.

A temperature of –78.5°C can be obtained with solid carbon dioxide or dry ice. However, large chunks of dry ice do not provide uniform contact with a flask being cooled. A liquid such as isopropyl alcohol is mixed with small pieces of dry ice to provide an efficient cooling mixture. Acetone and ethanol can be used in place of isopropyl alcohol. Be careful when handling dry ice because it can inflict severe frostbite. Extremely low temperatures can be obtained with liquid nitrogen (–195.8°C).

PROBLEMS

1. What would be the preferred heating device(s) in each of the following situations?
   (a) Reflux a solvent with a 56°C boiling point
   (b) Reflux a solvent with a 110°C boiling point
   (c) Distillation of a substance that boils at 220°C
2. Obtain the boiling points for the following compounds by using a handbook (Chapter 4). In each case, suggest a heating device(s) that should be used for refluxing the substance.
   (a) Butyl benzoate
   (b) 1-Pentanol
   (c) 1-Chloropropane

3. What type of bath would you use to get a temperature of –10°C?
4. Obtain the melting point and boiling point for benzene and ammonia from a handbook (Chapter 4) and answer the following questions.
   (a) A reaction was conducted in benzene as the solvent. Because the reaction was very exothermic, the mixture was cooled in a salt-ice bath. This was a bad choice. Why?
   (b) What bath should be used for a reaction that is conducted in liquid ammonia as the solvent?
5. Criticize the following techniques:
   (a) Refluxing a mixture that contains diethyl ether using a Bunsen burner.
   (b) Refluxing a mixture that contains a large amount of toluene using a hot water bath.
   (c) Refluxing a mixture using the apparatus shown in Figure 6.6, but with an unclamped thermometer.
   (d) Using a mercury thermometer that is inserted into an aluminum block on a hot plate.
   (e) Running a reaction with *tert*-butyl alcohol (2-methyl-2-propanol) that is cooled to 0°C in an ice bath.

CHAPTER 7
Reaction Methods

The successful completion of an organic reaction requires the chemist to be familiar with a variety of laboratory methods. These methods include operating safely, assembling the apparatus, heating and stirring reaction mixtures, adding liquid reagents, maintaining anhydrous and inert conditions in the reaction, and collecting gaseous products. Several techniques that are used in bringing a reaction to a successful conclusion are treated in this chapter.

7.1 ASSEMBLING THE APPARATUS

Care must be taken when assembling the glass components into the desired apparatus. You should always remember that Newtonian physics applies to chemical apparatus, and unsecured pieces of glassware are certain to respond to gravity.

Assembling an apparatus in the correct manner requires that the individual pieces of glassware are connected to each other securely and the entire apparatus is held in the correct position. This can be accomplished by using **adjustable metal clamps** or a combination of adjustable metal clamps and **plastic joint clips.**

Two types of adjustable metal clamps are shown in Figure 7.1. Although these two types of clamps can usually be interchanged, the extension clamp is more commonly used to hold round-bottom flasks in place, and the three-finger clamp is frequently used to clamp condensers. Both types of clamps must be attached to a ring stand using a clamp holder, shown in Figure 7.1C.

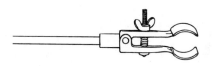

A. Extension clamp   B. Three-finger clamp   C. Clamp holder

Figure 7.1 Adjustable metal clamps.

A. Securing Macroscale Apparatus Assemblies

It is possible to assemble an apparatus using only adjustable metal clamps. An apparatus used to perform a distillation is shown in Figure 7.2. It is held together securely with three metal clamps. Because of the size of the apparatus and its geometry, the various clamps would likely be attached to three different ring stands. This apparatus would be somewhat difficult to assemble, because it is necessary to ensure that the individual pieces stay together while securing and adjusting the clamps required to hold the entire apparatus in place. In addition, one must be very careful not to bump any part of the apparatus or the ring stands after the apparatus is assembled.

A more convenient alternative is to use a combination of metal clamps and plastic joint clips. A plastic joint clip is shown in Figure 7.3A. These clips are very easy to use (they just clip on), will withstand temperatures up to 140°C, and are quite durable. They hold together two pieces of glassware that are connected by ground glass joints, as shown in Figure 7.3B. These clips come in different sizes to fit ground glass joints of different sizes and they are color coded for each size.

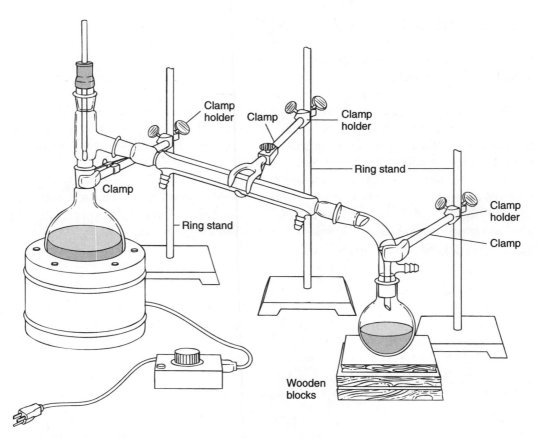

Figure 7.2 Distillation apparatus secured with metal clamps.

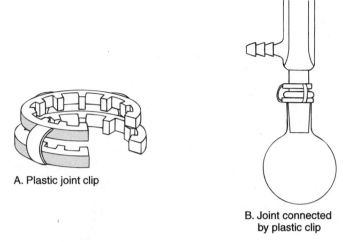

A. Plastic joint clip

B. Joint connected
by plastic clip

Figure 7.3 Plastic joint clip.

When used in combination with metal clamps, the plastic joint clips make it much easier to assemble most apparatus in a secure manner. There is less chance of dropping the glassware while assembling the apparatus, and once the apparatus is set up it is more secure. Figure 7.4 shows the same distillation apparatus held in place with both adjustable metal clamps and plastic joint clips.

To assemble this apparatus, first connect all of the individual pieces together using the plastic clips. The entire apparatus is then connected to the ring stands using the adjustable metal clamps. Note that only two ring stands are required and the wooden blocks are not needed.

B. Securing Microscale Apparatus Assemblies

The glassware in most microscale kits is made with standard-taper ground joints. The most common joint size is ⊤ 14/10. Some microscale glassware with ground-glass joints also has threads cast into the outside surface of the outer joints (see top of air condenser in Fig. 7.5). The threaded joint allows the use of a plastic screw cap with a hole in the top to fasten two pieces of glassware together securely. The plastic cap is slipped over the inner joint of the upper piece of glassware, followed by a rubber O-ring (see Fig. 7.5). The O-ring should be pushed down so that it fits snugly on top of the ground-glass joint. The inner ground-glass joint is then fitted into the outer joint of the bottom piece of glassware. The screw cap is tightened, without excessive force, to attach the entire apparatus firmly together. The O-ring provides an additional seal that makes this joint airtight. With this connecting system, it is unnecessary to use any type of grease to seal the joint. The O-ring *must be used* to obtain a good seal and to lessen the chances of breaking the glassware when you tighten the plastic cap.

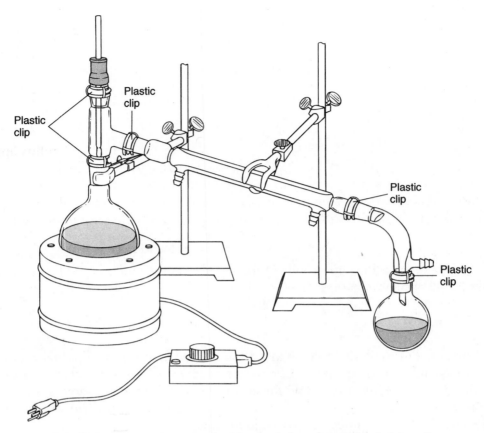

Figure 7.4 Distillation apparatus secured with metal clamps and plastic joint clips.

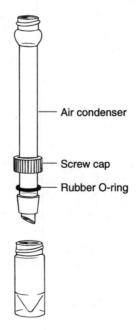

Figure 7.5 A microscale standard-taper joint assembly.

Microscale glassware connected together in this fashion can be assembled very easily. The entire apparatus is held together securely, and usually only one metal clamp is required to hold the apparatus onto a ring stand.

7.2 HEATING UNDER REFLUX

Often we wish to heat a mixture for a long time and to leave it untended. A **reflux apparatus** (see Fig. 7.6) allows such heating. The liquid is heated to a boil and the hot vapors are cooled and condensed as they rise into the water-jacketed condenser. Therefore, very little liquid is lost by evaporation and the mixture is kept at a constant temperature, the boiling point of the liquid. The liquid mixture is said to be **heating under reflux.**

Condenser. The **water-jacketed condenser** shown in Figure 7.6 consists of two concentric tubes with the outer cooling tube sealed onto the inner tube. The vapors rise within the inner tube and water circulates through the outer tube. The circulating water removes heat from the vapors and condenses them. Figure 7.6 also shows a typical microscale apparatus for heating small quantities of material under reflux (Fig. 7.6B).

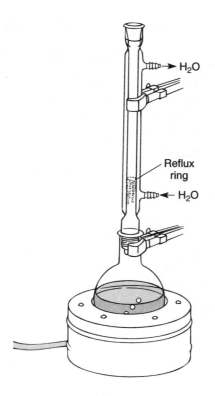

A. Reflux apparatus for macroscale reactions, using a heating mantle and water-jacketed condenser.

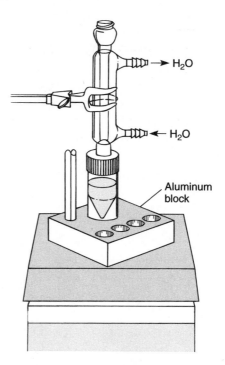

B. Reflux apparatus for microscale reactions, using a hot plate, aluminum block, and water-jacketed condenser.

Figure 7.6 Heating under reflux.

When using a water-jacketed condenser, the direction of the water flow should be such that the condenser will fill with cooling water. The water should enter the bottom of the condenser and leave from the top. The water should flow fast enough to withstand any changes in pressure in the water lines, but it should not flow any faster than absolutely necessary. An excessive flow rate greatly increases the chance of a flood, and high water pressure may force the hose from the condenser. Cooling water should be flowing before heating is begun! If the water is to remain flowing overnight, it is advisable to fasten the rubber tubing securely with wire to the condenser. If a flame is used as a source of heat, it is wise to use a wire gauze beneath the flask to provide an even distribution of heat from the flame. In most cases, a heating mantle, water bath, oil bath, aluminum block, sand bath, or steam bath is preferred over a flame.

Stirring.   When heating a solution, always use a magnetic stirrer or a boiling stone (see Sections 7.3 and 7.4) to keep the solution from "bumping" (see next section).

Rate of Heating.   If the heating rate has been correctly adjusted, the liquid being heated under reflux will travel only part way up the condenser tube before condensing. Below the condensation point, solvent will be seen running back into the flask; above it, the interior of the condenser will appear dry. The boundary between the two zones will be clearly demarcated, and a **reflux ring** or a ring of liquid will appear there. The reflux ring can be seen in Figure 7.6A. In heating under reflux, the rate of heating should be adjusted so that the reflux ring is no higher than a third to a half the distance to the top of the condenser. With microscale experiments, the quantities of vapor rising in the condenser frequently are so small that a clear reflux ring cannot be seen. In those cases, the heating rate must be adjusted so that the liquid boils smoothly but not so rapidly that solvent can escape the condenser. With such small volumes, the loss of even a small amount of solvent can affect the reaction. With macroscale reactions, the reflux ring is much easier to see, and one can adjust the heating rate more easily.

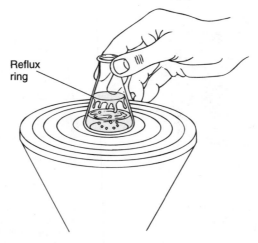

Reflux
ring

Figure 7.7 Tended reflux of small quantities on a steam cone.

Tended Reflux.   It is possible to heat small amounts of a solvent under reflux in an Erlenmeyer flask. By heating gently, the evaporated solvent will condense in the relatively cold neck of the flask and return to the solution. This technique (see Fig. 7.7) requires constant attention. The flask must be swirled frequently and removed from the heating source for a short period if the boiling becomes too vigorous. When heating is in progress, the reflux ring should not be allowed to rise into the neck of the flask.

7.3 STIRRING METHODS

When a solution is heated, there is a danger that it may become superheated. When this happens, very large bubbles sometimes erupt violently from the solution; this is called **bumping.** Bumping must be avoided because of the risk that material may be lost from the apparatus, that a fire might start, or that the apparatus may break.

Magnetic stirrers are used to prevent bumping because they produce turbulence in the solution. The turbulence breaks up the large bubbles that form in boiling solutions. An additional purpose for using a magnetic stirrer is to stir the reaction to ensure that all the reagents are thoroughly mixed. A magnetic stirring system consists of a magnet that is rotated by an electric motor. The rate at which this magnet rotates can be adjusted by a potentiometric control. A small magnet, which is coated with a nonreactive material such as Teflon or glass, is placed in the flask. The magnet within the flask rotates in response to the rotating magnetic field caused by the motor-driven magnet. The result is that the inner magnet stirs the solution as it rotates. A very common type of magnetic stirrer includes the stirring system within a hot plate. This type of hot plate-stirrer permits one to heat the reaction and stir it simultaneously.

For macroscale apparatus, magnetic stirring bars of various sizes and shapes are available. For microscale apparatus, a **magnetic spin vane** is often used. It is designed to contain a tiny bar magnet and to have a shape that conforms to the conical bottom of a reaction vial. A small Teflon-coated magnetic stirring bar works well with very small round-bottom boiling flasks. Small stirring bars of this type (often sold as "disposable" stirring bars) can be obtained very cheaply. A variety of magnetic stirring bars is illustrated in Figure 7.8.

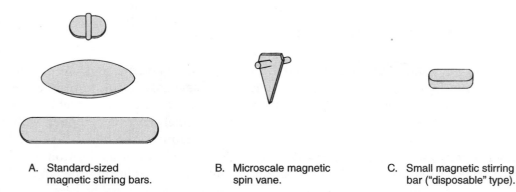

A. Standard-sized
    magnetic stirring bars.

B. Microscale magnetic
    spin vane.

C. Small magnetic stirring
    bar ("disposable" type).

Figure 7.8 Magnetic stirring bars.

There is also a variety of simple techniques that may be used to stir a liquid mixture in a centrifuge tube or conical vial. A thorough mixing of the components of a liquid can be achieved by repeatedly drawing the liquid into a Pasteur pipet and then ejecting the liquid back into the container by pressing sharply on the dropper bulb. Liquids can also be stirred effectively by placing the flattened end of a spatula into the container and twirling it rapidly.

7.4   BOILING STONES

A **boiling stone,** also known as a **boiling chip** or **Boileezer,** is a small lump of porous material that produces a steady stream of fine air bubbles when it is heated in a solvent. This stream of bubbles and the turbulence that accompanies it break up the large bubbles of gases in the liquid. In this way, it reduces the tendency of the liquid to become superheated, and it promotes the smooth boiling of the liquid. The boiling stone decreases the chances for bumping.

Two common types of boiling stones are carborundum and marble chips. Carborundum boiling stones are more inert, and the pieces are usually quite small, suitable for most applications. If available, carborundum boiling stones are preferred for most purposes. Marble chips may dissolve in strong acid solutions, and the pieces are larger. The advantage of marble chips is that they are cheaper.

Because boiling stones act to promote the smooth boiling of liquids, you should always make certain that a boiling stone has been placed in a liquid *before* heating is begun. If you wait until the liquid is hot, it may have become superheated. Adding a boiling stone to a superheated liquid will cause all the liquid to try to boil at once. The liquid, as a result, would erupt entirely out of the flask or froth violently.

As soon as boiling ceases in a liquid containing a boiling stone, the liquid is drawn into the pores of the boiling stone. When this happens, the boiling stone no longer can produce a fine stream of bubbles; it is spent. You may have to add a new boiling stone if you have allowed boiling to stop for a long period.

Wooden applicator sticks are used in some applications. They function in the same manner as boiling stones. Occasionally, glass beads are used. Their presence also causes sufficient turbulence in the liquid to prevent bumping.

7.5   ADDITION OF LIQUID REAGENTS

Liquid reagents and solutions are added to a reaction by several means, some of which are shown in Figure 7.9. The most common type of assembly for macroscale experiments is shown in Figure 7.9A. In this apparatus, a separatory funnel is attached to the sidearm of a Claisen head adapter. The separatory funnel must be equipped with a standard-taper, ground-glass joint to be used in this manner. The liquid is stored in the separatory funnel (which is called an **addition funnel** in this application) and is added to the reaction. The rate of addition is controlled by adjusting the stopcock. When it is being used as an addition funnel, the upper opening must be kept open to

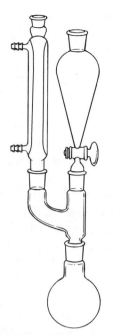

A. Macroscale equipment, using a
separatory funnel as an addition funnel.

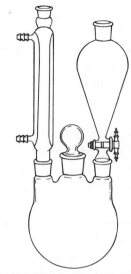

B. Macroscale, for larger amounts.

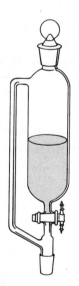

C. A pressure equalizing addition funnel.

D. Addition with a hypodermic
syringe inserted through a
rubber septum.

Figure 7.9 Methods for adding liquid reagents to a reaction.

the atmosphere. If the upper hole is stoppered, a vacuum will develop in the funnel and will prevent the liquid from passing into the reaction vessel. Because the funnel is open to the atmosphere, there is a danger that atmospheric moisture can contaminate the liquid reagent as it is being added. To prevent this outcome, a drying tube (see Section 7.6) may be attached to the upper opening of the addition funnel. The drying tube allows the funnel to maintain atmospheric pressure without allowing the passage of water vapor into the reaction. For reactions that are particularly sensitive to moisture, it is also advisable to attach a second drying tube to the top of the condenser.

Another macroscale assembly, suitable for larger amounts of material, is shown in Figure 7.9B. Drying tubes may also be used with this apparatus to prevent contamination from atmospheric moisture.

Figure 7.9C shows an alternative type of addition funnel that is useful for reactions that must be maintained under an atmosphere of inert gas. This is the **pressure-equalizing addition funnel.** With this glassware, the upper opening is stoppered. The sidearm allows the pressure above the liquid in the funnel to be in equilibrium with the pressure in the rest of the apparatus, and it allows the inert gas to flow over the top of the liquid as it is being added.

With either type of macroscale addition funnel, you can control the rate of addition of the liquid by carefully adjusting the stopcock. Even after careful adjustment, changes in pressure can occur, causing the flow rate to change. In some cases, the stopcock can become clogged. It is important, therefore, to monitor the addition rate carefully and to refine the adjustment of the stopcock as needed to maintain the desired rate of addition.

A fourth method, shown in Figure 7.9D, is suitable for use in microscale and some macroscale experiments where the reaction should be kept isolated from the atmosphere. In this approach, the liquid is kept in a hypodermic syringe. The syringe needle is inserted through a rubber septum, and the liquid is added dropwise from the syringe. The septum seals the apparatus from the atmosphere, which makes this technique useful for reactions that are conducted under an atmosphere of inert gas or where anhydrous conditions must be maintained. The drying tube is used to protect the reaction mixture from atmospheric moisture.

7.6 DRYING TUBES

With certain reactions, atmospheric moisture must be prevented from entering the reaction vessel. A **drying tube** can be used to maintain anhydrous conditions within the apparatus. Two types of drying tubes are shown in Figure 7.10. The typical drying tube is prepared by placing a small, loose plug of glass wool or cotton into the constriction at the end of the tube nearest the ground-glass joint or hose connection. The plug is tamped gently with a glass rod or piece of wire to place it in the correct position. A drying agent, typically calcium sulfate ("Drierite") or calcium chloride (see Chapter 12, Section 12.9, page 178), is poured on top of the plug to the approximate depth shown in Figure 7.10. Another loose plug of glass wool or cotton is placed on

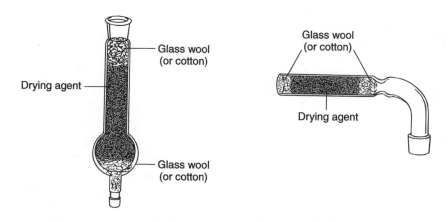

A. Macroscale drying tube.

B. Microscale drying tube.

Figure 7.10 Drying tubes.

top of the drying agent to prevent the solid material from falling out of the drying tube. The drying tube is then attached to the flask or condenser.

Air that enters the apparatus must pass through the drying tube. The drying agent absorbs any moisture from air passing through it, so that air entering the reaction vessel has had the water vapor removed from it.

7.7 REACTIONS CONDUCTED UNDER AN INERT ATMOSPHERE

Some reactions are very sensitive to oxygen and water vapor present in air and require an inert atmosphere in order to obtain satisfactory results. The usual reactions in which it is desirable to exclude air often include organometallic reagents, such as organomagnesium or organolithium reagents, where water vapor and oxygen (air) react with these compounds. The most common inert gases available in a laboratory are nitrogen and argon, which are available in gas cylinders. Nitrogen is probably the gas most often used to carry out reactions under an inert atmosphere, although argon has a distinct advantage because it is more dense than air. This allows the argon to push air away from the reaction mixture.

When laboratories are not equipped with individual gas lines to benches or hoods, it is very useful to supply nitrogen or argon to the reaction apparatus using a balloon assembly (shown in Fig. 7.11). Your instructor will provide you with the apparatus.

Construct the balloon assembly by cutting off the top of a 3-mL disposable plastic syringe. Attach a small balloon snugly to the top of the syringe, securing it with a small rubber band that has been doubled to hold the balloon securely to the body of the syringe. Attach a needle to the syringe. Fill the balloon with the inert gas through the needle using a piece of rubber tubing attached to the gas source. When the

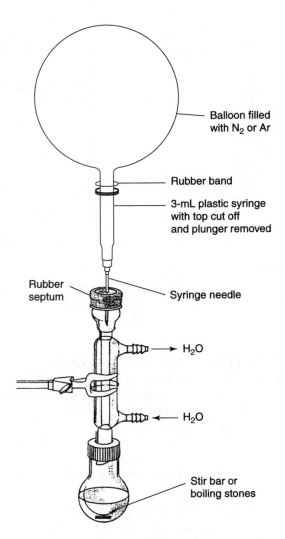

Balloon filled
with N_2 or Ar

Rubber band

3-mL plastic syringe
with top cut off
and plunger removed

Rubber
septum

Syringe needle

H_2O

H_2O

Stir bar or
boiling stones

Figure 7.11 Conducting a reaction under an inert atmosphere using a balloon assembly.

balloon has been inflated to 2–3 inches in diameter, quickly pinch off the neck of the balloon while removing the gas source. Now push the needle into a rubber stopper to keep the balloon inflated. It is possible to keep an assembly like this filled with inert gas for several days without the balloon deflating.

Before you start the reaction, you may need to dry your apparatus thoroughly in an oven. Add all reagents carefully to avoid water. The following instructions are based on the assumption that you are using an apparatus consisting of a round-bottom flask equipped with a condenser. Attach a rubber septum to the top of your condenser. Now flush the air out of the apparatus with the inert gas. It is best not to use the balloon assembly for this purpose, unless you are using argon (see next paragraph). Instead, remove the round-bottom flask from the apparatus, and, with the help of your instructor, flush it with the inert gas using a Pasteur pipet to bubble the gas through the solvent and reaction mixture in the flask. In this way, you can remove air from the reaction assembly prior to attaching the balloon assembly. Quickly reattach the flask to the apparatus. Pinch off the neck of the balloon between your fingers, remove the rubber stopper, and insert the needle into the rubber septum. The reaction apparatus is now ready for use.

When argon is employed as an inert gas, you can use the balloon assembly to remove air from the reaction apparatus in the following way. Insert the balloon assembly into the rubber septum as described above. Also insert a second needle (no syringe attached) through the septum. The pressure from the balloon will force argon down the reflux condenser (argon is more dense than air) and push the less dense air out through the second syringe needle. When the apparatus has been thoroughly flushed with argon, remove the second needle. Nitrogen does not work as well with this method because it is less dense than air and it will be difficult to remove the air that is in contact with the reaction mixture in the round-bottom flask.

For reactions conducted at room temperature, you can remove the condenser shown in Figure 7.11. Attach the rubber septum directly to the round-bottom flask, and insert the needle of an argon-filled balloon assembly through the rubber septum. To flush the air out of the reaction flask, insert a second syringe needle into the rubber septum. Any air present in the flask will be flushed out through this second syringe needle, and the air will be replaced with argon. Now remove the second needle and you have a reaction mixture free of air.

7.8 CAPTURING NOXIOUS GASES

Many organic reactions involve the production of a noxious gaseous product. The gas may be corrosive, such as hydrogen chloride, hydrogen bromide, or sulfur dioxide, or it may be toxic, such as carbon monoxide. The safest way to avoid exposure to these gases is to conduct the reaction in a ventilated hood where the gases can be safely drawn away by the ventilation system.

In many instances, however, it is quite safe and efficient to conduct the experiment on the laboratory bench, away from the hood. This is particularly true when the gases are soluble in water. Some techniques for capturing noxious gases are presented in this section.

A. External Gas Traps

One approach to capturing gases is to prepare a trap that is separate from the reaction apparatus. The gases are carried from the reaction to the trap by means of tubing. There are several variations on this type of trap. With macroscale reactions, a trap using an inverted funnel placed in a beaker of water is used. A piece of glass tubing, inserted through a thermometer adapter attached to the reaction apparatus, is connected to flexible tubing. The tubing is attached to a conical funnel. The funnel is clamped in place inverted over a beaker of water. The funnel is clamped so that its lip *almost touches* the water surface, but is not placed below the surface of the water. With this arrangement, water cannot be sucked back into the reaction if the pressure in the reaction vessel changes suddenly. This type of trap can also be used in microscale applications. An example of the inverted funnel type of gas trap is shown in Figure 7.12.

One method that works well for macroscale and microscale experiments is to place a thermometer adapter into the opening in the reaction apparatus. A Pasteur pipet is inserted upside-down through the adapter, and a piece of flexible tubing is fitted over the narrow tip. It might be helpful to break the Pasteur pipet before using it for this purpose, so that only the narrow tip and a short section of the barrel are used. The other end of the flexible tubing is placed through a large plug of moistened glass wool in a test tube. The water in the glass wool absorbs the water-soluble gases. This method is shown in Figure 7.13.

B. Drying-Tube Method

Some macroscale and most microscale experiments have the advantage that the amounts of gases produced are very small. Hence, it is easy to trap them and prevent them from escaping into the laboratory room. You can take advantage of the water solubility of corrosive gases such as hydrogen chloride, hydrogen bromide, and sulfur dioxide. A simple technique is to attach the drying tube (see Fig. 7.10) to the top of the reaction flask or condenser. The drying tube is filled with moistened glass wool. The moisture in the glass wool absorbs the gas, preventing its escape. To prepare this type of gas trap, fill the drying tube with glass wool and then add water dropwise to the glass

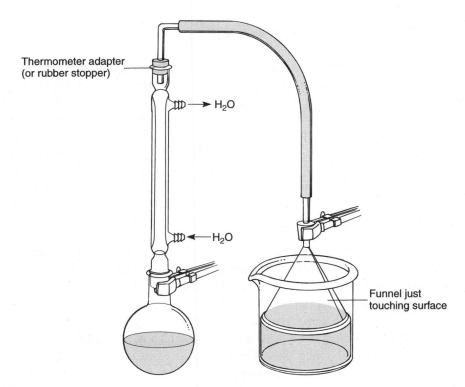

Figure 7.12 An inverted-funnel gas trap.

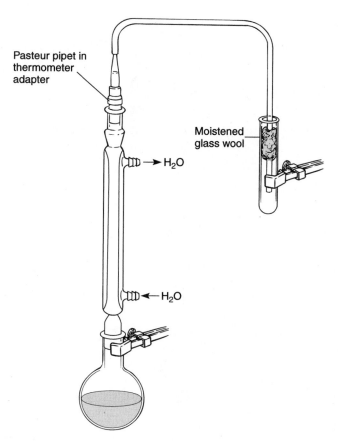

Figure 7.13 An external gas trap.

wool until it has been moistened to the desired degree. Moistened cotton can also be used, although cotton will absorb so much water that it is easy to plug the drying tube.

When using glass wool in a drying tube, moisture from the glass wool must not be allowed to drain from the drying tube into the reaction. It is best to use a drying tube that has a constriction between the part where the glass wool is placed and the neck, where the joint is attached (see Fig. 7.10B). The constriction acts as a partial barrier preventing the water from leaking into the neck of the drying tube. Make certain not to make the glass wool too moist. When it is necessary to use the drying tube shown in Figure 7.10A as a gas trap and it is essential that water not be allowed to enter the reaction flask, the modification shown in Figure 7.14 should be used. The rubber tubing between the thermometer adapter and the drying tube should be heavy enough to prevent crimping.

C. Removal of Noxious Gases Using an Aspirator

An aspirator can be used to remove noxious gases from the reaction. The simplest approach is to clamp a disposable Pasteur pipet so that its tip is placed well into the condenser atop the reaction flask. An inverted funnel clamped over the apparatus can also

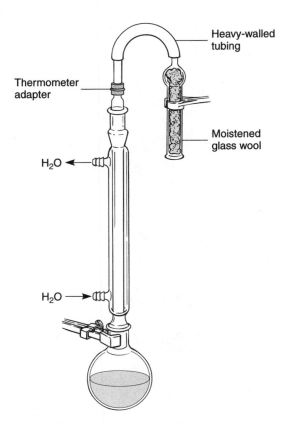

Figure 7.14 A drying tube used to capture evolved gases.

be used. The pipet or funnel is attached to an aspirator with flexible tubing. A trap should be placed between the pipet or funnel and the aspirator. As gases are liberated from the reaction, they rise into the condenser. The vacuum draws the gases away from the apparatus. Both types of systems are shown in Figure 7.15. In the special case in which the noxious gases are soluble in water, connecting a water aspirator to the pipet or funnel removes the gases from the reaction and traps them in the flowing water without the need for a separate gas trap.

7.9   COLLECTING GASEOUS PRODUCTS

In Section 7.8, means for removing unwanted gaseous products from the reaction system were examined. Some experiments produce gaseous products that you must collect and analyze. Methods to collect gaseous products are all based on the same principle. The gas is carried through tubing from the reaction to the opening of a flask or a test tube, which has been filled with water and is inverted in a container of water. The gas is allowed to bubble into the inverted collection tube (or flask). As the collection tube fills with gas, the water is displaced into the water container. If the col-

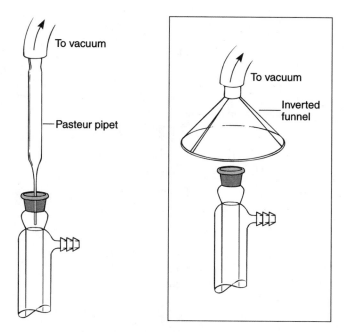

Figure 7.15 Removal of noxious gases under vacuum. (The inset shows an alternate assembly, using an inverted funnel in place of the Pasteur pipet.)

lection tube is graduated, as in a graduated cylinder or a centrifuge tube, you can monitor the quantity of gas produced in the reaction.

If the inverted gas collection tube is constructed from a piece of glass tubing, a rubber septum can be used to close the upper end of the container. This type of collection tube is shown in Figure 7.16. A sample of the gas can be removed using a gastight syringe equipped with a needle. The gas that is removed can be analyzed by gas chromatography (see Chapter 22).

In Figure 7.16, a piece of glass tubing is attached to the free end of the flexible hose. This piece of glass tubing sometimes makes it easier to fix the open end in the proper position in the opening of the collection tube or flask. The other end of the flexible tubing is attached to a piece of glass tubing or a Pasteur pipet that has been inserted into a thermometer adapter.

7.10 EVAPORATION OF SOLVENTS

In many experiments, it is necessary to remove excess solvent from a solution. An obvious approach is to allow the container to stand unstoppered in the hood for several hours until the solvent has evaporated. This method is generally not practical, however, and a quicker, more efficient means of evaporating solvents must be used.

▨▨▨ **CAUTION** You must always evaporate solvents in the hood.

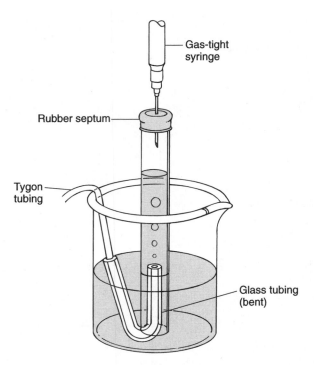

Figure 7.16 A gas collection tube, with rubber septum.

A. Large-Scale Methods

This can be done by evaporating the solvent from an open Erlenmeyer flask (Fig. 7.17A and B). Such an evaporation must be conducted in a hood, since many solvent vapors are toxic or flammable. A boiling stone must be used. A gentle stream of air directed toward the surface of the liquid will remove vapors that are in equilibrium with the solution and accelerate the evaporation. A Pasteur pipet connected by a short piece of rubber tubing to the compressed air line will act as a convenient air nozzle (Fig. 7.17A). A tube or an inverted funnel connected to an aspirator may also be used (Fig. 7.17B). In this case, vapors are removed by suction. It is better to use an Erlenmeyer flask than a beaker for this procedure, since deposits of solid will usually build up on the sides of the beaker where the solvent evaporates. The refluxing action in an Erlenmeyer flask does not allow this buildup. If a hot plate is used as the heat source, care must be taken with flammable solvents to ensure against fires caused by "flashing," when solvent vapors come into contact with the hot plate surface.

It is also possible to remove low-boiling solvents under reduced pressure (Fig. 7.17C). In this method, the solution is placed in a filter flask, along with a wooden applicator stick or a short length of capillary tubing. The flask is stoppered, and the sidearm is connected to an aspirator (by a trap), as described in Chapter 8, Section 8.3, page 107. Under reduced pressure, the solvent begins to boil. The wooden stick or capillary tubing serves the same function as a boiling stone. By this method, solvent

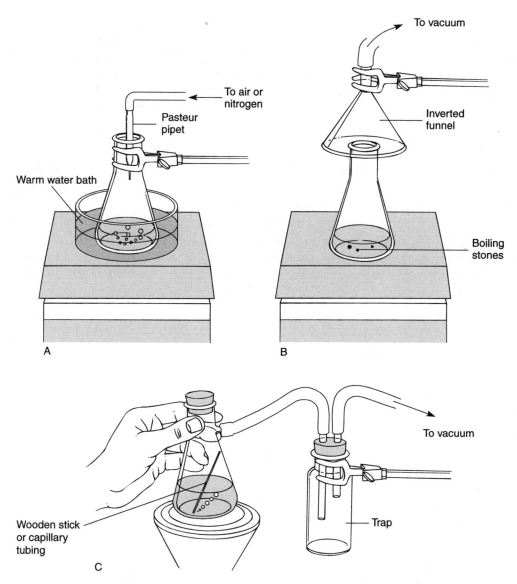

Figure 7.17 Evaporation of solvents (heat source can be varied among those shown).

can be evaporated from a solution without using much heat. This technique is often used when heating the solution might decompose thermally sensitive substances. The method has the disadvantage that when low-boiling solvents are used, solvent evaporation cools the flask below the freezing point of water. When this happens, a layer of frost forms on the outside of the flask. Since frost is insulating, it must be removed to keep evaporation proceeding at a reasonable rate. Frost is best removed by one of two methods: either the flask is placed in a bath of warm water (with constant swirling) or it is heated on the steam bath (again with swirling). Either method promotes efficient heat transfer.

Large amounts of a solvent should be removed by distillation (see Chapter 14). *Never evaporate ether solutions to dryness*, except on a steam bath or by the reduced-pressure method. The tendency of ether to form explosive peroxides is a serious potential hazard. If peroxides should be present, the large and rapid temperature increase in the flask once the ether evaporates could bring about the detonation of any residual peroxides. The temperature of a steam bath is not high enough to cause such a detonation.

B. Small-Scale Methods

A simple means of evaporating a small amount of solvent is to place a centrifuge tube in a warm water bath. The heat from the water bath will warm the solvent to a temperature at which it can evaporate within a short time. The heat from the water can be adjusted to provide the best rate of evaporation, but the liquid should not be allowed to boil vigorously. The evaporation rate can be increased by allowing a stream of dry air or nitrogen to be directed into the centrifuge tube (Fig. 7.18A). The moving gas stream will sweep the vapors from the tube and accelerate the evaporation. As an alternative, a vacuum can be applied above the tube to draw away solvent vapors.

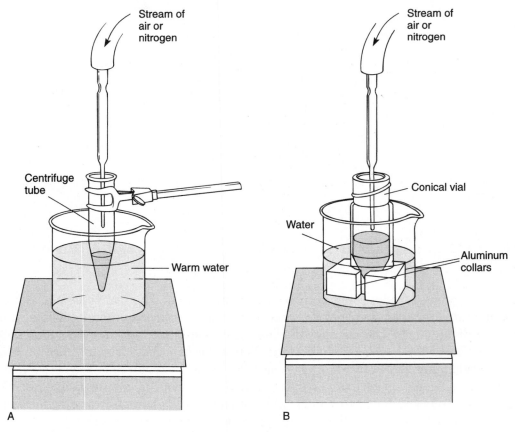

Figure 7.18 Evaporation of solvents (small-scale methods).

A convenient water bath suitable for microscale methods can be constructed by placing the aluminum collars, which are generally used with aluminum heating blocks, into a 150-mL beaker (Fig. 7.18B). In some cases, it may be necessary to round off the sharp edges of the collars with a file in order to allow them to fit properly into the beaker. Held by the aluminum collars, the conical vial will stand securely in the beaker. This assembly can be filled with water and placed on a hot plate for use in the evaporation of small amounts of solvent.

7.11 ROTARY EVAPORATOR

In the research laboratory, solvents are evaporated under reduced pressure using a **rotary evaporator.** This is a motor-driven device, which is designed for rapid evaporation of solvents, with heating, while minimizing the possibility of bumping. A vacuum is applied to the flask, and the motor spins the flask. The rotation of the flask spreads a thin film of the liquid over the surface of the glass. This accelerates evaporation. The rotation also agitates the solution sufficiently to reduce the problem of bumping. A water bath can be placed under the flask to warm the solution and increase the vapor pressure of the solvent. One can select the speed at which the flask is rotated and the temperature of the water bath to attain the desired evaporation rate. A rotary evaporator is shown in Figure 7.19.

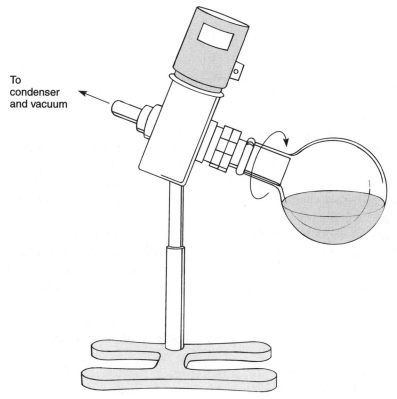

To condenser and vacuum

Figure 7.19 A rotary evaporator.

PROBLEMS

1. What is the best type of stirring device to use for stirring a reaction that takes place in
   (a) a conical vial?
   (b) a 10-mL round-bottom flask?
   (c) a 250-mL round-bottom flask?
2. Should you use a drying tube for the following reaction? Explain.

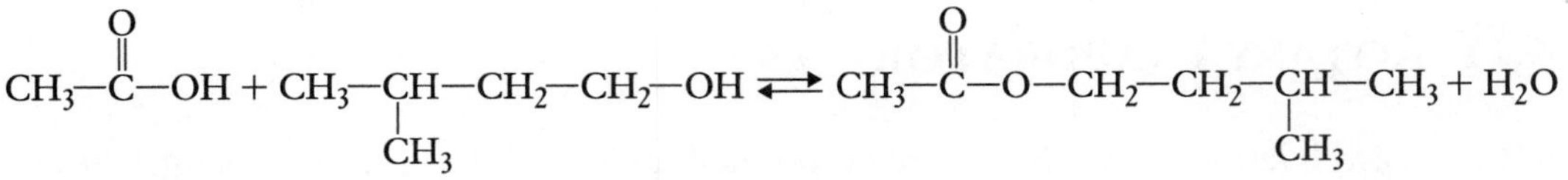

$$CH_3-\overset{\overset{\displaystyle O}{\|}}{C}-OH + CH_3-\underset{\underset{\displaystyle CH_3}{|}}{CH}-CH_2-CH_2-OH \rightleftharpoons CH_3-\overset{\overset{\displaystyle O}{\|}}{C}-O-CH_2-CH_2-\underset{\underset{\displaystyle CH_3}{|}}{CH}-CH_3 + H_2O$$

3. For which of the following reactions should you use a trap to collect noxious gases?

   (a) 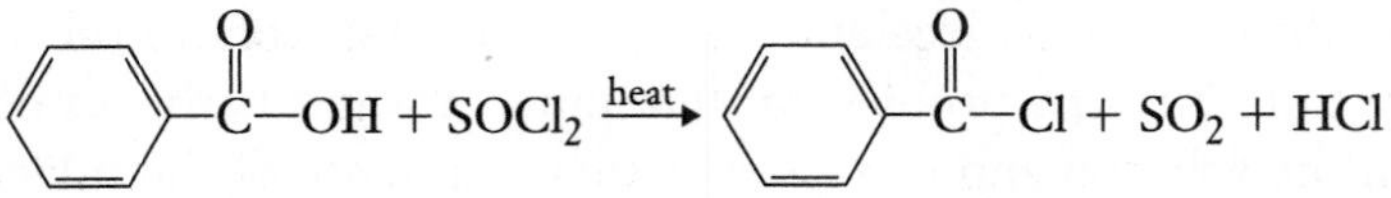
$$\text{C}_6\text{H}_5-\overset{\overset{\displaystyle O}{\|}}{C}-OH + SOCl_2 \xrightarrow{\text{heat}} \text{C}_6\text{H}_5-\overset{\overset{\displaystyle O}{\|}}{C}-Cl + SO_2 + HCl$$

   (b) 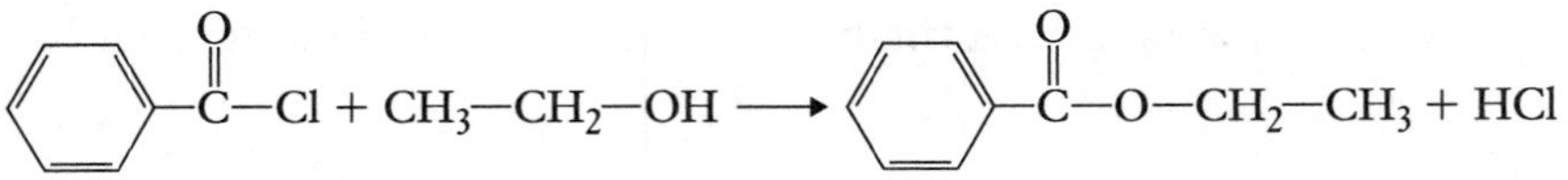
$$\text{C}_6\text{H}_5-\overset{\overset{\displaystyle O}{\|}}{C}-Cl + CH_3-CH_2-OH \longrightarrow \text{C}_6\text{H}_5-\overset{\overset{\displaystyle O}{\|}}{C}-O-CH_2-CH_3 + HCl$$

   (c) $C_{12}H_{22}O_{11} + H_2O \longrightarrow 4\,CH_3-CH_2-OH + 4\,CO_2$
       (Sucrose)

   (d) $CH_3-\underset{\underset{\displaystyle H}{|}}{C}=NH + H_2O \xrightarrow[\text{heat}]{\text{base}} CH_3-\underset{\underset{\displaystyle H}{|}}{C}=O + NH_3$

4. Criticize the following techniques:
   (a) A reflux is conducted with a stopper in the top of the condenser.
   (b) Water is passed through the reflux condenser at the rate of 1 gallon per minute.
   (c) No water hoses are attached to the condenser during a reflux.
   (d) A boiling stone is not added to the round-bottom flask until the mixture is boiling vigorously.
   (e) To save money, you decide to save your boiling stones for another experiment.
   (f) The reflux ring is located near the top of the condenser in a reflux setup.
   (g) A rubber O-ring is omitted when the water condenser is attached to a conical vial.
   (h) A gas trap is assembled with the funnel in Figure 7.12 completely submerged in the water in the beaker.
   (i) Powdered drying agent is used rather than granular material.
   (j) A reaction involving hydrogen chloride is conducted on the laboratory bench and not in a hood.
   (k) An air-sensitive reaction apparatus is set up as shown in Figure 7.6.
   (l) Air is used to evaporate solvent from an air-sensitive compound.

CHAPTER 8

Filtration

Filtration is a technique used for two main purposes. The first is to remove solid impurities from a liquid. The second is to collect a desired solid from the solution from which it was precipitated or crystallized. Several different kinds of filtration are commonly used: two general methods include gravity filtration and vacuum (or suction) filtration. Two techniques specific to the microscale laboratory are filtration with a filter-tip pipet and filtration with a Craig tube. The various filtration techniques and their applications are summarized in Table 8.1. These techniques are discussed in more detail in the following sections.

Table 8.1 Filtration Methods

Method	Application	Section
Gravity Filtration		
Filter cones	The volume of liquid to be filtered is about 10 mL or greater, and the solid collected in the filter is saved.	8.1A
Fluted filters	The volume of liquid to be filtered is greater than about 10 mL, and solid impurities are removed from a solution; often used in crystallization procedures.	8.1B
Filtering pipets	Used with volumes less than about 10 mL to remove solid impurities from a liquid.	8.1C
Vacuum Filtration		
Büchner funnels	Primarily used to collect a desired solid from a liquid when the volume is greater than about 10 mL; used frequently to collect the crystals obtained from crystallization.	8.3
Hirsch funnels	Used in the same way as Büchner funnels, except the volume of liquid is usually smaller (1–10 mL)	8.3
Filtering Media	Used to remove finely divided impurities.	8.4
Filter Tip Pipets	May be used to remove a small amount of solid impurities from a small volume (1–2 mL) of liquid; also useful for pipetting volatile liquids, especially in extraction procedures.	8.6
Craig Tubes	Used to collect a small amount of crystals resulting from crystallizations in which the volume of the solution is less than 2 mL.	8.7
Centrifugation	Although not strictly a filtration technique, centrifugation may be used to remove suspended impurities from a liquid (1–25 mL).	8.8

8.1 GRAVITY FILTRATION

The most familiar filtration technique is probably filtration of a solution through a paper filter held in a funnel, allowing gravity to draw the liquid through the paper. Because even a small piece of filter paper will absorb a significant volume of liquid, this technique is useful only when the volume of mixture to be filtered is greater than 10 mL. For many microscale procedures a more suitable technique, which also makes use of gravity, is to use a Pasteur (or disposable) pipet with a cotton or glass wool plug (called a filtering pipet).

A. Filter Cones

This filtration technique is most useful when the solid material being filtered from a mixture is to be collected and used later. The filter cone, because of its smooth sides, can easily be scraped free of collected solids. Because of the many folds, fluted filter paper, described in the next section, cannot be scraped easily. The filter cone is likely to be used in experiments only when a relatively large volume (greater than 10 mL) is being filtered and when a Büchner or Hirsch funnel (Section 8.3) is not appropriate.

The filter cone is prepared as indicated in Figure 8.1. It is then placed into a funnel of an appropriate size. With filtrations using a simple filter cone, solvent may form seals between the filter and the funnel and between the funnel and the lip of the receiving flask. When a seal forms, the filtration stops because the displaced air has no possibility of escaping. To avoid the solvent seal, you can insert a small piece of paper, a paper clip, or some other bent wire between the funnel and the lip of the flask to let the displaced air escape. As an alternative, you can support the funnel by a clamp fixed *above* the flask, rather than by placing it on the neck of the flask. A gravity filtration using a filter cone is shown in Figure 8.2.

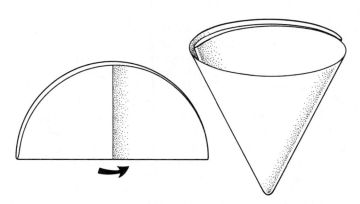

Figure 8.1 Folding a filter cone.

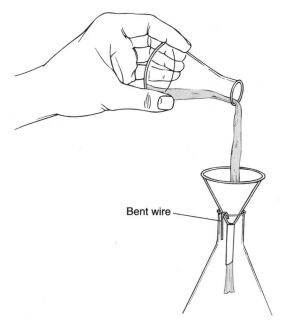

Bent wire

Figure 8.2 Gravity filtration with a filter cone.

B. Fluted Filters

This filtration method is also most useful when filtering a relatively large amount of liquid. Because a fluted filter is used when the desired material is expected to remain in solution, this filter is used to remove undesired solid materials, such as dirt particles, decolorizing charcoal, and undissolved impure crystals. A fluted filter is often used to filter a hot solution saturated with a solute during a crystallization procedure.

The technique for folding a fluted filter paper is shown in Figure 8.3. An advantage of a fluted filter is that it increases the speed of filtration, which occurs for two reasons. First, it increases the surface area of the filter paper through which the solvent seeps; second, it allows air to enter the flask along its sides to permit rapid pressure equalization. If pressure builds up in the flask from hot vapors, filtering slows down. This problem is especially pronounced with filter cones. The fluted filter tends to reduce this problem considerably, but it may be a good idea to clamp the funnel above the receiving flask or to use a piece of paper, paper clip, or wire between the funnel and the lip of the flask as an added precaution against solvent seals.

Filtration with a fluted filter is relatively easy to perform when the mixture is at room temperature. However, when it is necessary to filter a hot solution saturated with a dissolved solute, a number of steps must be taken to ensure that the filter does not become clogged by solid material accumulated in the stem of the funnel or in the

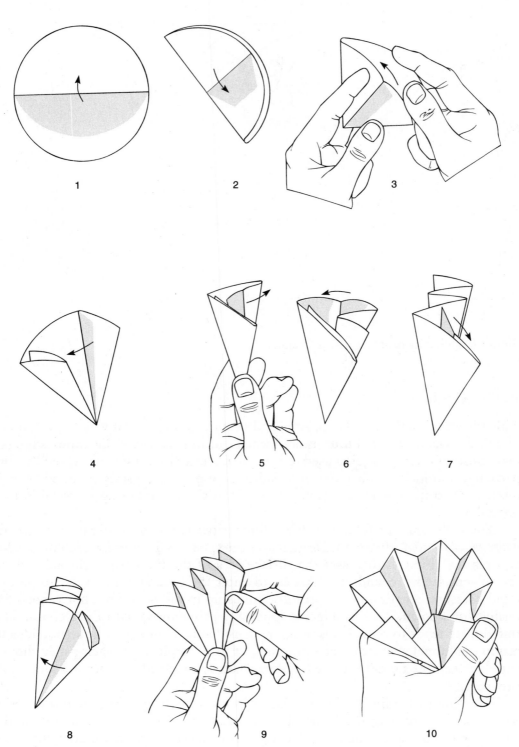

Figure 8.3 Folding a fluted filter paper, or origami at work in the organic chemistry laboratory.

filter paper. When the hot saturated solution comes in contact with a relatively cold funnel (or a cold flask, for that matter), the solution is cooled and may become super-saturated. If crystallization then occurs in the filter, either the crystals will fail to pass through the filter paper or they will clog the stem of the funnel.

To keep the filter from clogging, use one of the following four methods. The first is to use a short-stemmed or a stemless funnel. With these funnels, it is less likely that the stem of the funnel will become clogged by solid material. The second method is to keep the liquid to be filtered at or near its boiling point at all times. The third way is to preheat the funnel by pouring hot solvent through it before the actual filtration. This keeps the cold glass from causing instantaneous crystallization. And fourth, it is helpful to keep the **filtrate** (filtered solution) in the receiver hot enough to continue boiling *slightly* (by setting it on a hot plate, for example). The refluxing solvent heats the receiving flask and the funnel stem and washes them clean of solids. This boiling of the filtrate also keeps the liquid in the funnel warm.

C. Filtering Pipets

A filtering pipet is a microscale technique most often used to remove solid impurities from a liquid with a volume less than 10 mL. It is important that the mixture being filtered be at or near room temperature because it is difficult to prevent premature crystallization in a hot solution saturated with a solute.

To prepare this filtration device, a small piece of cotton is inserted into the top of a Pasteur (disposable) pipet and pushed down to the beginning of the lower constriction in the pipet, as shown in Figure 8.4. It is important that enough cotton is used to collect all the solid being filtered; however, the amount of cotton used should not be so large that the flow rate through the pipet is significantly restricted. For the same reason, the cotton should not be packed too tightly. The cotton plug can be pushed down gently with a long thin object such as a glass stirring rod or a wooden applicator stick. It is advisable to wash the cotton plug by passing about 1 mL of solvent (usually the same solvent that is to be filtered) through the filter.

In some cases, such as when filtering a strongly acidic mixture or when performing a very rapid filtration to remove dirt or impurities of large particle size from a solution, it may be better to use glass wool in place of the cotton. The disadvantage in using glass wool is that the fibers do not pack together as tightly, and small particles will pass through the filter more easily.

To conduct a filtration (with either a cotton or glass wool plug), the filtering pipet is clamped so that the filtrate will drain into an appropriate container. The mixture to be filtered is usually transferred to the filtering pipet with another Pasteur pipet. If a small volume of liquid is being filtered (less than 1 or 2 mL), it is advisable to rinse the filter and plug with a small amount of solvent after the last of the filtrate has passed through the filter. The rinse solvent is then combined with the original filtrate. If desired, the rate of filtration can be increased by gently applying pressure to the top of the pipet using a pipet bulb.

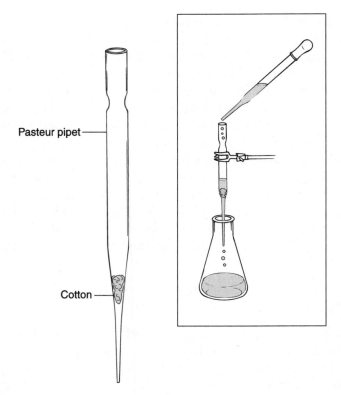

Pasteur pipet

Cotton

Figure 8.4 A filtering pipet.

Depending on the amount of solid being filtered and the size of the particles (small particles are more difficult to remove by filtration), it may be necessary to put the filtrate through a second filtering pipet. This should be done with a new filtering pipet rather than the one already used.

8.2 FILTER PAPER

Many kinds and grades of filter paper are available. The paper must be correct for a given application. In choosing filter paper, you should be aware of its various properties. **Porosity** is a measure of the size of the particles that can pass through the paper. Highly porous paper does not remove small particles from solution; paper with low porosity removes very small particles. **Retentivity** is a property that is the opposite of porosity. Paper with low retentivity does not remove small particles from the filtrate. The **speed** of filter paper is a measure of the time it takes a liquid to drain through the filter. Fast paper allows the liquid to drain quickly; with slow paper, it takes much longer to complete the filtration. Because all these properties are related, fast filter paper usually has a low retentivity and high porosity, and slow filter paper usually has high retentivity and low porosity.

Table 8.2 Some Common Qualitative Filter Paper Types and
Approximate Relative Speeds and Retentivities

Fine	High	Slow				
					Type (by number)	
Porosity ↓	Retentivity ↓	Speed ↓	Speed	E&D	S&S	Whatman
			Very slow	610	576	5
			Slow	613	602	3
			Medium	615	597	2
			Fast	617	595	1
			Very fast	—	604	4
Coarse	Low	Fast				

Table 8.2 compares some commonly available qualitative filter paper types and ranks them according to porosity, retentivity, and speed. Eaton-Dikeman (E&D), Schleicher and Schuell (S&S), and Whatman are the most common brands of filter paper. The numbers in the table refer to the grades of paper used by each company.

8.3 VACUUM FILTRATION

Vacuum, or suction, filtration is more rapid than gravity filtration and is most often used to collect solid products resulting from precipitation or crystallization. This technique is used primarily when the volume of liquid being filtered is more than 1–2 mL. With smaller volumes, use of the Craig tube (Section 8.7) is the preferred technique. In a vacuum filtration, a receiver flask with a sidearm, a **filter flask,** is used. For macroscale laboratory work, the most useful sizes of filter flasks range from 50 mL to 500 mL, depending on the volume of liquid being filtered. For microscale work, the most useful size is a 50-mL filter flask. The sidearm is connected by *heavy-walled* rubber tubing (see Chapter 16, Figure 16.2, page 246) to a source of vacuum. Thin-walled tubing will collapse under vacuum, due to atmospheric pressure on its outside walls, and will seal the vacuum source from the flask. Because this apparatus is unstable and can tip over easily, it must be clamped, as shown in Figure 8.5.

CAUTION It is essential that the filter flask be clamped.

Two types of funnels are useful for vacuum filtration, the Büchner funnel and the Hirsch funnel. The **Büchner funnel** is used for filtering larger amounts of solid from solution in macroscale applications. Büchner funnels are usually made from

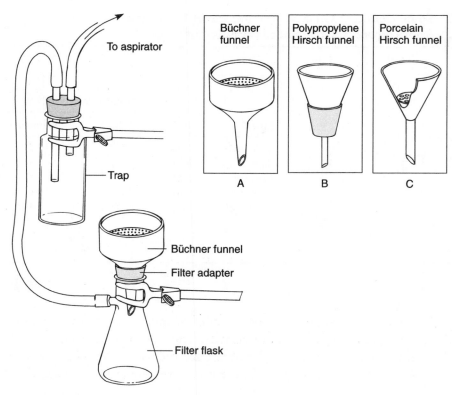

Figure 8.5 Vacuum filtration.

polypropylene or porcelain. A Büchner funnel (see Fig. 8.5) is sealed to the filter flask by a rubber stopper or a filter (Neoprene) adapter. The flat bottom of the Büchner funnel is covered with an unfolded piece of circular filter paper. To prevent the escape of solid materials from the funnel, you must be certain that the filter paper fits the funnel exactly. It must cover all the holes in the bottom of the funnel but not extend up the sides. Before beginning the filtration, it is advisable to moisten the paper with a small amount of solvent. The moistened filter paper adheres more strongly to the bottom of the funnel and prevents unfiltered mixture from passing around the edges of the filter paper.

The **Hirsch funnel,** which is shown in Figure 8.5B and C, operates on the same principle as the Büchner funnel, but it is usually smaller and its sides are sloped rather than vertical. The Hirsch funnel is used primarily in microscale experiments. The polypropylene Hirsch funnel (see Fig. 8.5B) is sealed to a 50-mL filter flask by a small section of Gooch tubing or a one-hole rubber stopper. This Hirsch funnel has a built-in adapter that forms a tight seal with some 25-mL filter flasks without the Gooch tubing. A polyethylene fritted disk fits into the bottom of the funnel. To prevent the holes in this disk from becoming clogged with solid material, the funnel should always be used with a circular filter paper that has the same diameter (1.27 cm) as the polyethylene disk. With a polypropylene Hirsch funnel, it is also important to moisten the paper with a small amount of solvent before beginning the filtration.

The porcelain Hirsch funnel is sealed to the filter flask with a rubber stopper or a Neoprene adapter. In this Hirsch funnel, the filter paper must also cover all the holes in the bottom but must not extend up the sides.

Because the filter flask is attached to a source of vacuum, a solution poured into a Büchner funnel or Hirsch funnel is literally "sucked" rapidly through the filter paper. For this reason, vacuum filtration is generally not used to separate fine particles such as decolorizing charcoal, since the small particles would likely be pulled through the filter paper. However, this problem can be alleviated when desired by the use of specially prepared filter beds (see Section 8.4).

8.4 FILTERING MEDIA

It is occasionally necessary to use specially prepared filter beds to separate fine particles when using vacuum filtration. Often, very fine particles either pass right through a paper filter or they clog it so completely that the filtering stops. This is avoided by using a substance called Filter Aid, or Celite. This material is also called **diatomaceous earth** because of its source. It is a finely divided inert material derived from the microscopic shells of dead diatoms (a type of phytoplankton that grows in the sea).

CAUTION Diatomaceous earth is a lung irritant. When using Filter Aid, take care not to breathe the dust.

Filter Aid will not clog the fiber pores of filter paper. It is **slurried,** mixed with a solvent to form a rather thin paste, and filtered through a Hirsch or Büchner funnel (with filter paper in place) until a layer of diatoms about 2–3 mm thick is formed on top of the filter paper. The solvent in which the diatoms were slurried is poured from the filter flask, and, if necessary, the filter flask is cleaned before the actual filtration is begun. Finely divided particles can now be suction filtered through this layer and will be caught in the Filter Aid. This technique is used for removing impurities, not for collecting a product. The filtrate (filtered solution) is the desired material in this procedure. If the material caught in the filter were the desired material, you would have to try to separate the product from all those diatoms! Filtration with Filter Aid is not appropriate when the desired substance is likely to precipitate or crystallize from solution.

In microscale work, it may sometimes be more convenient to use a column prepared with a Pasteur pipet to separate fine particles from a solution. The Pasteur pipet is packed with alumina or silica gel, as shown in Figure 8.6.

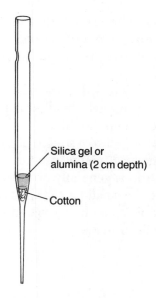

Figure 8.6 A Pasteur pipet with filtering media.

8.5 The Aspirator

The most common source of vacuum (approximately 10–20 mm Hg) in the laboratory is the water aspirator, or "water pump," illustrated in Figure 8.7. This device passes water rapidly past a small hole to which a sidearm is attached. The water pulls air in

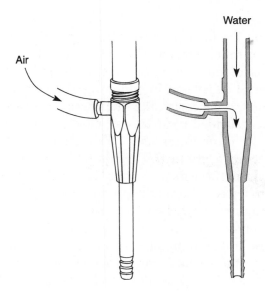

Figure 8.7 An aspirator.

through the sidearm. This phenomenon, called the Bernoulli effect, causes a reduced pressure along the side of the rapidly moving water stream and creates a partial vacuum in the sidearm.

> **NOTE:** The aspirator works most effectively when the water is turned on to the fullest extent.

A water aspirator can never lower the pressure beyond the vapor pressure of the water used to create the vacuum. Hence, there is a lower limit to the pressure (on cold days) of 9–10 mm Hg. A water aspirator does not provide as high a vacuum in the summer as in the winter, due to this water-temperature effect.

A trap must be used with an aspirator. One type of trap is illustrated in Figure 8.5. Another method for securing this type of trap is shown in Figure 8.8. This simple holder can be constructed from readily available material and can be placed anywhere on the laboratory bench. Although not often needed, a trap can prevent water from contaminating your experiment. If the water pressure in the laboratory drops suddenly, the pressure in the filter flask may suddenly become lower than the pressure in the water aspirator. This would cause water to be drawn from the aspirator stream into the filter flask and contaminate the filtrate or even the material in the filter. The trap stops this reverse flow. A similar flow will occur if the water flow at the aspirator is stopped before the tubing connected to the aspirator sidearm is disconnected.

> **NOTE:** Always disconnect the tubing before stopping the aspirator.

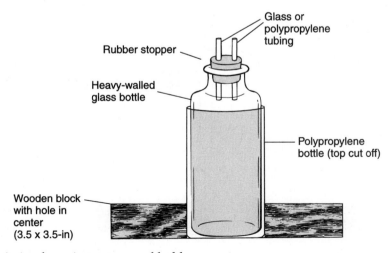

Figure 8.8 A simple aspirator trap and holder.

If a "backup" begins, disconnect the tubing as rapidly as possible before the trap fills with water. Some chemists like to fit a stopcock into the stopper on top of the trap. A three-hole stopper is required for this purpose. With a stopcock in the trap, the system can be vented before the aspirator is shut off. Then, water cannot back up into the trap.

Aspirators do not work well if too many people use the water line at the same time because the water pressure is lowered. Also, the sinks at the ends of the lab benches or the lines that carry away the water flow may have a limited capacity for draining the resultant water flow from too many aspirators. Care must be taken to avoid floods.

8.6 FILTER-TIP PIPET

The filter-tip pipet, illustrated in Figure 8.9, has two common uses. The first is to remove a small amount of solid, such as dirt or filter paper fibers, from a small volume of liquid (1–2 mL). It can also be helpful when using a Pasteur pipet to transfer a highly volatile liquid, especially during an extraction procedure (see Chapter 12, Section 12.5, page 171).

Preparing a filter-tip pipet is similar to preparing a filtering pipet, except that a much smaller amount of cotton is used. A *very tiny* piece of cotton is loosely shaped into a ball and placed into the large end of a Pasteur pipet. Using a wire with a diameter slightly smaller than the inside diameter of the narrow end of the pipet, the ball of cotton is pushed to the bottom of the pipet. If it becomes difficult to push the cotton, you have probably started with too much cotton; if the cotton slides through the narrow end with little resistance, you probably have not used enough.

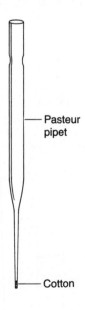

Figure 8.9 A filter-tip pipet.

To use a filter-tip pipet as a filter, the mixture is drawn up into the Pasteur pipet using a pipet bulb and then expelled. With this procedure, a small amount of solid will be captured by the cotton. However, very fine particles, such as activated charcoal, cannot be removed efficiently with a filter-tip pipet, and this technique is not effective in removing more than a trace amount of solid from a liquid.

Transferring many organic liquids with a Pasteur pipet can be a somewhat difficult procedure for two reasons. First, the liquid may not adhere well to the glass. Second, as you handle the Pasteur pipet, the temperature of the liquid in the pipet increases slightly, and the increased vapor pressure may tend to "squirt" the liquid out the end of the pipet. This problem can be particularly troublesome when separating two liquids during an extraction procedure. The purpose of the cotton plug in this situation is to slow the rate of flow through the end of the pipet so you can control the movement of liquid in the Pasteur pipet more easily.

8.7 CRAIG TUBES

The **Craig tube,** illustrated in Figure 8.10, is used primarily to separate crystals from a solution after a microscale crystallization procedure has been performed (Chapter 11, Section 11.4, page 150). Although it may not be a filtration procedure in the traditional sense, the outcome is similar. The outer part of the Craig tube is similar to a test tube, except that the diameter of the tube becomes wider part of the way up the tube, and the glass is ground at this point so that the inside surface is rough. The inner part (plug) of the Craig tube may be made of Teflon or glass. If this part is glass, the end of the plug is also ground. With either a glass or a Teflon inner plug, there is only a partial seal where the plug and the outer tube come together. Liquid may pass through, but solid will not. This is where the solution is separated from the crystals.

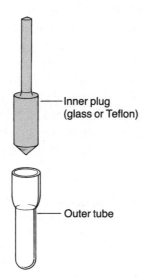

Inner plug
(glass or Teflon)

Outer tube

Figure 8.10 A Craig tube (2 mL).

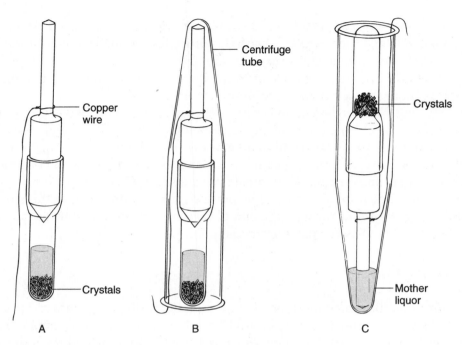

Figure 8.11 Separation with a Craig tube.

After crystallization has been completed in the outer Craig tube, replace the inner plug (if necessary) and connect a thin copper wire or strong thread to the narrow part of the inner plug, as indicated in Figure 8.11A. While holding the Craig tube in an upright position, a plastic centrifuge tube is placed over the Craig tube so that the bottom of the centrifuge tube rests on top of the inner plug, as shown in Figure 8.11B. The copper wire should extend just below the lip of the centrifuge tube and is now bent upward around the lip of the centrifuge tube. This apparatus is then turned over so that the centrifuge tube is in an upright position. The Craig tube is spun in a centrifuge (be sure it is balanced by placing another tube filled with water on the opposite side of the centrifuge) for several minutes until the **mother liquor** (solution from which the crystals grew) goes to the bottom of the centrifuge tube and the crystals collect on the end of the inner plug (see Figure 8.11C). Depending on the consistency of the crystals and the speed of the centrifuge, the crystals may spin down to the inner plug, or (if you are unlucky) they may remain at the other end of the Craig tube.[1] If

---

[1]**Note to the Instructor:** In some centrifuges, the bottom of the Craig tube may be very close to the center of the centrifuge when the Craig tube assembly is placed into the centrifuge. In this situation very little centrifugal force will be applied to the crystals and it likely that the crystals will not spin down. It may then be helpful to use an inner plug with a shorter stem. The stem on a Teflon inner plug can be easily cut off about 0.5 inch with a pair of wire cutters. Not only will this help to spin down the crystals to the inner plug, but also the centrifuge can be run at a lower speed, which can help prevent breakage of the Craig tube.

the latter situation occurs, it may be helpful to centrifuge the Craig tube longer, or, if this problem is anticipated, to stir the crystal and solution mixture with a spatula or stirring rod before centrifugation.

Using the copper wire, the Craig tube is then pulled out of the centrifuge tube. If the crystals collected on the end of the inner plug, it is now a simple procedure to remove the plug and scrape the crystals with a spatula onto a watch glass, a clay plate, or a piece of smooth paper. Otherwise, it will be necessary to scrape the crystals from the inside surface of the outer part of the Craig tube.

8.8 CENTRIFUGATION

Sometimes centrifugation is more effective in removing solid impurities than conventional filtration techniques. Centrifugation is particularly effective in removing suspended particles, which are so small that the particles would pass through most filtering devices. Another situation in which centrifugation may be useful is when the mixture must be kept hot to prevent premature crystallization while the solid impurities are removed.

Centrifugation is performed by placing the mixture in one or two centrifuge tubes (be sure to balance the centrifuge) and centrifuging for several minutes. The supernatant liquid is then decanted (poured off) or removed with a Pasteur pipet.

PROBLEM

1. In each of the following situations, what type of filtration device would you use?
   (a) Remove powdered decolorizing charcoal from 20 mL of solution.
   (b) Collect crystals obtained from crystallizing a substance from about 1 mL of solution.
   (c) Remove a very small amount of dirt from 1 mL of liquid.
   (d) Isolate 0.2 g of crystals from about 5 mL of solution after performing a crystallization.
   (e) Remove dissolved colored impurities from about 3 mL of solution.
   (f) Remove solid impurities from 5 mL of liquid at room temperature.

CHAPTER 9

Physical Constants of Solids: The Melting Point

9.1 PHYSICAL PROPERTIES

The physical properties of a compound are those properties that are intrinsic to a given compound when it is pure. A compound may often be identified simply by determining a number of its physical properties. The most commonly recognized physical properties of a compound include its color, melting point, boiling point, density, refractive index, molecular weight, and optical rotation. Modern chemists would include the various types of spectra (infrared, nuclear magnetic resonance, mass, and ultraviolet-visible) among the physical properties of a compound. A compound's spectra do not vary from one pure sample to another. In this chapter, we look at methods of determining the melting point. Boiling point and density of compounds are covered in Chapter 13. Refractive index, optical rotation, and spectra are considered separately in their own chapters.

Many reference books list the physical properties of substances. You should consult Chapter 4 for a complete discussion on how to find data for specific compounds. The most useful works for finding lists of values for the nonspectroscopic physical properties include

The Merck Index
The CRC Handbook of Chemistry and Physics
Lange's Handbook of Chemistry
Aldrich Handbook of Fine Chemicals

Complete citations for these references may be found in Chapter 30 (Guide to the Chemical Literature, page 507). Although the *CRC Handbook* has very good tables, it adheres strictly to IUPAC nomenclature. For this reason, it may be easier to use one of the other references, particularly *The Merck Index* or the *Aldrich Handbook of Fine Chemicals*, in your first attempt to locate information (see Chapter 4).

9.2 THE MELTING POINT

The melting point of a compound is used by the organic chemist not only to identify it, but also to establish its purity. A small amount of material is heated *slowly* in a special apparatus equipped with a thermometer or thermocouple, a heating bath or heating coil, and a magnifying eyepiece for observing the sample. Two temperatures are noted. The first is the point at which the first drop of liquid forms among the crystals; the second is the point at which the whole mass of crystals turns to a *clear* liquid. The melting point is recorded by giving this range of melting. You might say, for example, that the melting point of a substance is 51–54°C. That is, the substance melted over a 3-degree range.

The melting point indicates purity in two ways. First, the purer the material, the higher its melting point. Second, the purer the material, the narrower its melting point range. Adding successive amounts of an impurity to a pure substance generally causes its melting point to decrease in proportion to the amount of impurity. Looking at it another way, adding impurities lowers the freezing point. The freezing point, a colligative property, is simply the melting point (solid → liquid) approached from the opposite direction (liquid → solid).

Figure 9.1 is a graph of the usual melting-point behavior of mixtures of two substances, A and B. The two extremes of the melting range (the low and high temperature) are shown for various mixtures of the two. The upper curves indicate the temperatures at which all the sample has melted. The lower curves indicate the temperature at which melting is observed to begin. With pure compounds, melting is sharp and without any range. This is shown at the left- and right-hand edges of the graph. If you begin with pure A, the melting point decreases as impurity B is added. At some point, a minimum temperature, or **eutectic,** is reached, and the melting point begins to increase to that of substance B. The vertical distance between the lower and upper curves represents the melting range. Notice that for mixtures that contain relatively small amounts of impurity (<15%) and are not close to the eutectic, the melting range increases as the sample becomes less pure. The range indicated by the lines in Figure 9.1 represents the typical behavior.

We can generalize the behavior shown in Figure 9.1. Pure substances melt with a narrow range of melting. With impure substances, the melting range becomes wider, and the entire melting range is lowered. Be careful to note, however, that at the minimum point of the melting-point–composition curves, the mixture often forms a

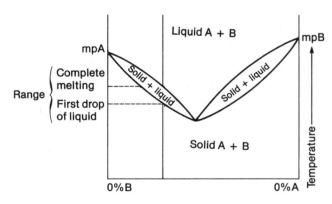

Figure 9.1 A melting-point–composition curve.

eutectic, which also melts sharply. Not all binary mixtures form eutectics, and some caution must be exercised in assuming that every binary mixture follows the previously described behavior. Some mixtures may form more than one eutectic, others might not form even one. In spite of these variations, both the melting point and its range are useful indications of purity, and they are easily determined by simple experimental methods.

9.3   MELTING-POINT THEORY

Figure 9.2 is a phase diagram describing the usual behavior of a two-component mixture (A + B) on melting. The behavior on melting depends on the relative amounts of A and B in the mixture. If A is a pure substance (no B), then A melts sharply at its melting point t_A. This is represented by point A on the left side of the diagram. When B is a pure substance, it melts at t_B; its melting point is represented by point B on the right side of the diagram. At either point A or point B, the pure solid passes cleanly, with a narrow range, from solid to liquid.

In mixtures of A and B, the behavior is different. Using Figure 9.2, consider a mixture of 80% A and 20% B on a mole-per-mole basis (that is, mole percentage). The melting point of this mixture is given by t_M at point M on the diagram. That is, adding B to A has lowered the melting point of A from t_A to t_M. It has also expanded the melting range. The temperature t_M corresponds to the **upper limit** of the melting range.

Lowering the melting point of A by adding impurity B comes about in the following way. Substance A has the lower melting point in the phase diagram shown, and if heated, it begins to melt first. As A begins to melt, solid B begins to dissolve in the liquid A that is formed. When solid B dissolves in liquid A, the melting point is depressed. To understand this, consider the melting point from the opposite direction. When a liquid at a high temperature cools, it reaches a point at which it solidi-

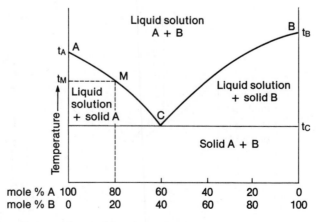

Figure 9.2 A phase diagram for melting in a two-component system.

fies, or "freezes." The temperature at which a liquid freezes is identical to its melting point. Recall that the freezing point of a liquid can be lowered by adding an impurity. Because the freezing point and the melting point are identical, lowering the freezing point corresponds to lowering the melting point. Therefore, as more impurity is added to a solid, its melting point becomes lower. There is, however, a limit to how far the melting point can be depressed. You cannot dissolve an infinite amount of the impurity substance in the liquid. At some point, the liquid will become saturated with the impurity substance. The solubility of B in A has an upper limit. In Figure 9.2, the solubility limit of B in liquid A is reached at point C, the **eutectic point.** The melting point of the mixture cannot be lowered below t_C, the melting temperature of the eutectic.

Now consider what happens when the melting point of a mixture of 80% A and 20% B is approached. As the temperature is increased, A begins to "melt." This is not really a visible phenomenon in the beginning stages; it happens before liquid is visible. It is a softening of the compound to a point at which it can begin to mix with the impurity. As A begins to soften, it dissolves B. As it dissolves B, the melting point is lowered. The lowering continues until all B is dissolved or until the eutectic composition (saturation) is reached. When the maximum possible amount of B has been dissolved, actual melting begins, and one can observe the first appearance of liquid. The initial temperature of melting will be below t_A. The amount below t_A at which melting begins is determined by the amount of B dissolved in A, but will never be below t_C. Once all B has been dissolved, the melting point of the mixture begins to rise as more A begins to melt. As more A melts, the semisolid solution is diluted by more A, and its melting point rises. While all this is happening, you can observe *both* solid and liquid in the melting-point capillary. Once all A has begun to melt, the composition of the mixture M becomes uniform and will reach 80% A and 20% B. At this point, the mixture finally melts sharply, giving a clear solution. The maximum melting point range will be $t_C - t_M$, because t_A is depressed by the impurity B that is present. The lower end of the melting range will always be t_C; however, melting will not always be observed at this temperature. An observable melting at t_C comes about only when a large amount of B is present. Otherwise, the amount of liquid formed at t_C will be too small to observe. Therefore, the melting behavior that is actually observed will have a smaller range, as shown in Figure 9.1.

9.4 MIXTURE MELTING POINTS

The melting point can be used as supporting evidence in identifying a compound in two different ways. Not only may the melting points of the two individual compounds be compared, but a special procedure called a **mixture melting point** may also be performed. The mixture melting point requires that an authentic sample of the same compound be available from another source. In this procedure, the two compounds (authentic and suspected) are finely pulverized and mixed together in equal quantities. Then the melting point of the mixture is determined. If there is a melting-point depression, or if the range of melting is expanded by a large amount, compared to the individual substances, you may conclude that one compound has acted as an impurity toward the other and that they are not the same compound. If there is no lowering of the melting point for the mixture (the melting point is identical with those of pure A and pure B), then A and B are almost certainly the same compound.

9.5   PACKING THE MELTING-POINT TUBE

Melting points are usually determined by heating the sample in a piece of thin-walled capillary tubing (1 mm × 100 mm) that has been sealed at one end. To pack the tube, press the open end gently into a *pulverized* sample of the crystalline material. Crystals will stick in the open end of the tube. The amount of solid pressed into the tube should correspond to a column no more than 1–2 mm high. To transfer the crystals to the closed end of the tube, drop the capillary tube, closed end first, down a ⅔-m length of glass tubing, which is held upright on the desk top. When the capillary tube hits the desk top, the crystals will pack down into the bottom of the tube. This procedure is repeated if necessary. Tapping the capillary on the desk top with fingers is not recommended, because it is easy to drive the small tubing into a finger if the tubing should break.

Some commercial melting-point instruments have a built-in vibrating device that is designed to pack capillary tubes. With these instruments, the sample is pressed into the open end of the capillary tube, and the tube is placed in the vibrator slot. The action of the vibrator will transfer the sample to the bottom of the tube and pack it tightly.

9.6   DETERMINING THE MELTING POINT—THE THIELE TUBE

There are two principal types of melting-point apparatus available: the Thiele tube and commercially available, electrically heated instruments. The Thiele tube, shown in Figure 9.3, is the simpler device and was once widely used. It is a glass tube designed to contain a heating oil (mineral oil or silicone oil) and a thermometer to which a capillary tube containing the sample is attached. The shape of the Thiele tube allows convection currents to form in the oil when it is heated. These currents maintain a uniform temperature distribution through the oil in the tube. The sidearm of the tube is designed to generate these convection currents and thus transfer the heat from the flame evenly and rapidly throughout the oil. The sample, which is in a capillary tube attached to the thermometer, is held by a rubber band or a thin slice of rubber tubing. It is important that this rubber band be above the level of the oil (allowing for expansion of the oil on heating), so that the oil does not soften the rubber and allow the capillary tubing to fall into the oil. If a cork or a rubber stopper is used to hold the thermometer, a triangular wedge should be sliced in it to allow pressure equalization.

The Thiele tube is usually heated by a microburner. During the heating, the rate of temperature increase should be regulated. Hold the burner by its cool base, and,

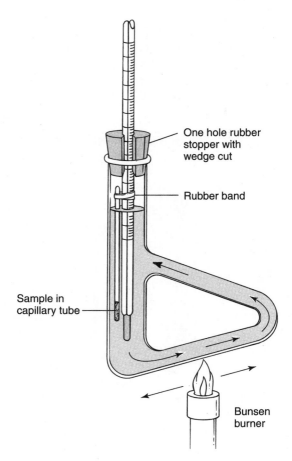

One hole rubber stopper with wedge cut

Rubber band

Sample in capillary tube

Bunsen burner

Figure 9.3 A Thiele tube.

using a low flame, move the burner slowly back and forth along the bottom of the arm of the Thiele tube. If the heating is too fast, remove the burner for a few seconds, and then resume heating. The rate of heating should be *slow* near the melting point (about 1°C per minute) to ensure that the temperature increase is not faster than the rate at which heat can be transferred to the sample being observed. At the melting point, it is necessary that the mercury in the thermometer and the sample in the capillary tube be at temperature equilibrium.

9.7 DETERMINING THE MELTING POINT— ELECTRICAL INSTRUMENTS

Three types of electrically heated melting-point instruments are illustrated in Figure 9.4. In each case, the melting-point tube is filled as described in Section 9.5 and placed in a holder located just behind the magnifying eyepiece. The apparatus is operated by

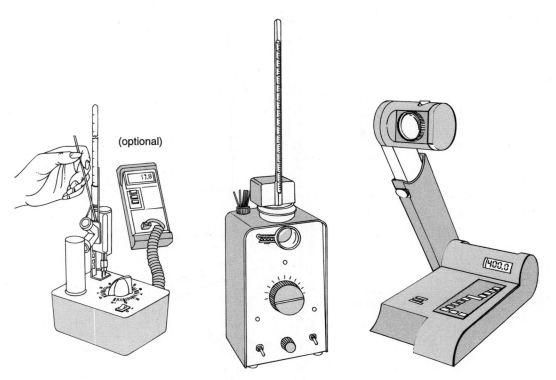

Figure 9.4 Melting-point apparatus.

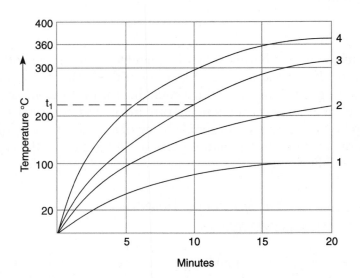

Figure 9.5 Heating-rate curves.

moving the switch to the ON position, adjusting the potentiometric control dial for the desired rate of heating, and observing the sample through the magnifying eyepiece. The temperature is read from a thermometer, or in the most modern instruments, from a digital display attached to a thermocouple. Your instructor will demonstrate and explain the type used in your laboratory.

Most electrically heated instruments do not heat or increase the temperature of the sample linearly. Although the rate of increase may be linear in the early stages of heating, it usually decreases and leads to a constant temperature at some upper limit. The upper-limit temperature is determined by the setting of the heating control. Thus, a family of heating curves is usually obtained for various control settings, as shown in Figure 9.5. The four hypothetical curves shown (1–4) might correspond to different control settings. For a compound melting at temperature t_1, the setting corresponding to curve 3 would be ideal. In the beginning of the curve, the temperature is increasing too rapidly to allow determination of an accurate melting point, but after the change in slope, the temperature increase will have slowed to a more usable rate.

If the melting point of the sample is unknown, you can often save time by preparing two samples for melting-point determination. With one sample, you can rapidly determine a crude melting-point value. Then repeat the experiment more carefully using the second sample. For the second determination, you already have an approximate idea of what the melting-point temperature should be and a proper rate of heating can be chosen.

When measuring temperatures above 150°C, thermometer errors can become significant. For an accurate melting point with a high melting solid, you may wish to apply a **stem correction** to the thermometer as described in Chapter 13, Section 13.4. An even better solution is to calibrate the thermometer as described in Section 9.9.

9.8 DECOMPOSITION, DISCOLORATION, SOFTENING, SHRINKAGE, AND SUBLIMATION

Many solid substances undergo some degree of unusual behavior before melting. At times it may be difficult to distinguish these types of behavior from actual melting. You should learn, through experience, how to recognize melting and how to distinguish it from decomposition, discoloration, and particularly softening and shrinkage.

Some compounds decompose on melting. This decomposition is usually evidenced by discoloration of the sample. Frequently, this decomposition point is a reliable physical property to be used in lieu of an actual melting point. Such decomposition points are indicated in tables of melting points by placing the symbol *d* immediately after the listed temperature. An example of a decomposition point is thiamine hydrochloride, whose melting point would be listed as 248°d, indicating that this substance melts with decomposition at 248°C. When decomposition is a result of reaction with the oxygen in air, it may be avoided by determining the melting point in a sealed, evacuated melting-point tube.

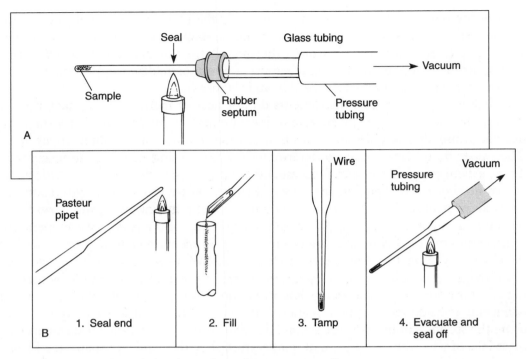

Figure 9.6 Evacuation and sealing of a melting-point capillary.

Figure 9.6 shows two simple methods of evacuating a packed tube. Method A uses an ordinary melting-point tube, while Method B constructs the melting-point tube from a disposable Pasteur pipet. Before using Method B, be sure to determine that the tip of the pipet will fit into the sample holder in your melting-point instrument.

Method A. In Method A, a hole is punched through a rubber septum using a large pin or a small nail, and the capillary tube is inserted from the inside, sealed end first. The septum is placed over a piece of glass tubing connected to a vacuum line. After evacuating the tube, the upper end of the tube may be sealed by heating and pulling it closed.

Method B. In Method B, the thin section of a 9-inch Pasteur pipet is used to construct the melting-point tube. Carefully seal the tip of the pipet using a flame. Be sure to hold the tip *upward* as you seal it. This will prevent water vapor from condensing inside the pipet. When the sealed pipet has cooled, the sample may be added through the open end using a microspatula. A small wire may be used to compress the sample into the closed tip. (If your melting-point apparatus has a vibrator, it may be used in place of the wire to simplify the packing.) When the sample is in place, the pipet is connected to the vacuum line with tubing and evacuated. The evacuated sample tube is sealed by heating it with a flame and pulling it closed.

Some substances begin to decompose *below* their melting points. Thermally unstable substances may undergo elimination reactions or anhydride formation reactions during heating. The decomposition products formed represent impurities in the original sample, so the melting point of the substance may be lowered due to their presence.

It is normal for many compounds to soften or shrink immediately before melting. Such behavior represents not decomposition but a change in the crystal structure or a mixing with impurities. Some substances "sweat," or release solvent of crystallization, before melting. These changes do not indicate the beginning of melting. Actual melting begins when the first drop of liquid becomes visible, and the melting range continues until the temperature is reached at which all the solid has been converted to the liquid state. With experience, you soon learn to distinguish between softening, or "sweating," and actual melting. If you wish, the temperature of the onset of softening or sweating may be reported as a part of your melting point range: 211°C (softens), 223–225°C (melts).

Some solid substances have such a high vapor pressure that they sublime at or below their melting points. In many handbooks, the sublimation temperature is listed along with the melting point. The symbols *sub, subl,* and sometimes *s* are used to designate a substance that sublimes. In such cases, the melting-point determination must be performed in a sealed capillary tube to avoid loss of the sample. The simplest way to accomplish sealing a packed tube is to heat the open end of the tube in a flame and pull it closed with a tweezers or forceps. A better way, although more difficult to master, is to heat the center of the tube in a small flame, rotating it about its axis, and keeping the tube straight, until the center collapses. If this is not done quickly, the sample may melt or sublime while you are working. With the smaller chamber, the sample will not be able to migrate to the cool top of the tube that may be above the viewing area. Figure 9.7 illustrates the method.

9.9 THERMOMETER CALIBRATION

When a melting-point or boiling-point determination has been completed, you expect to obtain a result that exactly duplicates the result recorded in a handbook or in the original literature. It is not unusual, however, to find a discrepancy of several

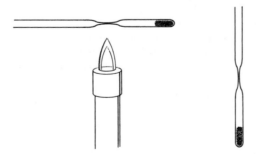

Figure 9.7 Sealing a tube for a substance that sublimes.

degrees from the literature value. Such a discrepancy does not necessarily indicate that the experiment was incorrectly performed or that the material is impure; rather, it may indicate that the thermometer used for the determination was slightly in error. Most thermometers do not measure the temperature with perfect accuracy.

To determine accurate values, you must calibrate the thermometer that is used. This calibration is done by determining the melting points of a variety of standard substances with the thermometer. A plot is drawn of the observed temperature vs. the published value of each standard substance. A smooth line is drawn through the points to complete the chart. A correction chart prepared in this way is shown in Figure 9.8. This chart is used to correct any melting point determined with that particular thermometer. Each thermometer requires its own calibration curve. A list of suitable standard substances for calibrating thermometers is provided in Table 9.1. The standard substances, of course, must be pure in order for the corrections to be valid.

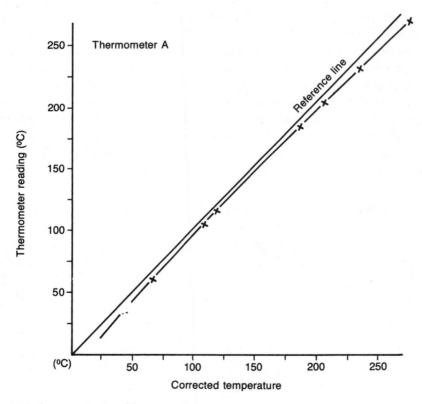

Figure 9.8 A thermometer calibration curve.

Table 9.1 Melting-Point Standards

Compound	Melting Point (°C)
Ice (solid–liquid water)	0
Acetanilide	115
Benzamide	128
Urea	132
Succinic acid	189
3,5-Dinitrobenzoic acid	205

PROBLEMS

1. Two substances, A and B, have the same melting point. How can you determine if they are the same without using any form of spectroscopy? Explain in detail.
2. Using Figure 9.5, determine which heating curve would be most appropriate for a substance with a melting point of about 150°C.
3. What steps can you take to determine the melting point of a substance that sublimes before it melts?
4. A compound melting at 134°C was suspected to be either aspirin (mp 135°C) or urea (mp 133°C). Explain how you could determine whether one of these two suspected compounds was identical to the unknown compound without using any form of spectroscopy.
5. An unknown compound gave a melting point of 230°C. When the molten liquid solidified, the melting point was redetermined and found to be 131°C. Give a possible explanation for this discrepancy.

CHAPTER 10

Solubility

The solubility of a **solute** (a dissolved substance) in a **solvent** (the dissolving medium) is the most important chemical principle underlying three basic techniques you will study in the organic chemistry laboratory: crystallization, extraction, and chromatography. In this chapter on solubility you will gain an understanding of the structural features of a substance that determine its solubility in various solvents. This understanding will help you to predict solubility behavior and to understand the techniques that are based on this property. Understanding solubility behavior will also help you understand what is going on during a reaction, especially when there is more than one liquid phase present or when a precipitate is formed.

10.1 DEFINITION OF SOLUBILITY

Although we often describe solubility behavior in terms of a substance being **soluble** (dissolved) or **insoluble** (not dissolved) in a solvent, solubility can be described more precisely in terms of the *extent* to which a substance is soluble. Solubility may be expressed in terms of grams of solute per liter (g/L) or milligrams of solute per milliliter (mg/mL) of solvent. Consider the solubilities at room temperature for the following three substances in water:

Cholesterol   0.002 mg/mL
Caffeine   22 mg/mL
Citric acid   620 mg/mL

In a typical test for solubility, 40 mg of solute is added to 1 mL of solvent. Therefore, if you were testing the solubility of these three substances, cholesterol would be insoluble, caffeine would be partially soluble, and citric acid would be soluble. Note that a small amount (0.002 mg) of cholesterol would dissolve. It is very unlikely, however, that you would be able to observe this small amount dissolving, and you would report that cholesterol is insoluble. On the other hand, 22 mg (55%) of the caffeine would dissolve. It is likely that you would be able to observe this, and you would state that caffeine is partially soluble.

If the organic compound being dissolved in a solvent is a liquid, then it is sometimes more appropriate to say that the compound and the solvent are **miscible** (mix homogeneously in all proportions). Likewise, if the liquid organic compound is insoluble in the solvent, then they are **immiscible** (do not mix, and form two liquid phases).

10.2 PREDICTING SOLUBILITY BEHAVIOR

A major goal of this chapter is to explain how to predict whether a substance will be soluble in a given solvent. This is not always easy, even for an experienced chemist. However, guidelines will help you make a good guess about the solubility of a compound in a specific solvent. In discussing these guidelines, it is helpful to separate the types of solutions we will be looking at into two categories: (A) solutions in which both the solvent and the solute are covalent (molecular), and (B) ionic solutions, in which the solute ionizes and dissociates.

A. Solutions in Which the Solvent and Solute Are Molecular

A very useful generalization in predicting solubility is the widely used rule "Like dissolves like." This rule is most commonly applied to polar and nonpolar compounds. According to this rule, a polar solvent will dissolve polar (or ionic) compounds and a nonpolar solvent will dissolve nonpolar compounds.

The reason for this behavior involves the nature of intermolecular forces of attraction. Although we will not be focusing on the nature of these forces, it is helpful to know what they are called. The force of attraction between polar molecules is called **dipole–dipole interaction;** between nonpolar molecules, forces of attraction are called **van der Waals forces** (also called **London** or **dispersion forces**). In both cases these attractive forces can occur between molecules of the same compound or different compounds. Consult your lecture textbook for more information on these forces.

To apply the rule "Like dissolves like," you must first determine whether a substance is polar or nonpolar. The polarity of a compound is dependent on both the polarities of the individual bonds and the shape of the molecule. For most organic compounds, evaluating these factors can become quite complicated because of the complexities of the molecules. However, it is possible to make some reasonable predictions just by looking at the types of atoms that a compound possesses. As you read the following guidelines, it is important to understand that although we often describe compounds as being polar or nonpolar, polarity is a matter of degree, ranging from nonpolar to highly polar.

Guidelines for Predicting Polarity and Solubility

1. All hydrocarbons are nonpolar.
  Examples:

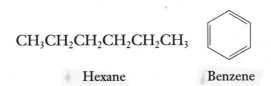

$$CH_3CH_2CH_2CH_2CH_2CH_3$$

        Hexane   Benzene

Hydrocarbons such as benzene are slightly more polar than hexane because of their pi (π) bonds, which allow for greater van der Waals or London attractive forces.

2. Compounds possessing the electronegative elements oxygen or nitrogen are polar. *Examples:*

$$\underset{\text{Acetone}}{CH_3\overset{\overset{\displaystyle O}{\|}}{C}CH_3} \qquad \underset{\text{Ethyl alcohol}}{CH_3CH_2OH} \qquad \underset{\text{Ethyl acetate}}{CH_3\overset{\overset{\displaystyle O}{\|}}{C}OCH_2CH_3}$$

$$\underset{\text{Ethylamine}}{CH_3CH_2NH_2} \qquad \underset{\text{Diethyl ether}}{CH_3CH_2OCH_2CH_3} \qquad \underset{\text{Water}}{H_2O}$$

The polarity of these compounds depends on the presence of polar C—O, C=O, OH, NH, and CN bonds. The compounds that are most polar are capable of forming hydrogen bonds (see guideline #6) and have NH or OH bonds. Although all these compounds are polar, the degree of polarity ranges from slightly polar to highly polar. This is due to the effect on polarity of the shape of the molecule and size of the carbon chain, and whether the compound can form hydrogen bonds.

3. The presence of halogen atoms, even though their electronegativities are relatively high, does not alter the polarity of an organic compound in a significant way. Therefore, these compounds are only slightly polar. The polarities of these compounds are more similar to hydrocarbons, which are nonpolar, than to water, which is highly polar.
Examples:

$$CH_2Cl_2$$

Methylene chloride (dichloromethane)   Chlorobenzene

4. When comparing organic compounds within the same family, note that adding carbon atoms to the chain decreases the polarity. For example, methyl alcohol (CH_3OH) is more polar than propyl alcohol ($CH_3CH_2CH_2OH$). This is because hydrocarbons are nonpolar, and increasing the length of a carbon chain makes the compound more hydrocarbon like.

5. Compounds that contain four or fewer carbons and also contain oxygen or nitrogen are often soluble in water. Almost any functional group containing these elements will lead to water solubility for low-molecular-weight (up to C_4) compounds. Compounds having five or six carbons and containing one of these elements are often insoluble in water or have borderline solubility.

6. As mentioned earlier, the force of attraction between polar molecules is dipole–dipole interaction. A special case of dipole–dipole interaction is hydrogen bonding. Hydrogen bonding is a possibility when a compound possesses a hydrogen atom

bonded to a nitrogen, oxygen, or fluorine atom. The bond is formed by the attraction between this hydrogen atom and a nitrogen, oxygen, or fluorine atom in another molecule. Hydrogen bonding may occur between two molecules of the same compound or between molecules of different compounds:

Hydrogen bonding is the strongest type of dipole–dipole interaction. When hydrogen bonding between solute and solvent is possible, solubility is greater than one would expect for compounds of similar polarity that cannot form hydrogen bonds. Hydrogen bonding is very important in organic chemistry, and you should be alert for situations in which hydrogen bonding may occur.

7. Another factor that can affect solubility is the degree of branching of the alkyl chain in a compound. Branching of the alkyl chain in a compound lowers the intermolecular forces between the molecules. This is usually reflected in a greater solubility in water for the branched compound than for the corresponding straight-chain compound. This occurs simply because the molecules of the branched compounds are more easily separated from one another.

8. The solubility rule ("Like dissolves like") may be applied to organic compounds that belong to the same family. For example, 1-octanol (an alcohol) is soluble in the solvent ethyl alcohol. Most compounds within the same family have similar polarity. However, this generalization may not apply if there is a substantial difference in size between the two compounds. For example, cholesterol, an alcohol with a molecular weight (MW) of 386.64, is only slightly soluble in methanol (MW 32.04). The large hydrocarbon component of cholesterol negates the fact that they belong to the same family.

9. The stability of the crystal lattice also affects solubility. Other things being equal, the higher the melting point (the more stable the crystal), the less soluble the compound. For instance, *p*-nitrobenzoic acid (mp 242°C) is, by a factor of 10, less soluble in a fixed amount of ethanol than the *ortho* (mp 147°C) and *meta* (mp 141°C) isomers.

You can check your understanding of some of these guidelines by studying the list given in Table 10.1, which is given in order of increasing polarity. The structures of these compounds were given on pages 129–130.

This list can be used to make some predictions about solubility, based on the rule "Like dissolves like." Substances that are close to one another on this list will have similar polarities. Thus you would expect hexane to be soluble in methylene chloride, but not in water. Acetone should be soluble in ethyl alcohol. On the other hand, you

Table 10.1 Compounds in Increasing Order of Polarity

	Increasing Polarity
Aliphatic hydrocarbons	
Hexane (nonpolar)	
Aromatic hydrocarbons (π bonds)	
Benzene (nonpolar)	
Halocarbons	
Methylene chloride (slightly polar)	
Compounds with polar bonds	
Diethyl ether (slightly polar)	
Ethyl acetate (intermediate polarity)	
Acetone (intermediate polarity)	
Compounds with polar bonds and hydrogen bonding	
Ethyl alcohol (intermediate polarity)	
Methyl alcohol (intermediate polarity)	
Water (highly polar)	

might predict that ethyl alcohol would be insoluble in hexane. However, ethyl alcohol is soluble in hexane, because ethyl alcohol is somewhat less polar than methyl alcohol or water. This last example demonstrates that you must be careful in using the guidelines on polarity for predicting solubilities. Ultimately, solubility tests must be done to confirm predictions until you gain more experience.

The trend in polarities shown in Table 10.1 can be expanded by including more organic families. The list in Table 10.2 gives an approximate order for the decreasing polarity of organic functional groups. It may appear that there are some discrepancies between the information provided in these two tables. This is because Table 10.1 provides information about specific compounds, whereas the trend shown in Table 10.2 is for major organic families and is approximate.

Table 10.2 Solvents, in Decreasing Order of Polarity

Decreasing Polarity (Approximate)	
H_2O	Water
RCOOH	Organic acids (acetic acid)
$RCONH_2$	Amides (*N,N*-dimethylformamide)
ROH	Alcohols (methanol, ethanol)
RNH_2	Amines (triethylamine, pyridine)
RCOR	Aldehydes, ketones (acetone)
RCOOR	Esters (ethyl acetate)
RX	Halides ($CH_2Cl_2 > CHCl_3 > CCl_4$)
ROR	Ethers (diethyl ether)
ArH	Aromatics (benzene, toluene)
RH	Alkanes (hexane, petroleum ether)

B. Solutions in Which the Solute Ionizes and Dissociates

Many ionic compounds are highly soluble in water because of the strong attraction between ions and the highly polar water molecules. This also applies to organic compounds that can exist as ions. For example, sodium acetate consists of Na^+ and CH_3COO^- ions, which are highly soluble in water. Although there are some exceptions, you may assume that all organic compounds that are in the ionic form will be water soluble.

The most common way by which organic compounds become ions is in acid–base reactions. For example, carboxylic acids can be converted to water-soluble salts when they react with dilute aqueous NaOH:

$$\underset{\text{Water-insoluble carboxylic acid}}{CH_3CH_2CH_2CH_2CH_2CH_2\overset{\displaystyle O}{\overset{\displaystyle \|}{C}}OH} + NaOH\ (aq) \longrightarrow$$

$$\underset{\text{Water-soluble salt}}{CH_3CH_2CH_2CH_2CH_2CH_2\overset{\displaystyle O}{\overset{\displaystyle \|}{C}}O^-\ Na^+} + H_2O$$

The water-soluble salt can then be converted back to the original carboxylic acid (which is insoluble in water) by adding another acid (usually aqueous HCl) to the solution of the salt. The carboxylic acid precipitates out of solution.

Amines, which are organic bases, can also be converted to water-soluble salts when they react with dilute aqueous HCl:

Water-insoluble amine   Water-soluble salt

This salt can be converted back to the original amine by adding a base (usually aqueous NaOH) to the solution of the salt.

10.3 ORGANIC SOLVENTS

Organic solvents must be handled safely. Always remember that organic solvents are all at least mildly toxic and that many are flammable. You should become thoroughly familiar with the introductory chapter on laboratory safety.

READ: "Laboratory Safety," pages 4–21.

Table 10.3 Common Organic Solvents

Solvent	bp (°C)	Solvent	bp (°C)
Hydrocarbons		Ethers	
Pentane	36	**Ether** (diethyl)	35
Hexane	69	**Dioxane**[a]	101
Benzene[a]	80	**1,2-Dimethoxyethane**	83
Toluene	111	Others	
Hydrocarbon mixtures		Acetic acid	118
Petroleum ether	30–60	Acetic anhydride	140
Ligroin	60–90	**Pyridine**	115
Chlorocarbons		**Acetone**	56
Methylene chloride	40	**Ethyl acetate**	77
Chloroform[a]	61	Dimethylformamide	153
Carbon tetrachloride[a]	77	Dimethylsulfoxide	189
Alcohols			
Methanol	65		
Ethanol	78		
Isopropyl alcohol	82		

Note: **Boldface type** indicates flammability.
[a]Suspect carcinogen (see page 20).

The most common organic solvents are listed in Table 10.3 along with their boiling points. Solvents marked in boldface type will burn. Ether, pentane, and hexane are especially dangerous; if they are combined with the correct amount of air, they will explode.

The terms **petroleum ether** and **ligroin** are often confusing. Petroleum ether is a mixture of hydrocarbons with isomers of formulas C_5H_{12} and C_6H_{14} predominating. Petroleum ether is not an ether at all because there are no oxygen-bearing compounds in the mixture. In organic chemistry, an ether is usually a compound containing an oxygen atom to which two alkyl groups are attached. Figure 10.1 shows some of the hydrocarbons that appear commonly in petroleum ether. It also shows the structure of ether (diethyl ether). Use special care when instructions call for either **ether** or **petroleum ether;** the two must not become accidentally confused. Confusion is particularly easy when one is selecting a container of solvent from the supply shelf.

Ligroin, or high-boiling petroleum ether, is like petroleum ether in composition except that compared with petroleum ether, ligroin generally includes higher-boiling alkane isomers. Depending on the supplier, ligroin may have different boiling ranges. Whereas some brands of ligroin have boiling points ranging from about 60°C to about 90°C, other brands have boiling points ranging from about 60°C to about 75°C. The boiling point ranges of petroleum ether and ligroin are often included on the labels of the containers.

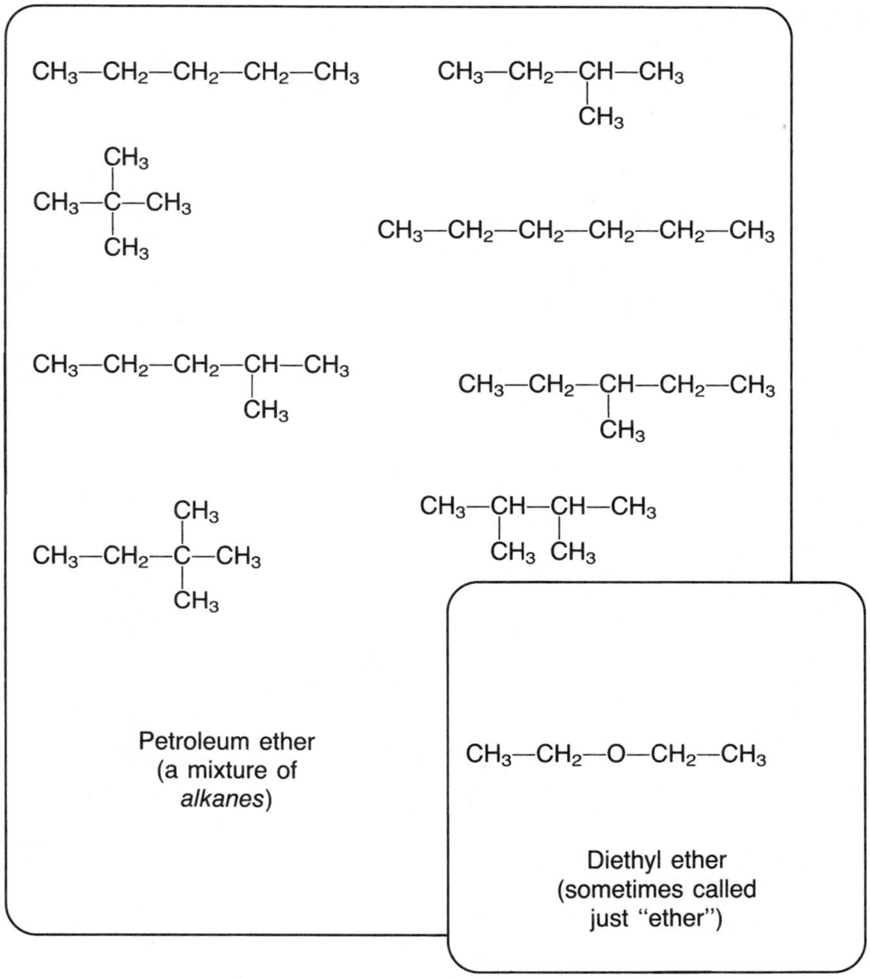

Figure 10.1 A comparison between "ether" (diethyl ether) and "petroleum ether."

PROBLEMS

1. For each of the following pairs of solutes and solvent, predict whether the solute would be soluble or insoluble. After making your predictions, you can check your answers by looking up the compounds in *The Merck Index* or the *CRC Handbook of Chemistry and Physics*. Generally, *The Merck Index* is the easier reference book to use. If the substance has a solubility greater than 40 mg/mL, you may conclude that it is soluble.

(a) Malic acid in water

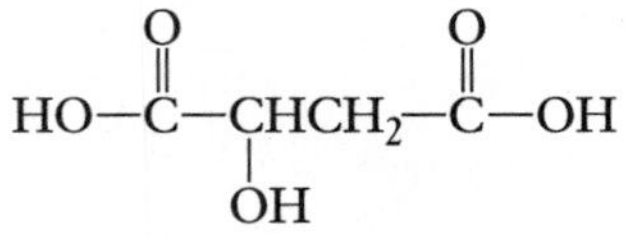

$$HO-\overset{\overset{O}{\|}}{C}-\underset{\underset{OH}{|}}{CH}CH_2-\overset{\overset{O}{\|}}{C}-OH$$

Malic acid

(b) Naphthalene in water

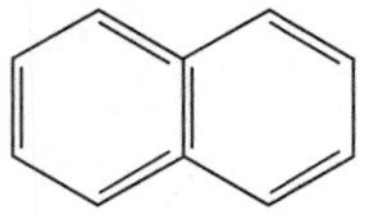

Naphthalene

(c) Amphetamine in ethyl alcohol

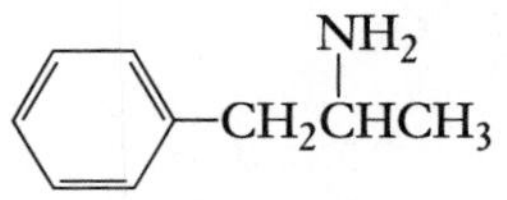

$$-CH_2\underset{\underset{}{}}{\overset{\overset{NH_2}{|}}{C}}HCH_3$$

Amphetamine

(d) Aspirin in water

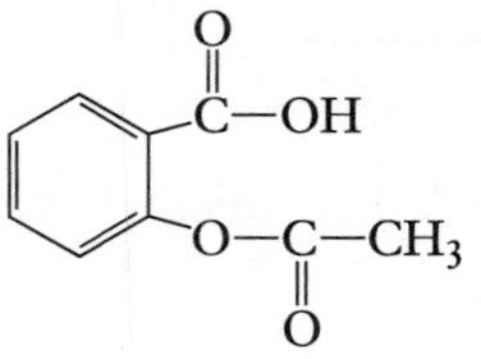

Aspirin

(e) Succinic acid in hexane (*note:* the polarity of hexane is similar to petroleum ether)

$$HO-\overset{\overset{O}{\|}}{C}-CH_2CH_2-\overset{\overset{O}{\|}}{C}-OH$$

Succinic acid

(f) Ibuprofen in diethyl ether

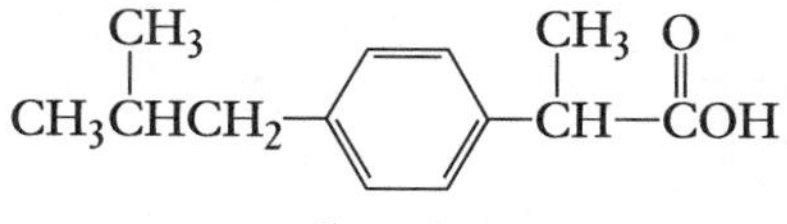

Ibuprofen

(g) 1-Decanol (*n*-decyl alcohol) in water

$$CH_3(CH_2)_8CH_2OH$$
1-Decanol

2. Predict whether the following pairs of liquids would be miscible or immiscible:
   (a) Water and methyl alcohol
   (b) Hexane and benzene
   (c) Methylene chloride and benzene
   (d) Water and toluene

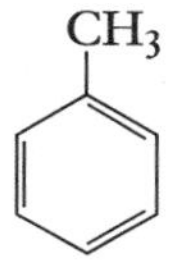

Toluene

   (e) Ethyl alcohol and isopropyl alcohol

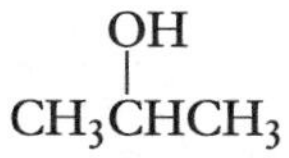

Isopropyl alcohol

3. Would you expect ibuprofen (see **1f**) to be soluble or insoluble in 1.0 *M* NaOH? Explain.
4. Thymol is very slightly soluble in water and very soluble in 1.0 *M* NaOH. Explain.

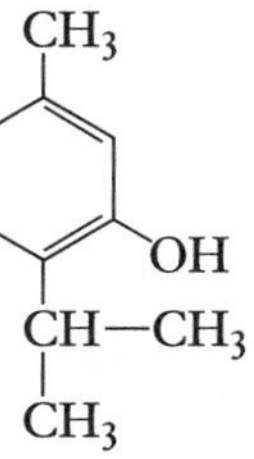

Thymol

5. Although tetrahydrocannabinol and methyl alcohol are both alcohols, tetrahydro-cannabinol is very slightly soluble in methyl alcohol at room temperature. Explain.

Tetrahydrocannabinol

6. What is the difference between
   (a) Ether and petroleum ether?
   (b) Ether and diethyl ether?
   (c) Ligroin and petroleum ether?

CHAPTER 11

Crystallization: Purification of Solids

In most organic chemistry experiments, the desired product is first isolated in an impure form. If this product is a solid, the most common method of purification is crystallization. The general technique involves dissolving the material to be crystallized in a *hot* solvent (or solvent mixture) and cooling the solution slowly. The dissolved material has a decreased solubility at lower temperatures and will separate from the solution as it is cooled. This phenomenon is called either **crystallization,** if the crystal growth is relatively slow and selective, or **precipitation,** if the process is rapid and nonselective. Crystallization is an equilibrium process and produces very pure material. A small seed crystal is formed initially, and it then grows layer by layer in a reversible manner. In a sense, the crystal "selects" the correct molecules from the solution. In precipitation, the crystal lattice is formed so rapidly that impurities are trapped within the lattice. Therefore, any attempt at purification with too rapid a process should be avoided. Because the impurities are usually present in much smaller amounts than the compound being crystallized, most of the impurities will remain in the solvent even when it is cooled. The purified substance can then be separated from the solvent and from the impurities by filtration.

In organic chemistry laboratory work, two methods are commonly used to perform crystallizations. The first method, which is carried out with an Erlenmeyer flask to dissolve the material and either a Büchner or Hirsch funnel to filter the crystals, is normally used when the weight of solid to be crystallized is more than 0.1 g. This technique, called **macroscale crystallization,** is discussed in Section 11.3. The second method is performed with a Craig tube and is used with smaller amounts of solid. Referred to as **microscale crystallization,** this technique is discussed in Section 11.4. The weight of solid to be crystallized, however, is not the only factor to consider when choosing a method for crystallization. Because the solubility of a substance in a given solvent must also be taken into account, the weight, 0.1 g, should not be adhered to rigidly in determining which method to use.

When the macroscale crystallization procedure described in Section 11.3 is used with a Hirsch funnel, the procedure is sometimes referred to as a **semimicroscale crystallization.** This procedure is commonly used in microscale work when the amount of solid is greater than 0.1 g.

Part A. Theory

11.1 SOLUBILITY

The first problem in performing a crystallization is selecting a solvent in which the material to be crystallized shows the desired solubility behavior. In an ideal case, the material should be sparingly soluble at room temperature and yet quite soluble at the boiling point of the solvent selected. The solubility curve should be steep, as can be seen in line A of Figure 11.1. A curve with a low slope (line B, Fig. 11.1) would not cause significant crystallization when the temperature of the solution was lowered. A solvent in which the material is very soluble at all temperatures (line C, Fig. 11.1) also would not be a suitable crystallization solvent. The basic problem in performing a crystallization is to select a solvent (or mixed solvent) that provides a steep solubility-vs.-temperature curve for the material to be crystallized. A solvent that allows the behavior shown in line A of Figure 11.1 is an ideal crystallization solvent. It should also be mentioned that solubility curves are not always linear, as they are depicted in Figure 11.1. Figure 11.1 represents an idealized form of solubility behavior. The solubility curve for sulfanilamide in 95% ethyl alcohol, shown in Figure 11.2, is typical of many organic compounds and shows what solubility behavior might look like for a real substance.

The solubility of organic compounds is a function of the polarities of both the solvent and the **solute** (dissolved material). A general rule is "Like dissolves like." If the solute is very polar, a very polar solvent is needed to dissolve it; if the solute is nonpolar, a nonpolar solvent is needed. Applications of this rule are discussed extensively in Chapter 10, Section 10.2, page 129 and in Section 11.5, page 153.

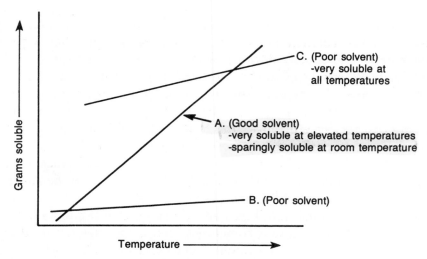

Figure 11.1 Graph of solubility vs. temperature.

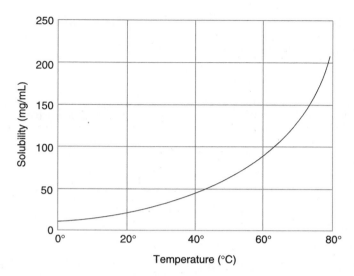

Figure 11.2 Solubility of sulfanilamide in 95% ethyl alcohol.

11.2 THEORY OF CRYSTALLIZATION

A successful crystallization depends on a large difference between the solubility of a material in a hot solvent and its solubility in the same solvent when it is cold. When the impurities in a substance are equally soluble in both the hot and the cold solvent, an effective purification is not easily achieved through crystallization. A material can be purified by crystallization when both the desired substance and the impurity have similar solubilities, but only when the impurity represents a small fraction of the total solid. The desired substance will crystallize on cooling, but the impurities will not.

For example, consider a case in which the solubilities of substance A and its impurity B are both 1 g/100 mL of solvent at 20°C and 10 g/100 mL of solvent at 100°C. In the impure sample of A, the composition is 9 g of A and 2 g of B. In the calculations for this example, it is assumed that the solubilities of both A and B are unaffected by the presence of the other substance. One hundred milliliters of solvent is used in each crystallization to make the calculations easier to understand. Normally, the minimum amount of solvent required to dissolve the solid would be used.

At 20°C, this total amount of material will not dissolve in 100 mL of solvent. However, if the solvent is heated to 100°C, all 11 g dissolve. The solvent has the capacity to dissolve 10 g of A *and* 10 g of B at this temperature. If the solution is cooled to 20°C, only 1 g of each solute can remain dissolved, so 8 g of A and 1 g of B crystallize, leaving 2 g of material in the solution. This crystallization is shown in Figure 11.3. The solution that remains after a crystallization is called the **mother liquor.** If the process is now repeated by treating the crystals with 100 mL of fresh

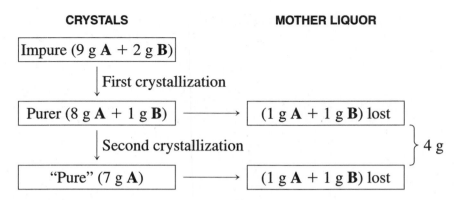

Figure 11.3 Purification of a mixture by crystallization.

solvent, 7 g of A will crystallize again, leaving 1 g of A and 1 g of B in the mother liquor. As a result of these operations, 7 g of pure A is obtained, but with the loss of 4 g of material (2 g of A plus 2 g of B). Again, this second crystallization step is illustrated in Figure 11.3. The final result illustrates an important aspect of crystallization—it is wasteful. Nothing can be done to prevent this waste; some A must be lost along with the impurity B for the method to be successful. Of course, if the impurity B were *more* soluble than A in the solvent, the losses would be reduced. Losses could also be reduced if the impurity were present in *much smaller* amounts than the desired material.

Note that in the preceding case, the method operated successfully because A was present in substantially larger quantity than its impurity B. If there had been a 50/50 mixture of A and B initially, no separation would have been achieved. In general, a crystallization is successful only if there is a *small* amount of impurity. As the amount of impurity increases, the loss of material must also increase. Two substances with nearly equal solubility behavior, present in equal amounts, cannot be separated. If the solubility behavior of two components present in equal amounts is different, however, a separation or purification is frequently possible.

In the preceding example, two crystallization procedures were performed. Normally this is not necessary; however, when it is, the second crystallization is more appropriately called **recrystallization.** As illustrated in this example, a second crystallization results in purer crystals, but the yield is lower.

In some experiments you will be instructed to cool the crystallizing mixture in an ice-water bath before collecting the crystals by filtration. Cooling the mixture increases the yield by decreasing the solubility of the substance; however, even at this reduced temperature, some of the product will be soluble in the solvent. It is not possible to recover all your product in a crystallization procedure even when the mixture is cooled in an ice-water bath. A good example of this is illustrated by the solubility curve for sulfanilamide shown in Figure 11.2. The solubility of sulfanilamide at 0°C is still significant, 14 mg/mL.

Part B. Macroscale Crystallization

11.3 MACROSCALE CRYSTALLIZATION

The crystallization technique described in this section is used when the weight of solid to be crystallized is more than 0.1 g. The four main steps in a macroscale crystallization are

1. Dissolving the solid
2. Removing insoluble impurities (when necessary)
3. Crystallization
4. Isolation of crystals

These steps are illustrated in Figure 11.4. An Erlenmeyer flask of an appropriate size must be chosen. It should be pointed out that a microscale crystallization with a Craig tube involves the same four steps, although the apparatus and procedures are somewhat different (see Section 11.4).

A. Dissolving the Solid

To minimize losses of material to the mother liquor, it is desirable to *saturate* the boiling solvent with solute. This solution, when cooled, will return the maximum possible amount of solute as crystals. To achieve this high return, the solvent is brought to its boiling point, and the solute is dissolved in the *minimum amount* (!) *of boiling solvent.* For this procedure, it is advisable to maintain a container of boiling solvent (on a hot plate). From this container, a small portion (about 1–2 mL) of the solvent is added to the Erlenmeyer flask containing the solid to be crystallized, and this mixture is heated while swirling occasionally until it resumes boiling.

> **▨▨▨ CAUTION** Do not heat the flask containing the solid until after you have added the first portion of solvent.

If the solid does not dissolve in the first portion of boiling solvent, then another small portion of boiling solvent is added to the flask. The mixture is swirled and heated again until it resumes boiling. If the solid dissolves, no more solvent is added. But if the solid has not dissolved, another portion of boiling solvent is added, as before, and the process is repeated until the solid dissolves. It is important to stress that the portions of solvent added each time are small, so that only the *minimum* amount of solvent necessary for dissolving the solid is added. It is also important to emphasize that the procedure requires the addition of solvent to solid. You must never add portions of solid to a fixed quantity of boiling solvent. By this latter method, it may be impossible to determine when saturation has been achieved. This entire procedure should be performed fairly rapidly or you may lose

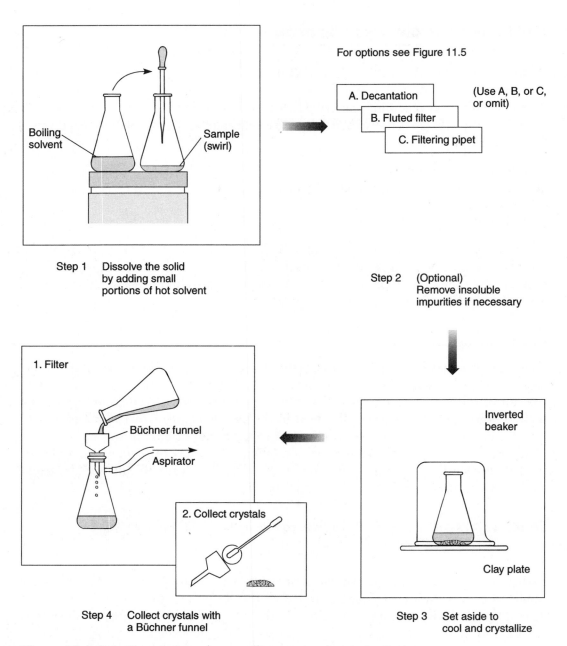

Step 1   Dissolve the solid by adding small portions of hot solvent

Step 2   (Optional) Remove insoluble impurities if necessary

Step 4   Collect crystals with a Büchner funnel

Step 3   Set aside to cool and crystallize

Figure 11.4 Steps in a macroscale crystallization (no decolorization).

solvent through evaporation nearly as quickly as you are adding it and this procedure will take a very long time. This is most likely to happen when using highly volatile solvents such as methyl alcohol or ethyl alcohol. The time from the first addition of solvent until the solid dissolves completely should not be longer than 15–20 minutes.

Comments on This Procedure for Dissolving the Solid

1. One of the most common mistakes is to add too much solvent. This can happen most easily if the solvent is not hot enough or if the mixture is not stirred sufficiently. If too much solvent is added, the percent recovery will be reduced; it is even possible that no crystals will form when the solution is cooled. If too much solvent is added, you must evaporate the excess by heating the mixture. A nitrogen or air stream directed into the container will accelerate the evaporation process (see Chapter 7, Section 7.10, page 95).

2. It is very important not to heat the solid until you have added some solvent. Otherwise, the solid may melt and possibly form an oil or decompose, and it may not crystallize easily (see page 155).

3. It is also important to use an Erlenmeyer flask rather than a beaker for performing the crystallization. A beaker should not be used because the large opening allows the solvent to evaporate too rapidly and allows dust particles to get in too easily.

4. In some experiments a specified amount of solvent for a given weight of solid will be recommended. In these cases, you should use the amount specified rather than the minimum amount of solvent necessary to dissolve the solid. The amount of solvent recommended has been selected to provide the optimum conditions for good crystal formation.

5. Occasionally, you may encounter an impure solid that contains small particles of insoluble impurities, pieces of dust, or paper fibers that will not dissolve in the hot crystallizing solvent. A common error is to add too much of the hot solvent in an attempt to dissolve these small particles, not realizing that they are insoluble. In such cases, you must be careful not to add too much solvent.

6. It is sometimes necessary to decolorize the solution by adding activated charcoal or by passing the solution through a column containing alumina or silica gel (see Section 11.7, and Chapter 19, Section 19.15, page 307). A decolorization step should be performed only if the mixture is *highly* colored and it is clear that the color is due to impurities and not to the actual color of the substance being crystallized. If decolorization is necessary, it should be accomplished before the following filtration step.

B. Removing Insoluble Impurities

It is necessary to use one of the following three methods only if insoluble material remains in the hot solution or if decolorizing charcoal has been used.

CAUTION Indiscriminate use of the procedure can lead to needless loss of your product.

Decantation is the easiest method of removing solid impurities and should be considered first. If filtration is required, a filtering pipet is used when the volume of liquid to be filtered is less than 10 mL (see Chapter 8, Section 8.1C, page 105), and you should use gravity filtration through a fluted filter when the volume is 10 mL or greater (see Chapter 8, Section 8.1B, page 103). These three methods are illustrated in Figure 11.5, and each is discussed below.

Decantation. If the solid particles are relatively large in size or they easily settle to the bottom of the flask, it may be possible to separate the hot solution from the impurities by carefully pouring off the liquid, leaving the solid behind. This is accomplished most easily by holding a glass stirring rod along the top of the flask and tilting the flask so that the liquid pours out along one end of the glass rod into another container. A technique similar in principle to decantation, which may be easier to perform with smaller amounts of liquid, is to use a **preheated Pasteur pipet** to remove the hot solution. With this method, it may be helpful to place the tip of the pipet against the bottom of the flask when removing the last portion of solution. The small space between the tip of the pipet and the inside surface of the flask prevents solid material from being drawn into the pipet. An easy way to preheat the pipet is to draw up a small portion of hot *solvent* (not the *solution* being transferred) into the pipet and expel the liquid. Repeat this process several times.

Fluted Filter. This method is the most effective way to remove solid impurities when the volume of liquid is greater than 10 mL or when decolorizing charcoal has been used (see Chapter 8, Section 8.1B, page 103) and Section 11.7. You should first add a small amount of extra solvent to the hot mixture. This action helps prevent crystal formation in the filter paper or the stem of the funnel during the filtration. The funnel is then fitted with a fluted filter and installed at the top of the Erlenmeyer flask to be used for the actual filtration. It is advisable to place a small piece of wire between the funnel and the mouth of the flask to relieve any increase in pressure caused by hot filtrate.

The Erlenmeyer flask containing the funnel and fluted paper is placed on top of a hot plate (low setting). The liquid to be filtered is brought to its boiling point and poured through the filter in portions. (If the volume of the mixture is less than 10 mL, it may be more convenient to transfer the mixture to the filter with a preheated Pasteur pipet.) It is necessary to keep the solutions in both flasks at their boiling temperatures to prevent premature crystallization. The refluxing action of the filtrate keeps the funnel warm and reduces the chance that the filter will clog with crystals that may have formed during the filtration. With low-boiling solvents, be aware that some solvent may be lost through evaporation. Consequently, extra solvent must be added to make up for this loss. If crystals begin to form in the filter during filtration, a minimum amount of boiling solvent is added to redissolve the crystals and to allow the solution to pass through the funnel. If the volume of liquid being filtered is less than 10 mL, a small amount of hot solvent should be used to rinse the filter after all the filtrate has been collected. The rinse solvent is then combined with the original filtrate.

After the filtration, it may be necessary to remove extra solvent by evaporation until the solution is once again saturated at the boiling point of the solvent (see Chapter 7, Section 7.10, page 95).

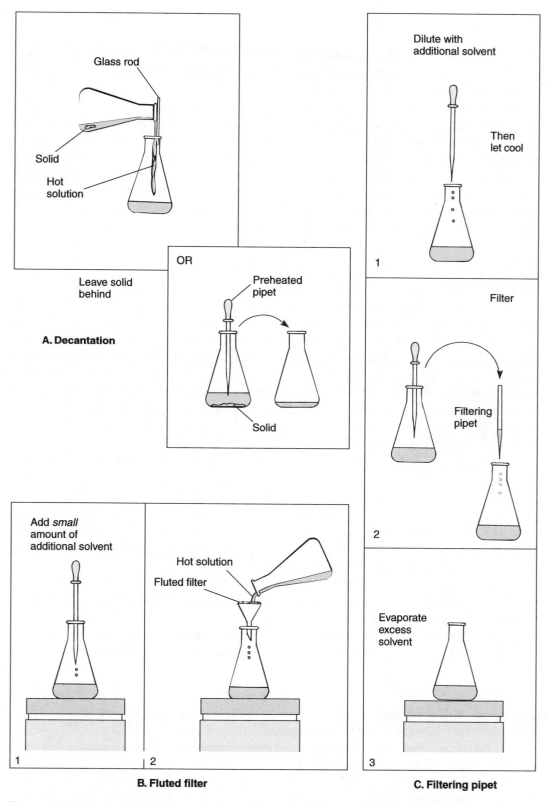

Figure 11.5 Methods for removing insoluble impurities in a macroscale crystallization.

Filtering Pipet. If the volume of solution after dissolving the solid in hot solvent is less than 10 mL, gravity filtration with a filtering pipet may be used to remove solid impurities. However, using a filtering pipet to filter a hot solution saturated with solute can be difficult without premature crystallization. The best way to prevent this from occurring is to add enough solvent to dissolve the desired product at room temperature (be sure not to add too much solvent) and perform the filtration at room temperature, as described in Chapter 8, Section 8.1C, page 105. After filtration, the excess solvent is evaporated by boiling until the solution is saturated at the boiling point of the mixture (see Chapter 7, Section 7.10, page 95). If powdered decolorizing charcoal was used, it will probably be necessary to perform two filtrations with a filtering pipet to remove all of the charcoal, or else a fluted filter can be used.

C. Crystallization

An Erlenmeyer flask, not a beaker, should be used for crystallization. The large open top of a beaker makes it an excellent dust catcher. The narrow opening of the Erlenmeyer flask reduces contamination by dust and allows the flask to be stoppered if it is to be set aside for a long period. Mixtures set aside for long periods must be stoppered after cooling to room temperature to prevent evaporation of solvent. If all the solvent evaporates, no purification is achieved, and the crystals originally formed become coated with the dried contents of the mother liquor. Even if the time required for crystallization to occur is relatively short, it is advisable to cover the top of the Erlenmeyer flask with a small watch glass or inverted beaker to prevent evaporation of solvent while the solution is cooling to room temperature.

The chances of obtaining pure crystals are improved if the solution cools to room temperature slowly. When the volume of solution is 10 mL or less, the solution is likely to cool more rapidly than is desired. This can be prevented by placing the flask on a surface that is a poor heat conductor and covering the flask with a beaker to provide a layer of insulating air. Appropriate surfaces include a clay plate or several pieces of filter paper on top of the laboratory bench. It may also be helpful to use a clay plate that has been warmed slightly on a hot plate or in an oven.

After crystallization has occurred, it is sometimes desirable to cool the flask in an ice-water bath. Because the solute is less soluble at lower temperatures, this will increase the yield of crystals.

If a cooled solution does not crystallize, it will be necessary to induce crystallization. Several techniques are described in Section 11.8A.

D. Isolation of Crystals

After the flask has been cooled, the crystals are collected by vacuum filtration through a Büchner (or Hirsch) funnel (see Chapter 8, Section 8.3, page 107, and Fig. 8.5). The crystals should be washed with a small amount of *cold* solvent to remove any mother liquor adhering to their surface. Hot or warm solvent will dissolve some of the crystals. The crystals should then be left for a short time (usually 5–10 minutes) in the funnel, where air, as it passes, will dry them free of most of the solvent. It is often wise to cover the Büchner funnel with an oversized filter paper or towel during this air drying. This precaution prevents accumulation of dust in the crystals. When the crystals are nearly dry, they should be gently scraped off (so paper fibers are not removed with the crystals) the filter paper onto a watch glass or clay plate for further drying (see Section 11.9).

The four steps in a macroscale crystallization are summarized in Figure 11.6.

A. Dissolving the Solid

1. Find a solvent with a steep solubility-vs-temperature characteristic. (Done by trial and error using small amounts of material or by consulting a handbook.)
2. Heat the desired solvent to its boiling point.
3. Dissolve the solid in a **minimum** of boiling solvent in a flask.
4. If necessary, add decolorizing charcoal or decolorize the solution on a silica gel or alumina column.

B. Removing Insoluble Impurities

1. Decant or remove the solution with a Pasteur pipet, or
2. Filter the hot solution through a fluted filter, a filtering pipet, or a filter tip pipet to remove insoluble impurities or charcoal.

> NOTE: If no decolorizing charcoal has been added or if there are no undissolved particles, Part B should be omitted.

C. Crystallizing

1. Allow the solution to cool.
2. If crystals appear, cool the mixture in an ice-water bath (if desired) and go to Part D. If crystals do not appear, go to the next step.
3. Inducing crystallization.
  (a) Scratch the flask with a glass rod.
  (b) Seed the solution with original solid, if available.
  (c) Cool the solution in an ice-water bath.
  (d) Evaporate excess solvent and allow the solution to cool again.

D. Collecting and Drying

1. Collect crystals by vacuum filtration using a Büchner funnel.
2. Rinse crystals with a small portion of **cold** solvent.
3. Continue suction until crystals are nearly dry.
4. Drying.
  (a) Air-dry the crystals, or
  (b) Place the crystals in a drying oven, or
  (c) Dry the crystals *in vacuo.*

Figure 11.6 Steps in a crystallization.

Part C. Microscale Crystallization

11.4 MICROSCALE CRYSTALLIZATION

In many microscale experiments, the amount of solid to be crystallized is small enough (generally less than 0.1 g) that a **Craig tube** (see Fig. 8.10, page 113) is the preferred method for crystallization. The main advantage of the Craig tube is that it minimizes the number of transfers of solid material, thus resulting in a greater yield of crystals. Also, separating the crystals from the mother liquor with the Craig tube is very efficient, and little time is required for drying the crystals. The steps involved are, in principle, the same as those performed when crystallization is accomplished with an Erlenmeyer flask and a Büchner (or Hirsch) funnel. The steps in a microscale crystallization using a Craig tube are illustrated in Figure 11.7.

A. Dissolving the Solid

In crystallizations in which a filtration step is not required to remove insoluble impurities such as dirt or activated charcoal, this first step can be performed directly in the Craig tube. Otherwise, use a small test tube. The solid is transferred to the Craig tube, and the appropriate solvent, contained in a test tube, is heated to boiling on an aluminum block. A small portion (several drops) of hot solvent is added to the Craig tube, which is subsequently heated on the aluminum block until the solution in the Craig tube starts to boil.

▨ CAUTION Do not heat the Craig tube containing the solid until after you have added the first portion of solvent.

The hot mixture should be stirred continuously with a microspatula using a twirling motion. Stirring not only helps to dissolve the solute but also prevents the boiling liquid from bumping. Additional portions of hot solvent are added until all the solid has dissolved. To obtain the maximum yield, it is important not to add too much solvent, although any excess solvent can be evaporated later. You should perform this procedure fairly rapidly. Otherwise you may lose solvent by evaporation nearly as quickly as you are adding it, and dissolving all the solid will take a long time. The time required to dissolve the solid should not be longer than 15 minutes.

For additional information about this procedure, see "Comments on this Procedure for Dissolving the Solid" on page 145.

B. Removing Insoluble Impurities

You should be alert for the presence of impurities that will not dissolve in the hot solvent, no matter how much solvent is added. If it appears that most of the solid has dissolved and the remaining solid has no tendency to dissolve, or if the liquid has been decolorized with charcoal, it will be necessary to remove the solid particles. Two methods are discussed.

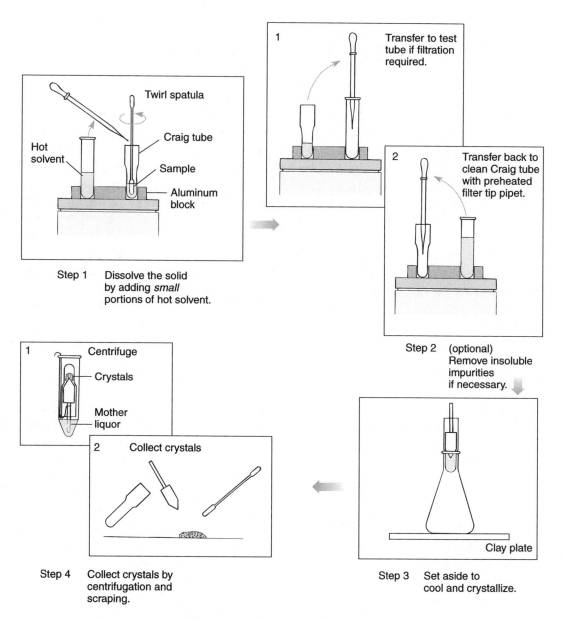

Figure 11.7 Steps in a microscale crystallization (no decolorization).

If the impurities are relatively large or concentrated in one part of the mixture, it may be possible to use a Pasteur pipet preheated with hot solvent to draw up the liquid without removing any solid. One way to do this is to expel the air from the pipet and then place the end of the pipet on the bottom of the tube, being careful not to trap any solid in the pipet. The small space between the pipet and the bottom of the tube should allow you to draw up the liquid without removing any solid.

When filtration is necessary, a preheated Pasteur pipet is used to transfer the mixture to a test tube. After making this transfer, the Craig tube is rinsed with a few drops of solvent, which are also added to the test tube. The Craig tube is then washed and dried. The test tube containing the mixture is also heated in an aluminum block. An additional 5 to 10 drops of solvent are added to the test tube to ensure that premature crystallization does not occur during the filtration step. To filter the mixture, take up the mixture in a filter-tip pipet (see Chapter 8, Section 8.6, page 112) that has been preheated with hot solvent, and quickly transfer the liquid to the clean Craig tube. Passing the liquid through the cotton plug in the filter-tip pipet should remove the solid impurities. *If this does not occur,* it may be necessary to add more solvent (to prevent crystallization) and filter the mixture through a filtering pipet (Chapter 8, Section 8.1C, page 105). In either case, once the filtered solution has been returned to the Craig tube, it will be necessary to evaporate some solvent until the solution is saturated near the boiling point of the liquid. This is most conveniently accomplished by placing the Craig tube in the aluminum block, and, while stirring rapidly using a microspatula (twirling is most effective), bringing the solution to a boil. When you begin to observe a trace of solid material coating the spatula just above the level of the liquid, the solution is near saturation, and evaporation should be stopped.

C. Crystallization

The hot solution is cooled slowly in the Craig tube to room temperature. Recall that slow cooling is important in the formation of pure crystals. When the volume of solution is 2 mL or less and the mass of glassware is relatively small, slow cooling is somewhat difficult to achieve. One method of increasing the cooling time is to insert the inner plug into the outer part of the Craig tube and place the Craig tube into a 10-mL Erlenmeyer flask. The layer of air in the flask will help insulate the hot solution as it cools. The Erlenmeyer flask is placed on a surface such as a clay plate (warmed slightly, if desired) or several pieces of paper. Another method is to fill a 10-mL Erlenmeyer flask with 8–10 mL of hot water at a temperature below the boiling point of the solvent. The assembled Craig tube is placed in the Erlenmeyer flask that is set on an appropriate surface. Be careful not to put so much water in the Erlenmeyer flask that the Craig tube floats. After crystallization at room temperature is complete, the Craig tube can be placed in an ice-water bath to maximize the yield.

If crystals have not formed after the solution has cooled to room temperature, it will be necessary to induce crystallization. Several techniques are described in Section 11.8.

A common occurrence with crystallizations using a Craig tube is to obtain a seemingly solid mass of very small crystals. This may not be a problem, but if there is very little mother liquor present or the crystals are impure, it may be necessary to repeat the crystallization. This situation may have resulted either because the cooling process occurred too rapidly, or because the solubility–temperature curve was so steep for a given solvent that very little mother liquor remained after the crystallization. In either case, you may want to repeat the crystallization to obtain a better (purer) yield of crystals. Three measures may be taken to avoid this problem. A small amount of extra solvent may be added before heating the mixture again and allowing it to cool. Second, the solution can be cooled more slowly. Finally, it may be helpful to try to induce crystallization *before* the solution has cooled to room temperature.

D. Isolation of Crystals

When the crystals have formed and the mixture has cooled in an ice-water bath (if desired), the Craig tube is placed in a centrifuge tube and the crystals are separated from the mother liquor by centrifugation (see Chapter 8, Section 8.7, page 113). The crystals are then scraped off the end of the inner plug or from inside the Craig tube onto a watch glass or piece of paper. Minimal drying will be necessary (see Section 11.9).

Part D. Additional Experimental Considerations: Macroscale and Microscale

11.5 SELECTING A SOLVENT

A solvent that dissolves little of the material to be crystallized when it is cold but a great deal of the material when it is hot is a good solvent for crystallization. Quite often, correct crystallization solvents are indicated in the experimental procedures that you will be following. When a solvent is not specified in a procedure, you can determine a good crystallization solvent by consulting a handbook or making an educated guess based on polarities, both discussed in this section. A third approach, involving experimentation, is discussed in Section 11.6.

With compounds that are well known, the correct crystallization solvent has already been determined through the experiments of earlier researchers. In such cases, the chemical literature can be consulted to determine which solvent should be used. Sources such as *The Merck Index* or the *CRC Handbook of Chemistry and Physics* may provide this information.

For example, consider naphthalene, which is found in *The Merck Index*. Under the entry for naphthalene it states: "Monoclinic prismatic plates from ether." This statement means that naphthalene can be crystallized from ether. It also gives the type of crystal structure. Unfortunately, the crystal structure may be given without reference to the solvent. Another way to determine the best solvent is by looking at solubility vs. temperature data. When this is given, a good solvent is one in which the solubility of the compound increases significantly as the temperature increases. Sometimes the solubility data will be given for only cold solvent and boiling solvent. This should provide enough information to determine whether this would be a good solvent for crystallization.

In most cases, however, the handbooks will state only whether a compound is soluble or not in a given solvent, usually at room temperature. Determining a good solvent for crystallization from this information can be somewhat difficult. The solvent in which the compound is soluble may or may not be an appropriate solvent for crystallization. Sometimes the compound may be too soluble in the solvent at all temperatures and you would recover very little of your product if this solvent was used for crystallization. It is possible that an appropriate solvent would be the one in which the compound is nearly insoluble at room temperature because the solubility vs. temperature curve is very steep. Although the solubility information may give you some ideas about what solvents to try, you will most likely need to determine a good crystallizing solvent by experimentation as described in Section 11.6.

When using *The Merck Index* or *Handbook of Chemistry and Physics* you should be aware that "alcohol" is frequently listed as a solvent. This generally refers to 95% or 100% ethyl alcohol. Since 100% (absolute) ethyl alcohol is more expensive than 95% ethyl alcohol, the cheaper grade is usually used in the chemistry laboratory. Another solvent frequently listed is benzene. Since benzene is a known carcinogen, it is rarely used in student laboratories. Toluene is a suitable substitute; the solubility behavior of a substance in benzene and toluene is so similar that you may assume any statement made about benzene also applies to toluene.

Another way to identify a solvent for crystallization is to consider the polarities of the compound and the solvents. Generally, you would look for a solvent that has a polarity somewhat similar to the compound to be crystallized. Consider the compound sulfanilamide, shown below. There are several polar bonds in sulfanilamide, the NH and the SO bonds. In addition, the NH_2 groups and the oxygen atoms in sulfanilamide can form hydrogen bonds. So although the benzene ring portion of sulfanilamide is nonpolar, sulfanilamide has an intermediate polarity because of the polar groups. A common organic solvent of intermediate polarity is 95% ethyl alcohol. Therefore it is likely that sulfanilamide would be soluble in 95% ethyl alcohol because they have similar polarities. (Note that the other 5% in 95% ethyl alcohol is usually a substance such as water or isopropyl alcohol, which does not alter the overall polarity of the solvent.) Although this kind of analysis is a good first step in determining an appropriate solvent for crystallization, without more information it is not enough to predict the shape of the solubility curve for the temperature vs. solubility data (see Figure 11.1, page 140). Therefore, knowing that sulfanilamide is soluble in 95% ethyl alcohol does not necessarily mean that this is a good solvent for crystallizing sulfanilamide. You would still need to test the solvent to see if it is appropriate. The solubility curve for sulfanilamide (see Figure 11.2, page 141) indicates that 95% ethyl alcohol is a good solvent for crystallizing this substance.

Sulfanilamide

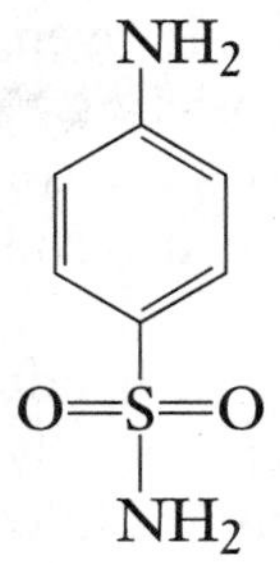

When choosing a crystallization solvent, do not select one whose boiling point is higher than the melting point of the substance (solute) to be crystallized. If the boiling point of the solvent is too high, the substance may come out of solution as a liquid rather than a crystalline solid. In such a case, the solid may **oil out.** Oiling out occurs when on cooling the solution to induce crystallization, the solute begins to come out of solution at a temperature above its melting point. The solute will then come out of solution as a liquid. Furthermore, as cooling continues, the substance may still not crystallize; rather, it will become a supercooled liquid. Oils may eventually solidify if the temperature is lowered, but often they will not actually crystallize. Instead, the solidified oil will be an amorphous solid or a hardened mass. In this case, purification of the substance will not have occurred as it does when the solid is crystalline. It can be very difficult to deal with oils when trying to obtain a pure substance. You must try to redissolve them and hope that the substance will crystallize with slow, careful cooling. During the cooling period, it may be helpful to scratch the glass container where the oil is present with a glass stirring rod that has not been fire polished. Seeding the oil as it cools with a small sample of the original solid is another technique that is sometimes helpful in working with difficult oils. Other methods of inducing crystallization are discussed in Section 11.8.

One additional criterion for selecting the correct crystallization solvent is the **volatility** of that solvent. Volatile solvents have low boiling points or evaporate easily. A solvent with a low boiling point may be removed from the crystals through evaporation without much difficulty. It will be difficult to remove a solvent with a high boiling point from the crystals without heating them under vacuum.

Table 11.1 lists common crystallization solvents. The solvents used most commonly are listed in the table first.

Table 11.1 Common Solvents for Crystallization

	Boils (°C)	Freezes (°C)	Soluble in H_2O	Flammability
Water	100	0	+	−
Methanol	65	a	+	+
95% Ethanol	78	a	+	+
Ligroin	60–90	a	−	+
Toluene	111	a	−	+
Chloroform[b]	61	a	−	−
Acetic acid	118	17	+	+
Dioxane[b]	101	11	+	+
Acetone	56	a	+	+
Diethyl ether	35	a	Slightly	++
Petroleum ether	30–60	a	−	++
Methylene chloride	41	a	−	−
Carbon tetrachloride[b]	77	a	−	−

[a]Lower than 0°C (ice temperature).
[b]Suspected carcinogen.

11.6 TESTING SOLVENTS FOR CRYSTALLIZATION

When the appropriate solvent is not known, select a solvent for crystallization by experimenting with various solvents and a very small amount of the material to be crystallized. Experiments are conducted on a small test-tube scale before the entire quantity of material is committed to a particular solvent. Such trial-and-error methods are common when trying to purify a solid material that has not been previously studied.

Procedure

1. Place about 0.05 g of the sample in a test tube.
2. Add about 0.5 mL of solvent at room temperature and stir the mixture by rapidly twirling a microspatula between your fingers. If all (or almost all) of the solid dissolves at room temperature, then your solid is *probably* too soluble in this solvent and little compound would be recovered if this solvent were used. Select another solvent.
3. If none (or very little) of the solid dissolves at room temperature, heat the tube carefully and stir with a spatula. (A hot water bath is perhaps better than an aluminum block because you can more easily control the temperature of the hot water bath. The temperature of the hot water bath should be slightly higher than the boiling point of the solvent.) Add more solvent dropwise, while continuing to heat and stir. Continue adding solvent until the solid dissolves, but do not add more than about 1.5 mL (total) of solvent. If all the solid dissolves, go to step 4. If all the solid has not dissolved by the time you have added 1.5 mL of solvent, this is probably not a good solvent. However, if most of the solid has dissolved at this point, you might try adding a little more solvent. Remember to heat and stir at all times during this step.
4. If the solid dissolves in about 1.5 mL or less of boiling solvent, then remove the test tube from the heat source, stopper the tube, and allow it to cool to room temperature. Then place it in an ice-water bath. If lots of crystals come out, this is most likely a good solvent. If crystals do not come out, scratch the sides of the tube with a glass stirring rod to induce crystallization. If crystals still do not form, this is probably not a good solvent.

Comments about This Procedure.
1. Selecting a good solvent is something of an art. There is no perfect procedure that can be used in all cases. You must think about what you are doing and use some common sense in deciding whether to use a particular solvent.
2. Do not heat the mixture above the melting point of your solid. This can occur most easily when the boiling point of the solvent is higher than the melting point of the solid. Normally, do not select a solvent that has a higher boiling point than the melting point of the substance. If you do, make certain that you do not heat the mixture beyond the melting point of your solid.

11.7 DECOLORIZATION

Small amounts of highly colored impurities may make the original crystallization solution appear colored; this color can often be removed by **decolorization,** either by using activated charcoal (often called Norit) or by passing the solution through a column packed with alumina or silica gel. A decolorizing step should be performed only if the color is due to impurities, not to the color of the desired product, and if the color is significant. Small amounts of colored impurities will remain in solution during crystallization, making the decolorizing step unnecessary. The use of activated charcoal is described separately for macroscale and microscale crystallizations, and then the column technique, which can be used with both crystallization techniques, is described.

A. Macroscale—Powdered Charcoal

As soon as the solute is dissolved in the minimum amount of boiling solvent, the solution is allowed to cool slightly and a small amount of Norit (powdered charcoal) is added to the mixture. The Norit adsorbs the impurities. When performing a crystallization in which the filtration is performed with a fluted filter, powdered Norit should be added because it has a larger surface area and can remove impurities more effectively. A reasonable amount of Norit is what could be held on the end of a microspatula, or about 0.01–0.02 g. If too much Norit is used, it will adsorb product as well as impurities. A small amount of Norit should be used, and its use should be repeated if necessary. (It is difficult to determine if the initial amount added is sufficient until after the solution is filtered, because the suspended particles of charcoal will obscure the color of the liquid.) Caution should be exercised so that the solution does not froth or erupt when the finely divided charcoal is added. The mixture is boiled with the Norit for several minutes and then filtered by gravity, using a fluted filter (see Section 11.3 and Chapter 8, Section 8.1B, page 103), and the crystallization is carried forward as described in Section 11.3.

The Norit preferentially adsorbs the colored impurities and removes them from the solution. The technique seems to be most effective with hydroxylic solvents. In using Norit, be careful not to breathe the dust. Normally, small quantities are used so that little risk of lung irritation exists.

B. Microscale—Pelletized Norit

If the crystallization is being performed in a Craig tube, it is advisable to use pelletized Norit. Although this is not as effective in removing impurities as powdered Norit, it is easier to remove, and the amount of pelletized Norit required is more easily determined because you can see the solution as it is being decolorized. Again, the Norit is added to the hot solution (the solution should not be boiling) after the solid has dissolved. This should be performed in a test tube rather than in a Craig tube. About 0.02 g is added, and the mixture is boiled for a minute or so to see if more Norit is required.

More Norit is added, if necessary, and the liquid is boiled again. It is important not to add too much pelletized Norit because the Norit will also adsorb some of the desired material, and it is possible that not all the color can be removed no matter how much is added. The decolorized solution is then removed with a preheated filter-tip pipet (see Chapter 8, Section 8.6, page 112) to filter the mixture and transferred to a Craig tube for crystallization as described in Section 11.4.

C. Decolorization on a Column

The other method for decolorizing a solution is to pass the solution through a column containing alumina or silica gel. The adsorbent removes the colored impurities while allowing the desired material to pass through (see Chapter 8, Fig. 8.6, page 110, and Chapter 19, Section 19.15, page 307). If this technique is used, it will be necessary to dilute the solution with additional solvent to prevent crystallization from occurring during the process. The excess solvent must be evaporated after the solution is passed through the column (Chapter 7, Section 7.10, page 95) and the crystallization procedure is continued as described in Sections 11.3 or 11.4.

11.8 INDUCING CRYSTALLIZATION

If a cooled solution does not crystallize, several techniques may be used to induce crystallization. Although identical in principle, the actual procedures vary slightly when performing macroscale and microscale crystallizations.

A. Macroscale

In the first technique, you should try scratching the inside surface of the flask vigorously with a glass rod that *has not been* fire polished. The motion of the rod should be vertical (in and out of the solution) and should be vigorous enough to produce an audible scratching. Such scratching often induces crystallization, although the effect is not well understood. The high-frequency vibrations may have something to do with initiating crystallization; or perhaps—a more likely possibility—small amounts of solution dry by evaporation on the side of the flask, and the dried solute is pushed into the solution. These small amounts of material provide "seed crystals," or nuclei, on which crystallization may begin.

A second technique that can be used to induce crystallization is to cool the solution in an ice bath. This method decreases the solubility of the solute.

A third technique is useful when small amounts of the original material to be crystallized are saved. The saved material can be used to "seed" the cooled solution. A small crystal dropped into the cooled flask often will start the crystallization—this is called **seeding.**

If all these measures fail to induce crystallization, it is likely that too much solvent was added. The excess solvent must then be evaporated (Chapter 7, Section 7.10, page 95) and the solution allowed to cool.

B. Microscale

The strategy is basically the same as described for macroscale crystallizations. Scratching vigorously with a glass rod *should be avoided*, however, because the Craig tube is fragile and expensive. Scratching *gently* is allowed.

Another measure is to dip a spatula or glass stirring rod into the solution and allow the solvent to evaporate so that a small amount of solid will form on the surface of the spatula or glass rod. When placed back into the solution, the solid will seed the solution. A small amount of the original material, if some was saved, may also be used to seed the solution.

A third technique is to cool the Craig tube in an ice-water bath. This method may also be combined with either of the previous suggestions.

If none of these measures is successful, it is possible that too much solvent is present, and it may be necessary to evaporate some of the solvent (Chapter 7, Section 7.10, page 95) and allow the solution to cool again.

11.9 DRYING CRYSTALS

The most common method of drying crystals involves placing them on a watch glass, a clay plate, or a piece of paper and allowing them to dry in air. An inverted beaker should be placed over the crystals to prevent accumulation of dust particles. The advantage of this method is that heat is not required, thus reducing the danger of decomposition or melting; however, exposure to atmospheric moisture may cause the hydration of strongly hygroscopic materials. A **hygroscopic** substance is a substance that absorbs moisture from the air.

Another method of drying crystals is to place the crystals on a watch glass, a clay plate, or a piece of absorbent paper in an oven. Although this method is simple, some possible difficulties deserve mention. Crystals that sublime readily should not be dried in an oven because they might vaporize and disappear. Care should be taken that the temperature of the oven does not exceed the melting point of the crystals. Remember that the melting point of crystals is lowered by the presence of solvent; allow for this melting-point depression when selecting a suitable oven temperature. Some materials decompose on exposure to heat, and they should not be dried in an oven. Finally, when many different samples are being dried in the same oven, crystals might be lost due to confusion or reaction with another person's sample. It is important to label the crystals when they are placed in the oven.

A third method, which requires neither heat nor exposure to atmospheric moisture, is drying *in vacuo*. Two procedures are illustrated in Figure 11.8.

Procedure A. In this method a desiccator is used. The sample is placed under vacuum in the presence of a drying agent. Two potential problems must be noted. The first deals with samples that sublime readily. Under vacuum, the likelihood of sublimation is increased. The second problem deals with the vacuum desiccator itself. Because the surface area of glass that is under vacuum is large, there is some danger that

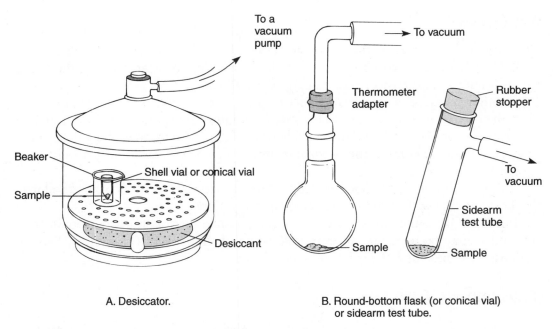

A. Desiccator.

B. Round-bottom flask (or conical vial) or sidearm test tube.

Figure 11.8 Methods for drying crystals in a vacuum.

the desiccator could implode. A vacuum desiccator should never be used unless it has been placed within a protective metal container (cage). If a cage is not available, the desiccator can be wrapped with electrical or duct tape. If you use an aspirator as a source of vacuum, you should use a water trap (see Fig. 8.5, p. 108).

Procedure B. This method can be accomplished with a round-bottom flask and a thermometer adapter equipped with a short piece of glass tubing, as illustrated in Figure 11.8B. In microscale work, the apparatus with the round-bottom flask can be modified by replacing the round-bottom flask with a conical vial. The glass tubing is connected by vacuum tubing to either an aspirator or a vacuum pump. A convenient alternative, using a sidearm test tube, is also shown in Figure 11.8B. With either apparatus, install a water trap when an aspirator is used.

11.10 MIXED SOLVENTS

Often the desired solubility characteristics for a particular compound are not found in a single solvent. In these cases, a mixed solvent may be used. You simply select a first solvent in which the solute is soluble and a second solvent, miscible with the first, in which the solute is relatively insoluble. The compound is dissolved in a minimum amount of the boiling solvent in which it is soluble. Following this, the second hot solvent is added to the boiling mixture, dropwise, until the mixture barely becomes cloudy. The cloudiness indicates precipitation. At this point, more of the first solvent

Table 11.2 Common Solvent Pairs for
Crystallization

Methanol–water	Ether–acetone
Ethanol–water	Ether–petroleum ether
Acetic acid–water	Toluene–ligroin
Acetone–water	Methylene chloride–methanol
Ether–methanol	Dioxane[a]–water

[a]Suspected carcinogen.

should be added. Just enough is added to clear the cloudy mixture. At that point, the solution is saturated, and as it cools, crystals should separate. Common solvent mixtures are listed in Table 11.2.

It is important not to add an excess of the second solvent or to cool the solution too rapidly. Either of these actions may cause the solute to oil out, or separate as a viscous liquid. If this happens, reheat the solution and add more of the first solvent.

PROBLEMS

1. Listed below are solubility–temperature data for an organic substance A dissolved in water.

Temperature (°C)	Solubility of A in 100 mL of Water
0	1.5 g
20	3.0 g
40	6.5 g
60	11.0 g
80	17.0 g

(a) Graph the solubility of A vs. temperature. Use the data given in the table. Connect the data points with a smooth curve.

(b) Suppose 0.1 g of A and 1.0 mL of water were mixed and heated to 80°C. Would all the substance A dissolve?

(c) The solution prepared in (b) is cooled. At what temperature will crystals of A appear?

(d) Suppose the cooling described in (c) were continued to 0°C. How many grams of A would come out of solution? Explain how you obtained your answer.

2. What would likely happen if a hot saturated solution were filtered by vacuum filtration using a Hirsch funnel? (*Hint:* The mixture will cool as it comes in contact with the Hirsch funnel.)

3. A compound you have prepared is reported in the literature to have a pale yellow color. When dissolving the substance in hot solvent to purify it by crystallization,

the resulting solution is yellow. Should you use decolorizing charcoal before allowing the hot solution to cool? Explain your answer.

4. After dissolving a crude product in 1.5 mL of hot solvent, the resulting solution is a dark brown color. Because the pure compound is reported in the literature to be colorless, it is necessary to perform a decolorizing procedure. Should you use pelletized Norit or powdered activated charcoal to decolorize the solution? Explain your answer.

5. While performing a crystallization, you obtain a light tan solution after dissolving your crude product in hot solvent. A decolorizing step is determined to be unnecessary, and there are no solid impurities present. Should you perform a filtration to remove impurities before allowing the solution to cool? Why or why not?

6. (a) Draw a graph of a cooling curve (temperature vs. time) for a solution of a solid substance that shows no supercooling effects. Assume that the solvent does not freeze.

   (b) Repeat the instructions in (a) for a solution for a solid substance that shows some supercooling behavior but eventually yields crystals if the solution is cooled sufficiently.

7. A solid substance A is soluble in water to the extent of 10 mg/mL of water at 25°C and 100 mg/mL of water at 100°C. You have a sample that contains 100 mg of A and an impurity B.

   (a) Assuming that 2 mg of B is present along with 100 mg of A, describe how you could purify A if B is completely insoluble in water. Your description should include the volume of solvent required.

   (b) Assuming that 2 mg of the impurity B is present along with 100 mg of A, describe how you could purify A if B had the same solubility behavior as A. Would one crystallization produce pure A? (Assume that the solubilities of both A and B are unaffected by the presence of the other substance).

   (c) Assume that 25 mg of the impurity B is present along with 100 mg of A. Describe how you could purify A if B had the same solubility behavior as A. Each time, use the minimum amount of water to just dissolve the solid. Would one crystallization produce absolutely pure A? How many crystallizations would be needed to produce pure A? How much A would have been recovered when the crystallizations had been completed?

8. An organic chemistry student dissolved 0.095 g of a crude product in 3.5 mL (the minimum amount required) of ethanol at 25°C. He cooled the solution in an ice-water bath for 15 minutes and obtained beautiful crystals. He filtered the crystals on a Hirsch funnel and rinsed them with about 0.5 mL of ice-cold ethanol. After drying, the weight of the crystals was found to be 0.005 g. Why was the recovery so low?

CHAPTER 12

Extractions, Separations, and Drying Agents

Part A. Theory

12.1 EXTRACTION

Transferring a solute from one solvent into another is called **extraction,** or more precisely, liquid–liquid extraction. The solute is extracted from one solvent into the other because the solute is more soluble in the second solvent than in the first. The two solvents must not be **miscible** (mix freely), and they must form two separate **phases** or layers, in order for this procedure to work. Extraction is used in many ways in organic chemistry. Many **natural products** (organic chemicals that exist in nature) are present in animal and plant tissues having high water content. Extracting these tissues with a water-immiscible solvent is useful for isolating the natural products. Often, diethyl ether (commonly referred to as "ether") is used for this purpose. Sometimes, alternative water-immiscible solvents such as hexane, petroleum ether, ligroin, and methylene chloride are used. For instance, caffeine, a natural product, can be extracted from an aqueous tea solution by shaking it successively with several portions of methylene chloride.

A generalized extraction process, using a specialized piece of glassware called a **separatory funnel,** is illustrated in Figure 12.1. The first solvent contains a mixture of black and white molecules (Fig. 12.1A). A second solvent that is not miscible with the first is added. After the separatory funnel is capped and shaken, the layers separate. In this example the second solvent (shaded) is less dense than the first, so it becomes the top layer (Fig. 12.1B). Because of differences in physical properties, the white molecules are more soluble in the second solvent, while the black molecules are more soluble in the first solvent. Most of the white molecules are in the upper layer, but there are some black molecules there, too. Likewise, most of the black molecules are in the lower layer. However, there are still a few white molecules in this lower phase. The lower phase may be separated from the upper phase by opening the stopcock at the bottom of the separatory funnel and allowing the lower layer to drain into a beaker (Fig. 12.1C). In this example, notice that it was not possible to effect a complete separation of the two types of molecules with a single extraction. This is a common occurrence in organic chemistry.

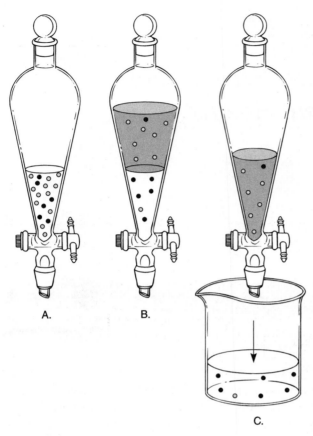

A. Solvent 1 contains a mixture of molecules (black and white).

B. After shaking with solvent 2 (shaded), most of the white molecules have been extracted into the new solvent. The white molecules are more soluble in the second solvent, while the black molecules are more soluble in the original solvent.

C. With removal of the lower phase, the black and white molecules have been partially separated.

Figure 12.1 The extraction process.

Many substances are soluble in both water and organic solvents. Water can be used to extract or "wash" water-soluble impurities from an organic reaction mixture. To carry out a "washing" operation, you add water and an immiscible organic solvent to the reaction mixture contained in a separatory funnel. After stoppering the funnel and shaking it, you allow the organic layer and the aqueous (water) layer to separate. A water wash removes highly polar and water-soluble materials, such as sulfuric acid, hydrochloric acid, or sodium hydroxide from the organic layer. The washing operation helps to purify the desired organic compound present in the original reaction mixture.

12.2 DISTRIBUTION COEFFICIENT

When a solution (solute A in solvent 1) is shaken with a second solvent (solvent 2) with which it is not miscible, the solute distributes itself between the two liquid phases. When the two phases have separated again into two distinct solvent layers, an equilibrium will have been achieved such that the ratio of the concentrations of the solute in each layer defines a constant. The constant, called the **distribution coefficient** (or partition coefficient) K, is defined by

$$K = \frac{C_2}{C_1}$$

where C_1 and C_2 are the concentrations at equilibrium, in grams per liter or milligrams per milliliter of solute A in solvent 1 and in solvent 2, respectively. This relationship is a ratio of two concentrations and is independent of the actual amounts of the two solvents mixed. The distribution coefficient has a constant value for each solute considered and depends on the nature of the solvents used in each case.

Not all the solute will be transferred to solvent 2 in a single extraction unless K is very large. Usually, it takes several extractions to remove all the solute from solvent 1. In extracting a solute from a solution, it is always better to use several small portions of the second solvent than to make a single extraction with a large portion. Suppose, as an illustration, a particular extraction proceeds with a distribution coefficient of 10. The system consists of 5.0 g of organic compound dissolved in 100 mL of water (solvent 1). In this illustration, the effectiveness of three 50-mL extractions with ether (solvent 2) is compared with one 150-mL extraction with ether. In the first 50-mL extraction, the amount extracted into the ether layer is given by the following calculation. The amount of compound remaining in the aqueous phase is given by x.

$$K = 10 = \frac{C_2}{C_1} = \frac{\left(\dfrac{5.0 - x}{50}\ \dfrac{\text{g}}{\text{mL ether}}\right)}{\left(\dfrac{x}{100}\ \dfrac{\text{g}}{\text{mL H}_2\text{O}}\right)}; \qquad 10 = \frac{(5.0 - x)(100)}{50x}$$

$$500x = 500 - 100x$$
$$600x = 500$$
$$x = 0.83 \text{ g remaining in the aqueous phase}$$
$$5.0 - x = 4.17 \text{ g in the ether layer}$$

As a check on the calculation, it is possible to substitute the value 0.83 g for x in the original equation and demonstrate that the concentration in the ether layer divided by the concentration in the water layer equals the distribution coefficient.

$$\frac{\left(\dfrac{5.0 - x}{50}\ \dfrac{\text{g}}{\text{mL ether}}\right)}{\left(\dfrac{x}{100}\ \dfrac{\text{g}}{\text{mL H}_2\text{O}}\right)} = \frac{\dfrac{4.17}{50}}{\dfrac{0.83}{100}} = \frac{0.083 \text{ g/mL}}{0.0083 \text{ g/mL}} = 10 = K$$

The second extraction with another 50-mL portion of fresh ether is performed on the aqueous phase, which now contains 0.83 g of the solute. The amount of solute extracted is given by the calculation shown in Figure 12.2. Also shown in the figure is a calculation for a third extraction with another 50-mL portion of ether. This third extraction will transfer 0.12 g of solute into the ether layer, leaving 0.02 g of solute remaining in the water layer. A total of 4.98 g of solute will be extracted into the combined ether layers, and 0.02 g will remain in the aqueous phase.

Figure 12.3 shows the result of a *single* extraction with 150 mL of ether. As shown there, 4.69 g of solute was extracted into the ether layer, leaving 0.31 g of compound in the aqueous phase. Three successive 50-mL ether extractions (Fig. 12.2) succeeded in removing 0.29 g more solute from the aqueous phase than using one 150-mL portion of either (Fig. 12.3). This differential represents 5.8% of the total material.

NOTE: Several extractions with smaller amounts of solvent are more effective than one extraction with a larger amount of solvent.

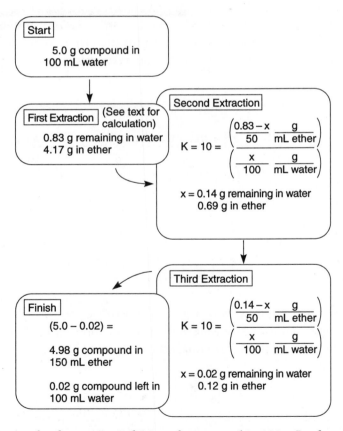

Figure 12.2 The result of extraction of 5.0 g of compound in 100 mL of water by three successive 50-mL portions of ether. Compare this result with that of Figure 12.3.

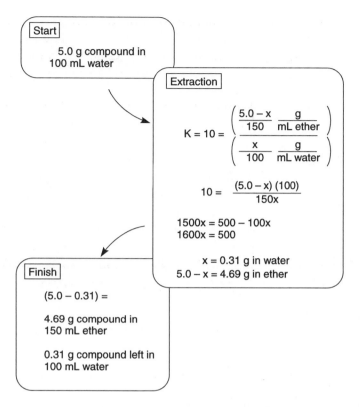

Figure 12.3 The result of extraction of 5.0 g of compound in 100 mL of water with one 150-mL portion of ether. Compare this result with that of Figure 12.2.

12.3 CHOOSING AN EXTRACTION METHOD AND A SOLVENT

Three types of apparatus are used for extractions: conical vials, centrifuge tubes, and separatory funnels (Fig. 12.4). Conical vials may be used with volumes of less than 4 mL; volumes of up to 10 mL may be handled in centrifuge tubes. A centrifuge tube equipped with a screw cap is particularly useful for extractions. Conical vials and centrifuge tubes are most often used in microscale experiments, although a centrifuge tube may also be used in some macroscale applications. The separatory funnel is used with larger volumes of liquid in macroscale experiments. The separatory funnel is discussed in Part B and the conical vial and centrifuge tube are discussed in Part C.

Most extractions consist of an aqueous phase and an organic phase. To extract a substance from an aqueous phase, an organic solvent that is not miscible with water must be used. Table 12.1 lists a number of the common organic solvents that are not miscible with water and are used for extractions.

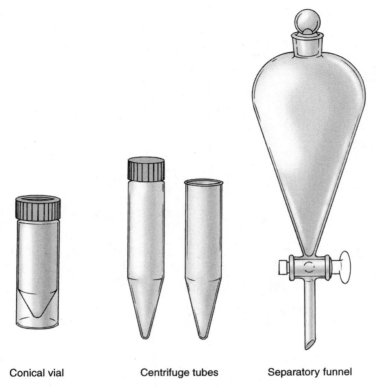

Conical vial   Centrifuge tubes   Separatory funnel

Figure 12.4 The apparatus used in extraction.

Solvents that have a density less than that of water (1.00 g/mL) will separate as the top layer when shaken with water. Solvents that have a density greater than that of water will separate into the lower layer. For instance, diethyl ether ($d = 0.71$ g/mL) when shaken with water will form the upper layer, whereas methylene chloride ($d = 1.33$ g/mL) will form the lower layer. When performing an extraction, slightly different methods are used to separate the lower layer (whether or not it is the aqueous layer or the organic layer) than to separate the upper layer.

Table 12.1 Densities of Common Extraction Solvents

Solvent	Density (g/mL)
Ligroin	0.67–0.69
Diethyl ether	0.71
Toluene	0.87
Water	1.00
Methylene chloride	1.330

Part B. Macroscale Extraction

12.4 THE SEPARATORY FUNNEL

A separatory funnel is illustrated in Figure 12.5. It is the piece of equipment used for carrying out extractions with medium to large quantities of material. To fill the separatory funnel, support it in an iron ring attached to a ring stand. Since it is easy to break a separatory funnel by "clanking" it against the metal ring, pieces of rubber tubing are often attached to the ring to cushion the funnel as shown in Figure 12.5. These are

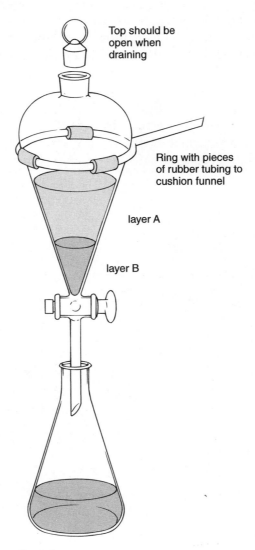

Figure 12.5 A separatory funnel.

short pieces of tubing, cut to a length of about 3 cm and slit open along their length. When slipped over the inside of the ring, they cushion the funnel in its resting place.

When beginning an extraction, the first step is to close the stopcock. (Don't forget!) Using a powder funnel (wide bore) placed in the top of the separatory funnel, fill it with both the solution to be extracted and the extraction solvent. Swirl the funnel gently by holding it by its upper neck, and then stopper it. Pick up the separatory funnel with two hands and hold it as shown in Figure 12.6. Hold the stopper in place firmly because the two immiscible liquids will build pressure when they mix, and this pressure may force the stopper out of the separatory funnel. To release this pressure, vent the funnel by holding it upside down (hold the stopper securely) and slowly open the stopcock. Usually the rush of vapors out of the opening can be heard. Continue shaking and venting until the "whoosh" is no longer audible. Now continue shaking the mixture gently for about 1 minute. This can be done by inverting the funnel in a rocking motion repeatedly or, if the formation of an emulsion is not a problem (see Section 12.10, page 181), by shaking the funnel more vigorously for less time.

NOTE: There is an art to shaking and venting a separatory funnel correctly, and it usually seems awkward to the beginner. The technique is best learned by observing a person, such as your instructor, who is thoroughly familiar with the separatory funnel's use.

Figure 12.6 The correct way of shaking and venting a separatory funnel.

When you have finished mixing the liquids, place the separatory funnel in the iron ring and remove the top stopper immediately. The two immiscible solvents separate into two layers after a short time, and they can be separated from one another by draining most of the lower layer through the stopcock.[1] Allow a few minutes to pass so that any of the lower phase adhering to the inner glass surfaces of the separatory funnel can drain down. Open the stopcock again and allow the remainder of the lower layer to drain until the interface between the upper and lower phases just begins to enter the bore of the stopcock. At this moment, close the stopcock and remove the remaining upper layer by pouring it from the top opening of the separatory funnel.

> **NOTE:** To minimize contamination of the two layers, the lower layer should always be drained from the bottom of the separatory funnel and the upper layer poured out from the top of the funnel.

When methylene chloride is used as the extracting solvent with an aqueous phase, it will settle to the bottom and be removed through the stopcock. The aqueous layer remains in the funnel. A second extraction of the remaining aqueous layer with fresh methylene chloride may be needed.

With a diethyl ether (ether) extraction of an aqueous phase, the organic layer will form on top. Remove the lower aqueous layer through the stopcock and pour the upper ether layer from the top of the separatory funnel. Pour the aqueous phase back into the separatory funnel and extract it a second time with fresh ether. The combined organic phases must be dried using a suitable drying agent (Section 12.9) before the solvent is removed.

For microscale procedures, a 60- or 125-mL separatory funnel is recommended. Because of surface tension, water has a difficult time draining from the bore of smaller funnels. Funnels larger than 125 mL are simply too large for microscale experiments, and a good deal of material is lost in "wetting" their surfaces.

Part C. Microscale Extraction

12.5 THE CONICAL VIAL—SEPARATING THE LOWER LAYER

Before using a conical vial for an extraction, make sure that the capped conical vial does not leak when shaken. To do this, place some water in the conical vial, place the Teflon liner in the cap, and screw the cap securely onto the conical vial. Shake the vial vigorously and check for leaks. Conical vials that are used for extractions must not be chipped on the edge of the vial or they will not seal adequately. If there is a leak, try

---

[1]A common error is to try to drain the separatory funnel without removing the top stopper. Under this circumstance, the funnel will not drain, because a partial vacuum is created in the space above the liquid.

tightening the cap or replacing the Teflon liner with another one. Sometimes it helps to use the silicone rubber side of the liner to seal the conical vial. Some laboratories are supplied with Teflon stoppers that fit into the 5-mL conical vials. You may find that this stopper eliminates leakage.

When shaking the conical vial, do it gently at first in a rocking motion. When it is clear that an emulsion will not form (see Section 12.10, page 181), you can shake it more vigorously.

In some cases, adequate mixing can be achieved by spinning your microspatula for at least 10 minutes in the conical vial. Another technique of mixing involves drawing the mixture up into a Pasteur pipet and squirting it rapidly back into the vial. Repeat this process for at least 5 minutes to obtain an adequate extraction.

The 5-mL conical vial is the most useful piece of equipment for carrying out extractions on a microscale level. In this section, we consider the method for removing the lower layer. A concrete example would be the extraction of a desired product from an aqueous layer using methylene chloride ($d = 1.33$ g/mL) as the extraction solvent. Methods for removal of the upper layer are discussed in the next section.

> **NOTE:** Always place a conical vial in a small beaker to prevent the vial from falling over.

Removing the Lower Layer.   Suppose that we extract an aqueous solution with methylene chloride. This solvent is more dense than water and will settle to the bottom of the conical vial. Use the following procedure, which is illustrated in Figure 12.7, to remove the lower layer.

1. Place the aqueous phase containing the dissolved product into a 5-mL conical vial (Fig. 12.7A).
2. Add about 1 mL of methylene chloride, cap the vial, and shake the mixture gently at first in a rocking motion and then more vigorously when it is clear that an emulsion will not form. Vent or unscrew the cap slightly to release the pressure in the vial. Allow the phases to separate completely so that you can detect two distinct layers in the vial. The organic phase will be the lower layer in the vial (Fig. 12.7B). If necessary, tap the vial with your finger or stir the mixture gently if some of the organic phase is suspended in the aqueous layer.
3. Prepare a Pasteur filter-tip pipet (Chapter 8, Section 8.6, page 112) using a 5¾-inch pipet. Attach a 2-mL rubber bulb to the pipet, depress the bulb, and insert the pipet into the vial so that the tip touches the bottom (Fig. 12.7C). The filter-tip pipet gives you better control in removing the lower layer. In some cases, however, you may be able to use a Pasteur pipet (no filter tip), but considerably more care must be taken to avoid losing liquid from the pipet during the transfer operation. With experience, you should be able to judge how much to squeeze the bulb to draw in the desired volume of liquid.

A. The aqueous solution contains the desired product.

B. Methylene chloride is used to extract the aqueous phase.

C. The Pasteur filter-tip pipet is placed in the vial.

D. The lower organic layer is removed from the aqueous phase.

E. The organic layer is transferred to a dry test tube or conical vial. The aqueous layer remains in the original extraction vial.

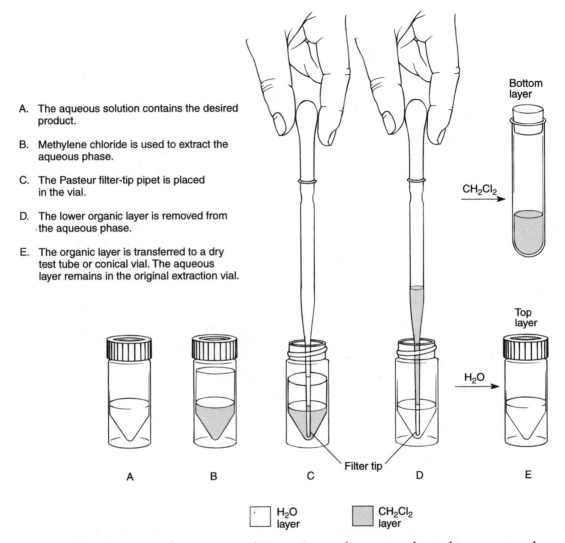

H_2O layer CH_2Cl_2 layer

Figure 12.7 Extraction of an aqueous solution using a solvent more dense than water: methylene chloride.

4. Slowly draw the lower layer (methylene chloride) into the pipet in such a way that you exclude the aqueous layer and any emulsion (Section 12.10) that might be at the interface between the layers (Fig. 12.7D). Be sure to keep the tip of the pipet squarely in the V at the bottom of the vial.

5. Transfer the withdrawn organic phase into a *dry* test tube or another *dry* conical vial if one is available. It is best to have the test tube or vial located next to the extraction vial. Hold the vials in the same hand between your index finger and thumb, as shown in Figure 12.8. This avoids messy and disastrous transfers. The aqueous layer (upper layer) is left in the original conical vial (Fig. 12.7E).

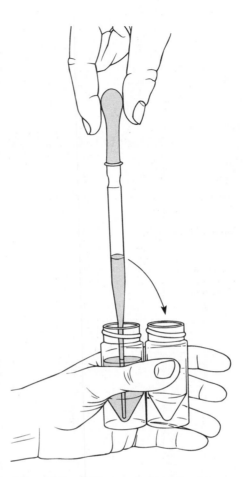

Figure 12.8 Method for holding vials while transferring liquids.

In performing an actual extraction in the laboratory, you would extract the aqueous phase with a second 1-mL portion of fresh methylene chloride to achieve a more complete extraction. Steps 2–5 would be repeated, and the organic layers from both extractions would be combined. In some cases, you may need to extract a third time with yet another 1-mL portion of methylene chloride. Again, the methylene chloride would be combined with the other extracts. The overall process would use three 1-mL portions of methylene chloride to transfer the product from the water layer into methylene chloride. Sometimes you will see the statement "extract the aqueous phase with three 1-mL portions of methylene chloride" in an experimental procedure. This statement describes in a shorter fashion the process described previously. Finally, the methylene chloride extracts will contain some water and must be dried with a drying agent as indicated in Section 12.9.

NOTE: If an organic solvent has been extracted with water, it should be dried with a drying agent (see Section 12.9) before proceeding.

In this example, we extracted water with the heavy solvent methylene chloride and removed it as the lower layer. If you were extracting a light solvent (for instance, diethyl ether) with water, and you wished to keep the water layer, the water would be the lower layer and would be removed using the same procedure. You would not dry the water layer, however.

12.6 THE CONICAL VIAL—SEPARATING THE UPPER LAYER

In this section, we consider the method used when you wish to remove the upper layer. A concrete example would be the extraction of a desired product from an aqueous layer using diethyl ether ($d = 0.71$ g/mL) as the extraction solvent. Methods for removing the lower layer were discussed previously.

> **NOTE:** Always place a conical vial in a small beaker to prevent the vial from falling over.

Removing the Upper Layer. Suppose we extract an aqueous solution with diethyl ether (ether). This solvent is less dense than water and will rise to the top of the conical vial. Use the following procedure, which is illustrated in Figure 12.9, to remove the upper layer.

1. Place the aqueous phase containing the dissolved product in a 5-mL conical vial (Fig. 12.9A).
2. Add about 1 mL of ether, cap the vial, and shake the mixture vigorously. Vent or unscrew the cap slightly to release the pressure in the vial. Allow the phases to separate completely so that you can detect two distinct layers in the vial. The ether phase will be the upper layer in the vial (Fig. 12.9B).
3. Prepare a Pasteur filter-tip pipet (Chapter 8, Section 8.6, page 112) using a 5¾-inch pipet. Attach a 2-mL rubber bulb to the pipet, depress the bulb, and insert the pipet into the vial so that the tip touches the bottom. The filter-tip pipet gives you better control in removing the lower layer. In some cases, however, you may be able to use a Pasteur pipet (no filter tip), but considerably more care must be taken to avoid losing liquid from the pipet during the transfer operation. With experience, you should be able to judge how much to squeeze the bulb to draw in the desired volume of liquid. Slowly draw the lower *aqueous* layer into the pipet. Be sure to keep the tip of the pipet squarely in the V at the bottom of the vial (Fig. 12.9C).
4. Transfer the withdrawn aqueous phase into a test tube or another conical vial for temporary storage. It is best to have the test tube or vial located next to the extraction vial. This avoids messy and disastrous transfers. Hold the vials in the same hand between your index finger and thumb as shown in Figure 12.8. The ether layer is left behind in the conical vial (Fig. 12.9D).

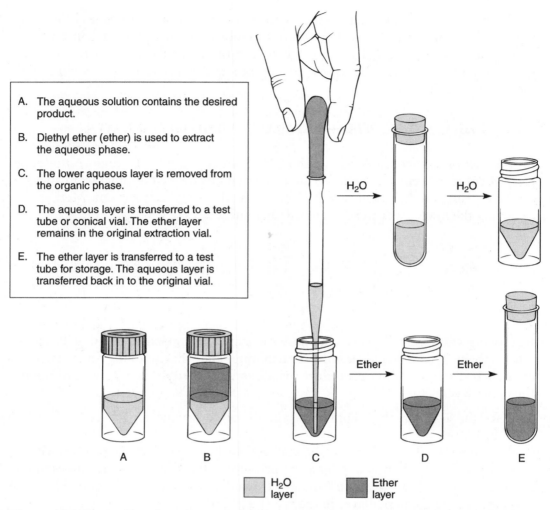

A. The aqueous solution contains the desired product.

B. Diethyl ether (ether) is used to extract the aqueous phase.

C. The lower aqueous layer is removed from the organic phase.

D. The aqueous layer is transferred to a test tube or conical vial. The ether layer remains in the original extraction vial.

E. The ether layer is transferred to a test tube for storage. The aqueous layer is transferred back in to the original vial.

H_2O

H_2O

Ether

Ether

A   B   C   D   E

H_2O layer

Ether layer

Figure 12.9 Extraction of an aqueous solution using a solvent less dense than water: diethyl ether.

5. The ether phase remaining in the original conical vial should be transferred with a Pasteur pipet into a test tube for storage and the aqueous phase returned to the original conical vial (Fig. 12.9E).

In performing an actual extraction, you would extract the aqueous phase with another 1-mL portion of fresh ether to achieve a more complete extraction. Steps 2–5 would be repeated, and the organic layers from both extractions would be combined in the test tube. In some cases, you may need to extract the aqueous layer a third time with yet another 1-mL portion of ether. Again, the ether would be combined with the other two layers. This overall process uses three 1-mL portions of ether to transfer the product from the water layer into ether. The ether extracts contain some water and must be dried with a drying agent as indicated in Section 12.9.

12.7 THE SCREW-CAP CENTRIFUGE TUBE

If you require an extraction that uses a larger volume than a conical vial can accommodate (about 4 mL), a centrifuge tube can often be used. A commonly available size of centrifuge tube has a volume of about 15 mL, and is supplied with a screw cap. In performing an extraction with a screw-capped centrifuge tube, the same procedures outlined for the conical vial (Sections 12.5 and 12.6) are used. As is the case for a conical vial, the tapered bottom of the centrifuge tube makes it easy to withdraw the lower layer with a Pasteur pipet.

> **NOTE:** A centrifuge tube has a large advantage over other methods of extraction. If an emulsion (Section 12.10) forms, you can use a centrifuge to aid in the separation of the layers.

You should check the capped centrifuge tube for leaks by filling it with water and shaking it vigorously. If it leaks, try replacing the cap with a different one. A **vortex mixer,** if available, provides an alternative to shaking the tube. In fact, a vortex mixer works well with a variety of containers, including small flasks, test tubes, conical vials, and centrifuge tubes. You start the mixing action on a vortex mixer by holding the test tube or other container on one of the neoprene pads. The unit mixes the sample by high-frequency vibration.

Part D. Additional Experimental Considerations: Macroscale and Microscale

12.8 HOW DO YOU DETERMINE WHICH ONE IS THE ORGANIC LAYER?

A common problem encountered during an extraction is trying to determine which of the two layers is the organic layer and which is the aqueous (water) layer. The most common situation is when the aqueous layer is on the bottom in the presence of an upper organic layer consisting of ether, ligroin, petroleum ether, or hexane (see densities in Table 12.1). However, the aqueous layer will be on the top when you use methylene chloride as a solvent (again, see Table 12.1). Although a laboratory procedure may frequently identify the expected relative positions of the organic and aqueous layers, sometimes their actual positions are reversed. Surprises usually occur in situations in which the aqueous layer contains a high concentration of sulfuric acid or a dissolved ionic compound, such as sodium chloride. Dissolved substances greatly increase the density of the aqueous layer, which may lead to the aqueous layer being found on the bottom even when coexisting with a relatively dense organic layer such as methylene chloride.

> **NOTE:** Always keep both layers until you have actually isolated the desired compound or until you are certain where your desired substance is located.

To determine if a particular layer is the aqueous one, add a few drops of water to the layer. Observe closely as you add the water to see where it goes. If the layer is water, then the drops of added water will dissolve in the aqueous layer and increase its volume. If the added water forms droplets or a new layer, however, you can assume that the suspected aqueous layer is actually organic. You can use a similar procedure to identify a suspected organic layer. This time, try adding more of the solvent, such as methylene chloride. The organic layer should increase in size, without separation of a new layer, if the tested layer is actually organic.

When performing an extraction procedure on the microscale level, you can use the following approach to identify the layers. When both layers are present, it is always a good idea to think carefully about the volumes of materials that you have added to the conical vial. You can use the graduations on the vial to help determine the volumes of the layers in the vial. If, for example, you have 1 mL of methylene chloride in a vial and you add 2 mL of water, you should expect the water to be on top because it is less dense than methylene chloride. As you add the water, *watch to see where it goes*. By noting the relative volumes of the two layers, you should be able to tell which is the aqueous layer and which is the organic layer. This approach can also be used when performing an extraction procedure using a centrifuge tube. Of course, you can always test to see which layer is the aqueous layer by adding one or two drops of water, as described previously.

12.9 DRYING AGENTS

After an organic solvent has been shaken with an aqueous solution, it will be "wet"; that is, it will have dissolved some water even though its miscibility with water is not great. The amount of water dissolved varies from solvent to solvent; diethyl ether represents a solvent in which a fairly large amount of water dissolves. To remove water from the organic layer, use a **drying agent**. A drying agent is an *anhydrous* inorganic salt that acquires waters of hydration when exposed to moist air or a wet solution:

$$
\underset{\substack{\text{Anhydrous}\\\text{drying agent}}}{\overset{\text{Insoluble}}{Na_2SO_4(s)}} + \text{Wet Solution }(nH_2O) \longrightarrow \underset{\substack{\text{Hydrated}\\\text{drying agent}}}{\overset{\text{Insoluble}}{Na_2SO_4 \cdot nH_2O\ (s)}} + \text{Dry Solution}
$$

The insoluble drying agent is placed directly into the solution, where it acquires water molecules and becomes hydrated. If enough drying agent is used, all of the water can be removed from a wet solution, making it "dry" or free of water.

The following anhydrous salts are commonly used: sodium sulfate, magnesium sulfate, calcium chloride, calcium sulfate (Drierite), and potassium carbonate. These salts vary in their properties and applications. For instance, not all will absorb the same amount of water for a given weight, nor will they dry the solution in the same extent. **Capacity** refers to the amount of water a drying agent absorbs per unit weight. Sodium and magnesium sulfates absorb a large amount of water (high capacity), but magnesium sulfate dries a solution more completely. **Completeness** refers to a compound's effectiveness in removing all the water from a solution by the time equilibrium has been reached. Magnesium ion, a strong Lewis acid, sometimes causes rearrangements of compounds such as epoxides. Calcium chloride is a good drying agent but cannot be used with many compounds containing oxygen or nitrogen because it forms complexes. Calcium chloride absorbs methanol and ethanol in addition to water, so it is useful for removing these materials when they are present as impurities. Potassium carbonate is a base and is used for drying solutions of basic substances, such as amines. Calcium sulfate dries a solution completely but has a low capacity.

Anhydrous sodium sulfate is the most widely used drying agent. The granular variety is recommended because it is easier to remove the dried solution from it than from the powdered variety. Sodium sulfate is mild and effective. It will remove water from most common solvents, with the possible exception of diethyl ether, in which case a prior drying with saturated salt solution may be advised (see page 181). Sodium sulfate must be used at room temperature to be effective; it cannot be used with boiling solutions. Table 12.2 compares the various common drying agents.

Macroscale.   To dry a large amount of solution, you should add enough granular anhydrous sodium sulfate to give a 1–3 mm layer on the bottom of the flask, depending on the volume of the solution. Dry the solution for at least 15 minutes, occasionally swirling the flask. The mixture is dry if it appears clear and shows the common signs of a dry solution given in Table 12.3.

If the solution remains cloudy after treatment with the first batch of drying agent, add more drying agent and repeat the drying procedure. If the drying agent clumps badly, with no drying agent that will flow freely when the flask is swirled, you should transfer the solution to a clean, dry flask and add a fresh portion of drying agent. If droplets of water are present, you should also transfer the solution to a new container and add a fresh quantity of drying agent. You should not add more drying agent if a puddle (water layer) forms. Instead, separate the layers, using a separatory funnel if necessary, and then add fresh drying agent to the organic layer, placed in a new container. If the first batch of drying agent does not completely dry the solution, it is not unusual to have to transfer (decant) the solution to a clean, dry flask and add a fresh portion of drying agent.

When the solution is dry, the drying agent should be removed by using decantation (pouring carefully to leave the drying agent behind). With granular sodium sulfate, decantation is quite easy to perform because of the size of the drying agent particles. If a powdered drying agent, such as magnesium sulfate, is used, it may be necessary to use a gravity filtration (Chapter 8, Section 8.1B, page 103) to remove the

Table 12.2 Common Drying Agents

	Acidity	Hydrated	Capacity[a]	Complete-ness[b]	Rate[c]	Use
Magnesium sulfate	Neutral	$MgSO_4 \cdot 7H_2O$	High	Medium	Rapid	General
Sodium sulfate	Neutral	$Na_2SO_4 \cdot 7H_2O$ $Na_2SO_4 \cdot 10H_2O$	High	Low	Medium	General
Calcium chloride	Neutral	$CaCl_2 \cdot 2H_2O$ $CaCl_2 \cdot 6H_2O$	Low	High	Rapid	Hydro carbons Halides
Calcium sulfate (Drierite)	Neutral	$CaSO_4 \cdot \frac{1}{2}H_2O$ $CaSO_4 \cdot 2H_2O$	Low	High	Rapid	General
Potassium carbonate	Basic	$K_2CO_3 \cdot 1\frac{1}{2}H_2O$ $K_2CO_3 \cdot 2H_2O$	Medium	Medium	Medium	Amines, esters, bases, ketones
Potassium hydroxide	Basic	—	—	—	Rapid	Amines only
Molecular sieves (3 or 4 Å)	Neutral	—	High	Extremely high	—	General

[a]Amount of water removed per given weight of drying agent.
[b]Refers to amount of H_2O still in solution at equilibrium with drying agent.
[c]Refers to rate of action (drying).

drying agent. The solvent is removed by distillation (Chapter 14, Section 14.3, page 206) or evaporation (Chapter 7, Section 7.10, page 95).

Microscale.   Before attempting to dry an organic layer, check closely to see that there are no visible signs of water. If you see droplets of water in the organic layer or water droplets clinging to the sides of the conical vial or test tube, transfer the organic layer with a *dry* Pasteur pipet to a *dry* container before adding any drying agent. Now add one spatulaful of granular anhydrous sodium sulfate (or other drying agent) from the V-grooved end of a microspatula into a solution contained in a conical vial or test

Table 12.3 Common Signs That Indicate a Solution Is Dry

1. There are no visible water droplets on the side of flask or suspended in solution.
2. There is not a separate layer of liquid or a "puddle."
3. The solution is clear, not cloudy. Cloudiness indicates water is present.
4. The drying agent (or a portion of it) flows freely on the bottom of the container when stirred or swirled and does not "clump" together as a solid mass.

tube. If all the drying agent "clumps," add another spatulaful of sodium sulfate. Dry the solution for at least 15 minutes. Stir the mixture occasionally with a spatula during that period. The mixture is dry if there are no visible signs of water and the drying agent flows freely in the container when stirred with a microspatula. The solution should not be cloudy. Add more drying agent if necessary. You should not add more drying agent if a "puddle" (water layer) forms or if drops of water are visible. Instead, you should transfer the organic layer to a dry container before adding fresh drying agent. When dry, use a *dry* Pasteur pipet or a *dry* filter-tip pipet (Chapter 8, Section 8.6, page 112) to remove the solution from the drying agent and transfer the solution to a *dry* conical vial. Rinse the drying agent with a small amount of fresh solvent and transfer this solvent to the vial containing the solution. Remove the solvent by evaporation using heat and a stream of air or nitrogen (Chapter 7, Section 7.10, page 95).

An alternative method of drying an organic phase is to pass it through a filtering pipet (Chapter 8, Section 8.1C, page 105) that has been packed with a small amount (ca. 2 cm) of drying agent. Again, the solvent is removed by evaporation.

Saturated Salt Solution.   At room temperature, diethyl ether (ether) dissolves 1.5% by weight of water, and water dissolves 7.5% of ether. Ether, however, dissolves a much smaller amount of water from a saturated aqueous sodium chloride solution. Hence, the bulk of water in ether, or ether in water, can be removed by shaking it with a saturated aqueous sodium chloride solution. A solution of high ionic strength is usually not compatible with an organic solvent and forces separation of it from the aqueous layer. The water migrates into the concentrated salt solution. The ether phase (organic layer) will be on top, and the saturated sodium chloride solution will be on the bottom ($d = 1.2$ g/mL). After removing the organic phase from the aqueous sodium chloride, dry the organic layer completely with sodium sulfate or with one of the other drying agents listed in Table 12.2.

12.10 EMULSIONS

An **emulsion** is a colloidal suspension of one liquid in another. Minute droplets of an organic solvent are often held in suspension in an aqueous solution when the two are mixed or shaken vigorously; these droplets form an emulsion. This is especially true if any gummy or viscous material was present in the solution. Emulsions are often encountered in performing extractions. Emulsions may require a long time to separate into two layers and are a nuisance to the organic chemist.

Fortunately, several techniques may be used to break a difficult emulsion once it has formed.

1. Often an emulsion will break up if it is allowed to stand for some time. Patience is important here. Gently stirring with a stirring rod or spatula may also be useful.
2. If one of the solvents is water, adding a saturated aqueous sodium chloride solution will help destroy the emulsion. The water in the organic layer migrates into the concentrated salt solution.

3. With microscale experiments, the mixture may be transferred to a centrifuge tube. The emulsion will often break during centrifugation. Remember to place another tube filled with water on the opposite side of the centrifuge to balance it. Both tubes should weigh the same.

4. Adding a very small amount of a water-soluble detergent may also help. This method has been used in the past for combating oil spills. The detergent helps to solubilize the tightly bound oil droplets.

5. Gravity filtration (see Chapter 8, Section 8.1, page 102) may help to destroy an emulsion by removing gummy polymeric substances. With large-scale reactions, you might try filtering the mixture through a fluted filter (Chapter 8, Section 8.1B, page 103) or a piece of cotton. With small-scale reactions, a filtering pipet may work (Chapter 8, Section 8.1C, page 105). In many cases, once the gum is removed, the emulsion breaks up rapidly.

6. If you are using a separatory funnel, you might try to use a gentle swirling action in the funnel to help break an emulsion. Gently stirring with a stirring rod may also be useful.

When you know through prior experience that a mixture may form a difficult emulsion, you should avoid shaking the mixture vigorously. When using conical vials for extractions, it may be better to use a magnetic spin vane for mixing and not shake the mixture at all. When using separatory funnels, extractions should be performed with gentle swirling instead of shaking or with several gentle inversions of the separatory funnel. Do not shake the separatory funnel vigorously in these cases. It is important to use a longer extraction period if the more gentle techniques described in this paragraph are being employed. Otherwise you will not transfer all the material from the first phase to the second one.

12.11 PURIFICATION AND SEPARATION METHODS

In nearly all synthetic experiments undertaken in the organic laboratory, a series of operations involving extractions is used after the actual reaction has been concluded. These extractions form an important part of the purification. Using them, the desired product is separated from unreacted starting materials or from undesired side products in the reaction mixture. These extractions may be grouped into three categories, depending on the nature of the impurities they are designed to remove.

The first category involves extracting or "washing" an organic mixture with water. Water washes are designed to remove highly polar materials, such as inorganic salts, strong acids or bases, and low-molecular-weight, polar substances including alcohols, carboxylic acids, and amines. Many organic compounds containing fewer than five carbons are water soluble. Water extractions are also used immediately following extractions of a mixture with either acid or base to ensure that all traces of acid or base have been removed.

The second category concerns extraction of an organic mixture with a dilute acid, usually 1–2 M hydrochloric acid. Acid extractions are intended to remove basic impurities, especially such basic impurities as organic amines. The bases are converted to their corresponding cationic salts by the acid used in the extraction. If an amine is one of the reactants, or if pyridine or another amine is a solvent, such an extraction might be used to remove any excess amine present at the end of a reaction.

$$RNH_2 + HCl \longrightarrow RNH_3^+Cl^-$$
(water-soluble ammonium salt)

Cationic ammonium salts are usually soluble in the aqueous solution, and they are thus extracted from the organic material. A water extraction may be used immediately following the acid extraction to ensure that all traces of the acid have been removed from the organic material.

The third category is extraction of an organic mixture with a dilute base, usually 5% sodium bicarbonate, although extractions with dilute sodium hydroxide can also be used. Such basic extractions are intended to convert acidic impurities, such as organic acids, to their corresponding anionic salts. For example, in the preparation of an ester, a sodium bicarbonate extraction might be used to remove any excess carboxylic acid that is present.

$$RCOOH + NaHCO_3 \longrightarrow RCOO^-Na^+ + H_2O + CO_2$$
(pK_a ~ 5)   (water-soluble carboxylate salt)

Anionic carboxylate salts, being highly polar, are soluble in the aqueous phase. As a result, these acid impurities are extracted from the organic material into the basic solution. A water extraction may be used after the basic extraction to ensure that all the base has been removed from the organic material.

Occasionally, phenols may be present in a reaction mixture as impurities, and removing them by extraction may be desired. Because phenols, although they are acidic, are about 10^5 times less acidic than carboxylic acids, basic extractions may be used to separate phenols from carboxylic acids by a careful selection of the base. If sodium bicarbonate is used as a base, carboxylic acids are extracted into the aqueous base, but phenols are not. Phenols are not sufficiently acidic to be deprotonated by the weak base bicarbonate. Extraction with sodium hydroxide, on the other hand, extracts both carboxylic acids and phenols into the aqueous basic solution, because hydroxide ion is a sufficiently strong base to deprotonate phenols.

Mixtures of acidic, basic, and neutral compounds are easily separated by extraction techniques. One such example is shown in Figure 12.10.

Organic acids or bases that have been extracted can be regenerated by neutralizing the extraction reagent. This would be done if the organic acid or base were a product

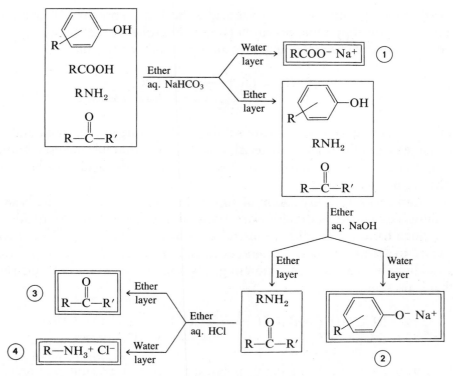

Figure 12.10 Separating a four-component mixture by extraction.

of a reaction rather than an impurity. For example, if a carboxylic acid has been extracted with aqueous base, the compound can be regenerated by acidifying the extract with 6 *M* HCl until the solution becomes *just* acidic, as indicated by litmus or pH paper. When the solution becomes acidic, the carboxylic acid will separate from the aqueous solution. If the acid is a solid at room temperature, it will precipitate and can be purified by filtration and crystallization. If the acid is a liquid, it will form a separate layer. In this case, it would usually be necessary to extract the mixture with ether or methylene chloride. After removing the organic layer and drying it, the solvent can be evaporated to yield the carboxylic acid.

In the example shown in Figure 12.10, you also need to perform a drying step at (3) before isolating the neutral compound. When the solvent is ether, you should first extract the ether solution with saturated aqueous sodium chloride to remove much of the water. The ether layer is then dried over a drying agent such as anhydrous sodium sulfate. If the solvent were methylene chloride, it would not be necessary to do the step with saturated sodium chloride.

When performing acid–base extractions, it is common practice to extract a mixture several times with the appropriate reagent. For example, if you were extracting a carboxylic acid from a mixture, you might extract the mixture three times with 2-mL portions of 1 *M* NaOH. In most published experiments, the procedure will specify the volume and concentration of extracting reagent and the number of times to do the extractions. If this information is not given, you must devise your own procedure. Using a carboxylic acid as an example, if you know the identity of the acid and the approximate amount present, you can actually calculate how much sodium hydroxide is needed. Because the carboxylic acid (assuming it is monoprotic) will react with sodium hydroxide in a 1:1 ratio, you would need the same number of moles of sodium hydroxide as there are moles of acid. To ensure that all the carboxylic acid is extracted, you should use about a *two-fold* excess of the base. From this you could calculate the number of milliliters of base needed. This should be divided into two or three equal portions, one portion for each extraction. In a similar fashion, you could calculate the amount of 5% sodium bicarbonate required to extract an acid or the amount of 1 *M* HCl required to extract a base. If the amount of organic acid or base is not known, then the situation is more difficult. A guideline that sometimes works is to do two or three extractions so that the total volume of the extracting reagent is approximately equal to the volume of the organic layer. To test this procedure, neutralize the aqueous layer from the last extraction. If a precipitate or cloudiness results, perform another extraction and test again. When no precipitate forms, you know that all the organic acid or base has been removed.

For some applications of acid–base extraction, an additional step, called **back-washing** or **back extraction,** is added to the scheme shown in Figure 12.10. Consider the first step, in which the carboxylic acid is extracted by sodium bicarbonate. This aqueous layer may contain some unwanted neutral organic material from the original mixture. To remove this contamination, the aqueous layer is backwashed with an organic solvent such as ether or methylene chloride. After shaking the mixture and allowing the layers to separate, the organic layer is removed and discarded. This technique may also be used when an amine is extracted with hydrochloric acid. The resulting aqueous layer is backwashed with an organic solvent to remove unwanted neutral material.

Part E. Continuous Extraction Methods

12.12 CONTINUOUS SOLID–LIQUID EXTRACTION

The technique of liquid–liquid extraction was described in Sections 12.1–12.8. In this section, solid–liquid extraction is described. Solid–liquid extraction is often used to extract a solid natural product from a natural source, such as a plant. A solvent is chosen that selectively dissolves the desired compound but that leaves behind the undesired insoluble solid. A continuous solid–liquid extraction apparatus, called a Soxhlet extractor, is commonly used in a research laboratory (Fig. 12.11).

As shown in Figure 12.11, the solid to be extracted is placed in a thimble made from filter paper, and the thimble is inserted into the central chamber. A low-boiling solvent, such as diethyl ether, is placed in the round-bottom distilling flask and is heated to reflux. The vapor rises through the left sidearm into the condenser where it

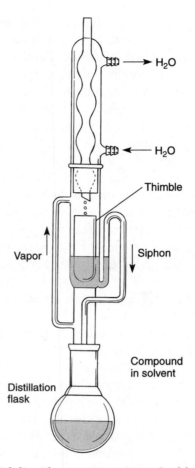

Figure 12.11 Continuous solid–liquid extraction using a Soxhlet extractor.

liquefies. The condensate (liquid) drips into the thimble containing the solid. The hot solvent begins to fill the thimble and extracts the desired compound from the solid. Once the thimble is filled with solvent, the sidearm on the right acts as a siphon, and the solvent, which now contains the dissolved compound, drains back into the distillation flask. The vaporization, condensation, extraction, siphoning process is repeated hundreds of times, and the desired product is concentrated in the distillation flask. The product is concentrated in the flask, because it has a boiling point higher than that of the solvent or because it is a solid.

12.13 CONTINUOUS LIQUID–LIQUID EXTRACTION

When a product is very soluble in water, it is often difficult to extract using the techniques described in Sections 12.4–12.7, because of an unfavorable distribution coefficient. In this case, you need to extract the aqueous solution numerous times with fresh batches of an immiscible organic solvent to remove the desired product from water. A less labor-intensive technique involves the use of a continuous liquid–liquid extraction apparatus. One type of extractor, used with solvents that are less dense than water, is shown in Figure 12.12. Diethyl ether is usually the solvent of choice.

The aqueous phase is placed in the extractor, which is then filled with diethyl ether up to the sidearm. The round-bottom distillation flask is partially filled with ether. The ether is heated to reflux in the round-bottom flask, and the vapor is liquefied in the water-cooled condenser. The ether drips into the central tube, passes through the porous sintered glass tip, and flows through the aqueous layer. The solvent extracts the desired compound from the aqueous phase, and the ether is recycled back into the round-bottom flask. The product is concentrated in the flask. The extraction is rather inefficient and must be placed in operation for at least 24 hours to remove the compound from the aqueous phase.

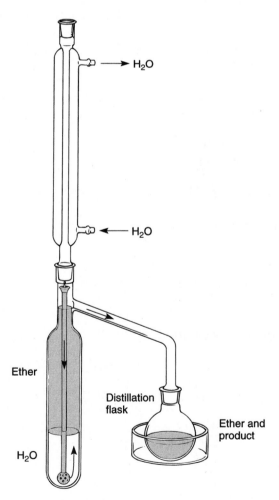

Figure 12.12 Continuous liquid–liquid extraction using a solvent less dense than water.

PROBLEMS

1. Suppose solute A has a distribution coefficient of 1.0 between water and diethyl ether. Demonstrate that if 100 mL of a solution of 5.0 g of A in water was extracted with two 25-mL portions of ether, a smaller amount of A would remain in the water than if the solution were extracted with one 50-mL portion of ether.

2. Write an equation to show how you could recover the parent compounds from their respective salts (1, 2, and 4) shown in Figure 12.10.

3. Aqueous hydrochloric acid was used *after* the sodium bicarbonate and sodium hydroxide extractions in the separation scheme shown in Figure 12.10. Is it possible to use this reagent earlier in the separation scheme so as to achieve the same overall result? If so, explain where you would perform this extraction.

4. Using aqueous hydrochloric acid, sodium bicarbonate, or sodium hydroxide solutions, devise a separation scheme using the style shown in Figure 12.10 to separate the following two-component mixtures. All the substances are soluble in ether. Also indicate how you would recover each of the compounds from their respective salts.

   (a) Give two different methods for separating this mixture.

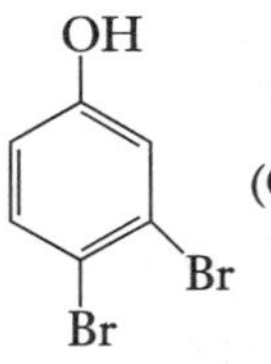

   $(CH_3CH_2CH_2CH_2)_3N$

   (b) Give two different methods for separating this mixture.

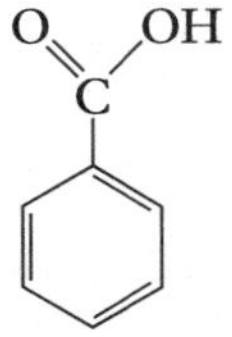

   $CH_3CH_2CH_2CH_2CH_2CH_2OH$

   (c) Give one method for separating this mixture.

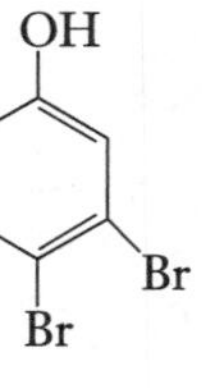

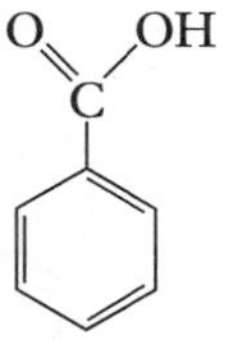

5. Solvents other than those in Table 12.1 may be used for extractions. Determine the relative positions of the organic layer and the aqueous layer in a conical vial or separatory funnel after shaking each of the following solvents with an aqueous phase. Find the densities for each of these solvents in a handbook (see Chapter 4, page 42).

   (a) 1,1,1-Trichloroethane

   (b) Hexane

6. A student prepares ethyl benzoate by the reaction of benzoic acid with ethanol using a sulfuric acid catalyst. The following compounds are found in the crude reaction mixture: ethyl benzoate (major component), benzoic acid, ethanol, and sulfuric acid. Using a handbook, obtain the solubility properties in water for each of these compounds (see Chapter 4, page 42). Indicate how you would remove benzoic acid, ethanol, and sulfuric acid from ethyl benzoate. At some point in the purification, you should also use an aqueous sodium bicarbonate solution.

7. Calculate the weight of water that could be removed from a wet organic phase using 50.0 mg of magnesium sulfate. Assume that it gives the hydrate listed in Table 12.2.

8. Explain exactly what you would do when performing the following laboratory instructions:

   (a) "Wash the organic layer with 1.0 mL of 5% aqueous sodium bicarbonate."

   (b) "Extract the aqueous layer three times with 1-mL portions of methylene chloride."

9. Just prior to drying an organic layer with a drying agent, you notice water droplets in the organic layer. What should you do next?

10. What should you do if there is some question about which layer is the organic one during an extraction procedure?

11. Saturated aqueous sodium chloride ($d = 1.2$ g/mL) is added to the following mixtures in order to dry the organic layer. Which layer is likely to be on the bottom in each case?

    (a) Sodium chloride layer or a layer containing a high-density organic compound dissolved in methylene chloride ($d = 1.4$ g/mL)

    (b) Sodium chloride layer or a layer containing a low-density organic compound dissolved in methylene chloride ($d = 1.1$ g/mL)

CHAPTER 13

Physical Constants of Liquids: The Boiling Point and Density

Part A. Boiling Points and Thermometer Correction

13.1 THE BOILING POINT

As a liquid is heated, the vapor pressure of the liquid increases to the point at which it just equals the applied pressure (usually atmospheric pressure). At this point, the liquid is observed to boil. The normal boiling point is measured at 760 mm Hg (760 torr) or 1 atm. At a lower applied pressure, the vapor pressure needed for boiling is also lowered, and the liquid boils at a lower temperature. The relation between applied pressure and temperature of boiling for a liquid is determined by its vapor pressure–temperature behavior. Figure 13.1 is an idealization of the typical vapor pressure–temperature behavior of a liquid.

Because the boiling point is sensitive to pressure, it is important to record the barometric pressure when determining a boiling point if the determination is being conducted at an elevation significantly above or below sea level. Normal atmospheric variations may affect the boiling point, but they are usually of minor importance.

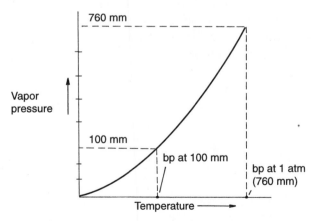

Figure 13.1 The vapor pressure–temperature curve for a typical liquid.

However, if a boiling point is being monitored during the course of a vacuum distillation (Chapter 16) that is being performed with an aspirator or a vacuum pump, the variation from the atmospheric value will be especially marked. In these cases, it is quite important to know the pressure as accurately as possible.

As a rule of thumb, the boiling point of many liquids drops about 0.5°C for a 10-mm decrease in pressure when in the vicinity of 760 mm Hg. At lower pressures, a 10°C drop in boiling point is observed for each halving of the pressure. For example, if the observed boiling point of a liquid is 150°C at 10 mm pressure, then the boiling point would be about 140°C at 5 mm Hg.

A more accurate estimate of the change in boiling point with a change of pressure can be made by using a nomograph. In Figure 13.2, a nomograph is given and a method is described for using it to obtain boiling points at various pressures when the boiling point is known at some other pressure.

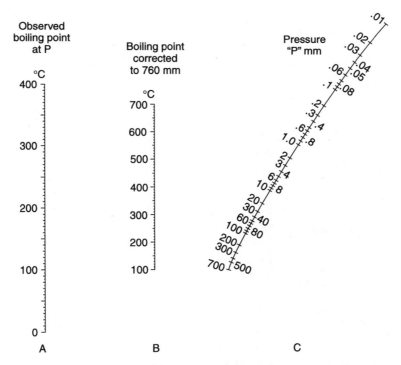

Figure 13.2 Pressure–temperature alignment nomograph. How to use the nomograph: Assume a reported boiling point of 100°C (column A) at 1 mm. To determine the boiling point at 18 mm, connect 100°C (column A) to 1 mm (column C) with a transparent plastic rule and observe where this line intersects column B (about 280°C). This value would correspond to the normal boiling point. Next, connect 280°C (column B) with 18 mm (column C) and observe where this intersects column A (151°C). The approximate boiling point will be 151°C at 18 mm. (Reprinted courtesy of MC/B Manufacturing Chemists, Inc.)

13.2 DETERMINING THE BOILING POINT—MICROSCALE METHODS

Two experimental methods of determining boiling points are easily available. When you have large quantities of material, you can simply record the boiling point (or boiling range) as viewed on a thermometer while performing a simple distillation (see Chapter 14). With smaller amounts of material, you can carry out a microscale or semimicroscale determination of the boiling point by using the apparatus shown in Figure 13.3.

Semimicroscale Method.   To carry out the semimicroscale determination, a piece of 5-mm glass tubing sealed at one end is attached to a thermometer with a rubber band or a thin slice of rubber tubing. The liquid whose boiling point is being determined is introduced with a Pasteur pipet into this piece of tubing, and a short piece of melting-point capillary (sealed at one end) is dropped in with the open end down. The whole unit is then placed in a Thiele tube. The rubber band should be placed above

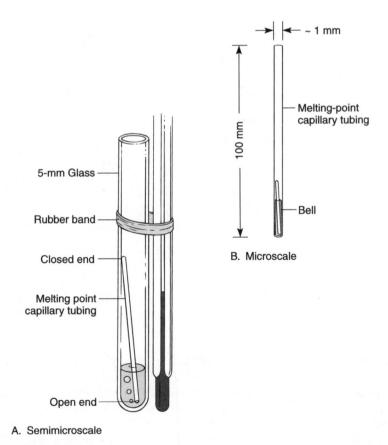

A. Semimicroscale

B. Microscale

Figure 13.3 Boiling-point determinations.

the level of the oil in the Thiele tube; otherwise the band may soften in the hot oil. When positioning the band, keep in mind that the oil will expand when heated. Next the Thiele tube is heated in the same fashion as described in Chapter 9, Section 9.6, page 120 for determining a melting point. Heating is continued until a rapid and continuous stream of bubbles emerges from the inverted capillary. At this point, you should stop heating. Soon, the stream of bubbles slows down and stops. When the bubbles stop, the liquid enters the capillary tube. The moment at which the liquid enters the capillary tube corresponds to the boiling point of the liquid, and the temperature is recorded.

Microscale Method.   In microscale experiments, there often is too little product available to use the semimicroscale method described above. However, the method can be scaled down in the following manner. The liquid is placed in a 1-mm melting-point capillary tube to a depth of about 4–6 mm. Use a syringe or a Pasteur pipet that has had its tip drawn thinner to transfer the liquid into the capillary tube. It may be necessary to use a centrifuge to transfer the liquid to the bottom of the tube. Next, prepare an appropriately-sized inverted capillary, or **bell.**

The easiest way to prepare a bell is to use a commercial micropipet, such as a 10-μL Drummond "microcap." These are available in vials of 50 or 100 microcaps and are very inexpensive. To prepare the bell, cut the microcap in half with a file or scorer and then seal one end by inserting it a small distance into a flame, turning it on its axis until the opening closes.

If microcaps are not available, a piece of 1-mm open-end capillary tubing (same size as a melting-point capillary) can be rotated along its axis in a flame while being held horizontally. Use your index fingers and thumbs to rotate the tube; do not change the distance between your two hands while rotating. When the tubing is soft, it is removed from the flame and pulled to a thinner diameter. When pulling, keep the tube straight by *moving both your hands and your elbows outward* by about 4 inches. Hold the pulled tube in place a few moments until it cools. Using the edge of a file or your fingernail, break out the thin center section. Seal one end of the thin section in the flame; then break it to a length that is about one and one-half times the height of your sample liquid (6–9 mm). Be sure the break is done squarely. Invert the bell (open end down), and place it in the capillary tube containing the sample liquid. Push the bell to the bottom with a fine copper wire if it adheres to the side of the capillary tube. A centrifuge may be used if you prefer. Figure 13.4 shows the construction method for the bell and the final assembly.

Place the microscale assembly in a standard melting-point apparatus (or a Thiele tube if an electrical apparatus is not available) to determine the boiling point. Heating is continued until a rapid and continuous stream of bubbles emerges from the inverted capillary. At this point, stop heating. Soon, the stream of bubbles slows down and stops. When the bubbles stop, the liquid enters the capillary tube. The moment at which the liquid enters the capillary tube corresponds to the boiling point of the liquid, and the temperature is recorded.

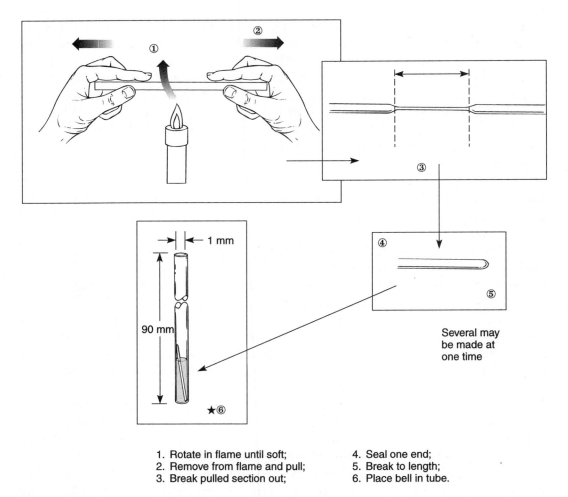

1. Rotate in flame until soft;
2. Remove from flame and pull;
3. Break pulled section out;
4. Seal one end;
5. Break to length;
6. Place bell in tube.

Figure 13.4 Construction of a microcapillary bell for microscale boiling-point determination.

Explanation of the Method. During the initial heating, the air trapped in the inverted bell expands and leaves the tube, giving rise to a stream of bubbles. When the liquid begins boiling, most of the air has been expelled; the bubbles of gas are due to the boiling action of the liquid. Once the heating is stopped, most of the vapor pressure left in the bell comes from the vapor of the heated liquid that seals its open end. There is always vapor in equilibrium with a heated liquid. If the temperature of the liquid is above its boiling point, the pressure of the trapped vapor will either exceed or equal the atmospheric pressure. As the liquid cools, its vapor pressure decreases. When the vapor pressure drops just below atmospheric pressure (just below the boiling point), the liquid is forced into the capillary tube.

Difficulties. Three problems are common to this method. The first arises when the liquid is heated so strongly that it evaporates or boils away. The second arises when the liquid is not heated above its boiling point before heating is discontinued. If the heating is stopped at any point below the actual boiling point of the sample, the liquid enters the bell *immediately*, giving an apparent boiling point that is too low. Be sure you observe a continuous stream of bubbles, too fast for individual bubbles to be distinguished, before lowering the temperature. Also, be sure the bubbling action decreases slowly before the liquid enters the bell. If your melting-point apparatus has fine enough control and fast response, you can actually begin heating again and force the liquid out of the bell before it becomes completely filled with the liquid. This allows a second determination to be performed on the same sample. The third problem is that the bell may be so light that the bubbling action of the liquid causes the bell to move up the capillary tube. This problem can sometimes be solved by using a longer (heavier) bell or by sealing the bell so that a larger section of solid glass is formed at the sealed end of the bell.

When measuring temperatures above 150°C, thermometer errors can become significant. For an accurate boiling point with a high-boiling liquid, you may wish to apply a *stem correction* to the thermometer as described in Section 13.4, or to calibrate the thermometer as described in Chapter 9, Section 9.9, page 125.

13.3 DETERMINING BOILING POINTS—OTHER METHODS

With some liquids, it is difficult to obtain an accurate boiling point by using the inverted capillary methods described previously. In these difficult cases (provided enough material is available), it may be necessary to use a more direct method.

One method is to perform a simple distillation in which the internal temperature is monitored. If using microscale glassware, use the apparatus shown in Figure 14.1 (page 201). Use 5–7 mL of liquid for the distillation. If using macroscale glassware, use the apparatus shown in Figure 14.5 (page 206) with about 7 mL of liquid. Use the smallest available round-bottom flask with this method.

Another method is shown in Figure 13.5. With this method, the bulb of the thermometer can be immersed in vapor from the boiling liquid for a period long enough to allow it to equilibrate and give a good temperature reading. The apparatus in Figure 13.5 uses a conical vial and an air condenser, which are joined by a ground-glass joint. As a precaution, it is a good idea to assemble the apparatus *without* the threaded plastic cap and O-ring. The cap and O-ring will melt at high temperatures. Place the bulb of the thermometer as close as possible to the boiling liquid without actually touching it. You should use a small, inert Carborundum (black) boiling stone. Do not use a marble or calcium carbonate (white) boiling chip. The aluminum collars should be used with high-boiling liquids in order to increase the amount of heat delivered to the area of the ground-glass joint. The joint is made of thick glass, and it absorbs a great deal of heat. This makes it necessary to heat the apparatus for a long period, and quite strongly, to achieve the equilibrium condition that is required. The liquid must boil vigorously, such that you see a reflux ring above the ground-glass joint. The temperature reading on the thermometer must remain constant at its highest observed value. If the temperature

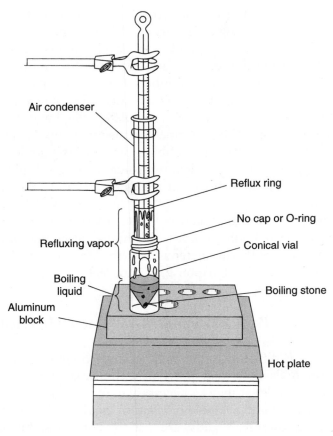

Figure 13.5 Other methods for determining the boiling point of a liquid.

continues to rise, the temperature of the thermometer has not yet reached the boiling point of the liquid. With high-boiling liquids (boiling points above approximately 150°C), it may not be possible to apply enough heat with this apparatus to reach the boiling point of the liquid. In this case you will need to use another method. One alternative is to replace the conical vial and air condenser in Figure 13.5 with a test tube. This method will work only if the test tube fits snugly in one of the holes of the aluminum block. Since the mass of glass in a test tube is less than in the conical vial–air condenser assembly, it will be possible to achieve a higher temperature with the test tube.

Another alternative is to attach an air condenser to a thin-walled conical vial. Support the conical vial on a wire screen and clamp the air condenser. Use a microburner to heat the liquid. As a precaution, it is a good idea to assemble the apparatus without the threaded plastic cap and O-ring. The cap and O-ring will melt at high temperatures. As mentioned previously, be sure to place the thermometer as close as possible to the boiling liquid and be sure to use a Carborundum boiling stone. The reflux ring that forms must be positioned above the ground-glass joint, and the temperature reading on the thermometer must remain constant at its highest observed value. If the temperature continues to rise, the temperature of the thermometer has not yet reached the boiling point of the liquid.

> **CAUTION:** You should use this method only if you have a high-boiling liquid and the method shown in Figure 13.5 does not work. Check with your instructor to make sure that you can use a microburner safely in your laboratory. You must watch carefully so that you do not heat so much as to cause the liquid to boil out of the air condenser, contact the flame, and cause a fire. Do not use a thick-walled conical vial with this method because it may crack when heated by a flame.

13.4 THERMOMETER STEM CORRECTIONS

Three types of thermometers are available: bulb immersion, stem immersion (partial immersion), and total immersion. **Bulb immersion** thermometers are calibrated by the manufacturer to give correct temperature readings when only the bulb (not the rest of the thermometer) is placed in the medium to be measured. **Stem immersion** thermometers are calibrated to give correct temperature readings when they are immersed to a specified depth in the medium to be measured. Stem immersion thermometers are easily recognized because the manufacturer always scores a mark, or immersion ring, completely around the stem at the specified depth of immersion. The immersion ring is normally found below any of the temperature calibrations. **Total immersion** thermometers are calibrated when the entire thermometer is immersed in the medium to be measured. The three types of thermometer are often marked on the back (opposite side from the calibrations) by the words *bulb*, *immersion*, or *total*, but this may vary from one manufacturer to another. Because total immersion thermometers are less expensive, they are most likely to be found in the laboratory.

Manufacturers design total immersion thermometers to read correctly only when they are immersed totally in the medium to be measured. The entire mercury thread must be covered. Because this situation is rare, a **stem correction** should be added to the observed temperature. This correction, which is positive, can be fairly large when high temperatures are being measured. Keep in mind, however, that if your thermometer has been calibrated for its desired use (such as described in Chapter 9, Section 9.9 for a melting-point apparatus), a stem correction should not be necessary for any temperature within the calibration limits. You are most likely to want a stem correction when you are performing a distillation. If you determine a melting point or boiling point using an uncalibrated, total immersion thermometer, you will also want to use a stem correction.

When you wish to make a stem correction for a total immersion thermometer, the following formula may be used. It is based on the fact that the portion of the mercury thread in the stem is cooler than the portion immersed in the vapor or the heated area around the thermometer. The mercury will not have expanded in the cool stem to the same extent as in the warmed section of the thermometer. The equation used is

$$(0.000154)(T - t_1)(T - t_2) = \text{correction to be added to } T \text{ observed}$$

1. The factor 0.000154 is a constant, the coefficient of expansion for the mercury in the thermometer.
2. The term $T - t_1$ corresponds to the length of the mercury thread not immersed in the heated area. It is convenient to use the temperature scale on the thermometer itself for this measurement rather than an actual length unit. T is the observed temperature, and t_1 is the *approximate* place where the heated part of the stem ends and the cooler part begins.
3. The term $T - t_2$ corresponds to the difference between the temperature of the mercury in the vapor T and the temperature of the mercury in the air outside the heated area (room temperature). The term T is the observed temperature, and t_2 is measured by hanging another thermometer so the bulb is close to the stem of the main thermometer.

Figure 13.6 shows how to apply this method for a distillation. By the formula just given, it can be shown that high temperatures are more likely to require a stem correction and that low temperatures need not be corrected. The following sample calculations illustrate this point.

Example 1	Example 2
$T = 200°C$	$T = 100°C$
$t_1 = 0°C$	$t_1 = 0°C$
$t_2 = 35°C$	$t_2 = 35°C$
$(0.000154)(200)(165) = 5.1°$ stem correction	$(0.000154)(100)(165) = 1.0°$ stem correction
$200°C + 5°C = 205°C$ corrected temperature	$100°C + 1°C = 101°C$ corrected temperature

Part B. Density

13.5 DENSITY

Density is defined as mass per unit volume and is generally expressed in units of grams per milliliter (g/mL) for a liquid and grams per cubic centimeter (g/cm^3) for a solid.

$$\text{Density} = \frac{\text{mass}}{\text{volume}} \quad \text{or} \quad D = \frac{M}{V}$$

In organic chemistry, density is most commonly used in converting the weight of liquid to a corresponding volume, or vice versa. It is often easier to measure a volume of a liquid than to weigh it. As a physical property, density is also useful for identifying liquids in much the same way that boiling points are used.

Although precise methods that allow the measurements of the densities of liquids at the microscale level have been developed, they are often difficult to perform. An approximate method for measuring densities can be found in using a 100-μL (0.100-mL) automatic pipet (Chapter 5, Section 5.6, page 61). Clean, dry, and preweigh one or more conical vials (including their caps and liners) and record their weights. Han-

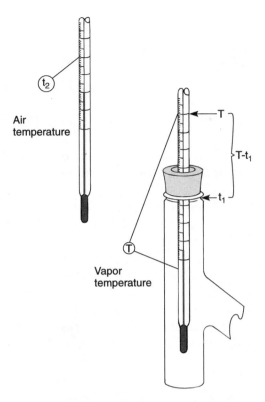

Figure 13.6 Measurement of a thermometer stem correction during distillation.

dle these vials with a tissue, to avoid getting your fingerprints on them. Adjust the automatic pipet to deliver 100μL and fit it with a clean, new tip. Use the pipet to deliver 100μL of the unknown liquid to each of your tared vials. Cap them so that the liquid does not evaporate. Reweigh the vials and use the weight of the 100μL of liquid delivered to calculate a density for each case. It is recommended that from three to five determinations be performed, that the calculations be performed to three significant figures, and that all the calculations be averaged to obtain the final result. This determination of the density will be accurate to within two significant figures. Table 13.1 compares some literature values with those that could be obtained by this method.

Table 13.1 Densities Determined by the
          Automatic Pipet Method (g/mL)

Substance	bp	lit	100 μL
Water	100	1.000	1.01
Hexane	69	0.660	0.66
Acetone	56	0.788	0.77
Dichloromethane	40	1.330	1.27
Diethyl ether	35	0.713	0.67

PROBLEMS

1. Using the temperature–pressure alignment chart in Figure 13.2, answer the following questions.
   (a) What is the normal boiling point (at 760 mm Hg) for a compound that boils at 150°C at 10 mm Hg pressure?
   (b) At what temperature would the compound in (a) boil if the pressure were 40 mm Hg?
   (c) A compound was distilled at atmospheric pressure and had a boiling point of 285°C. What would be the approximate boiling range for this compound at 15 mm Hg?
2. Calculate the corrected boiling point for nitrobenzene by using the method given in Section 13.4. The boiling point was determined using an apparatus similar to that shown in Figure 13.5. The observed boiling point was 205°C. The reflux ring in the air condenser just reached up to the 0°C mark on the thermometer. A second thermometer suspended alongside the test tube, at a slightly higher level than the one inside, gave a reading of 35°C.
3. Suppose that you had calibrated the thermometer in your melting-point apparatus against a series of melting-point standards. After reading the temperature and converting it using the calibration chart, should you also apply a stem correction? Explain.
4. The density of a liquid was determined by the automatic pipet method. A 100-μL automatic pipet was used. The liquid had a mass of 0.082 g. What was the density in grams per milliliter of the liquid?
5. During the microscale boiling-point determination of an unknown liquid, heating was discontinued at 154°C and the liquid immediately began to enter the inverted bell. Heating was begun again at once, and the liquid was forced out of the bell. Heating was again discontinued at 165°C, at which time a very rapid stream of bubbles emerged from the bell. On cooling, the rate of bubbling gradually diminished until the liquid reached a temperature of 161°C, and entered and filled the bell. Explain this sequence of events. What was the boiling point of the liquid?

CHAPTER 14

Simple Distillation

Distillation is the process of vaporizing a liquid, condensing the vapor, and collecting the condensate in another container. This technique is very useful for separating a liquid mixture when the components have different boiling points, or when one of the components will not distill. It is one of the principal methods of purifying a liquid. Four basic distillation methods are available to the chemist: simple distillation, fractional distillation, vacuum distillation (distillation at reduced pressure), and steam distillation. Fractional distillation will be discussed in Chapter 15. Vacuum distillation will be discussed in Chapter 16, and steam distillation will be covered in Chapter 18.

A typical modern distillation apparatus (semimicroscale) is shown in Figure 14.1. The liquid to be distilled is placed in the distilling flask and heated, usually by a hot

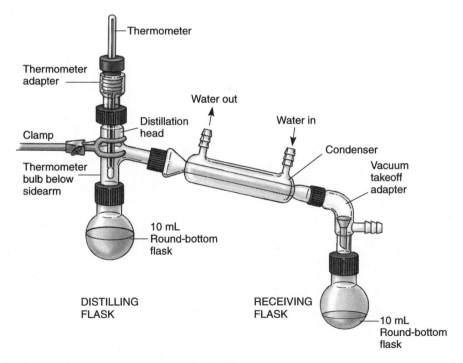

Figure 14.1 A typical apparatus for simple distillation.

plate with an aluminum block or a heating mantle. The heated liquid vaporizes and is forced upward past the thermometer and into the condenser. The vapor is condensed to liquid in the cooling condenser, and the liquid flows downward through the vacuum adapter (no vacuum is used) into the receiving flask.

14.1 THE EVOLUTION OF DISTILLATION EQUIPMENT

There are probably more types and styles of distillation apparatus than exist for any other technique in chemistry. Over the centuries, chemists have devised just about every conceivable design. The earliest known types of distillation apparatus were the

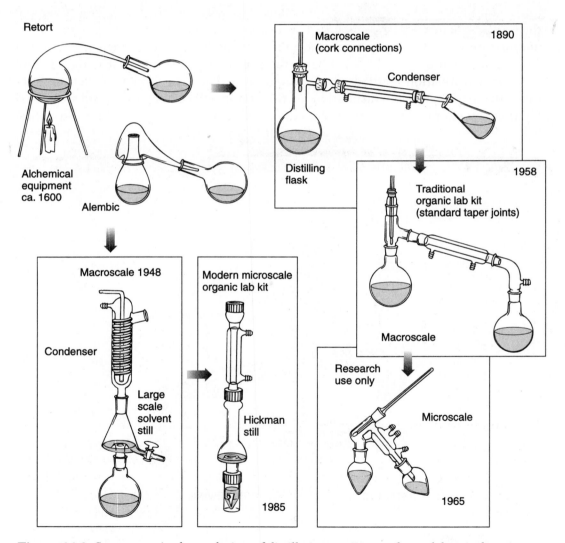

Figure 14.2 Some steps in the evolution of distillation equipment from alchemical equipment (dates represent approximate time of use).

alembic and the **retort** (Fig. 14.2). They were used by alchemists in the Middle Ages and the Renaissance, and probably even earlier by Arabic chemists. Most other distillation equipment has evolved as variations on these designs.

Figure 14.2 shows several stages in the evolution of distillation equipment as it relates to the organic laboratory. It is not intended to be a complete history; rather, it is representative. Until recently, equipment based on the retort design was common in the laboratory. Although the retort itself was still in use early in the twentieth century, it had evolved by that time into the distillation flask and water-cooled condenser combination. This early equipment was connected with drilled corks. By 1958, most introductory laboratories were beginning to use "organic lab kits" that included glassware connected by standard-taper glass joints. The original lab kits contained large ℑ 24/40 joints. Within a short time, they became smaller with ℑ 19/22 and even ℑ 14/20 joints. These later kits are still being used today in "macroscale" laboratory courses.

In the 1960s, researchers developed even smaller versions of these kits for working at the "microscale" level (in Fig. 14.2, see the box labeled 1965, research use only), but this glassware is generally too expensive to use in an introductory laboratory course. However, in the mid-1980s, several groups developed a different style of microscale distillation equipment based on the alembic design (see the box labeled 1985, Hickman still). This new microscale equipment has ℑ 14/10 standard-taper joints, threaded outer joints with screw cap connectors, and an internal O-ring for a compression seal. Microscale equipment similar to this is now used in many introductory courses. The advantages of this glassware are that there is less material used (lower cost), lower personal exposure to chemicals, and less waste generated. Because both types of equipment are in use today, after we describe macroscale equipment, we will also show the equivalent microscale distillation apparatus.

A third type of distillation, semimicroscale, will also be described in this chapter. This method more closely resembles the macroscale method of distillation, but the scale is smaller and glassware with ℑ 14/10 joints is used.

Part A. Theory

14.2 DISTILLATION THEORY

In the traditional distillation of a pure substance, vapor rises from the distillation flask and comes into contact with a thermometer that records its temperature. The vapor then passes through a condenser, which reliquefies the vapor and passes it into the receiving flask. The temperature observed during the distillation of a **pure substance** remains constant throughout the distillation so long as both vapor *and* liquid are present in the system (see Fig. 14.3A). When a **liquid mixture** is distilled, often the temperature does not remain constant but increases throughout the distillation. This is because the composition of the vapor that is distilling varies continuously during the distillation (see Fig. 14.3B).

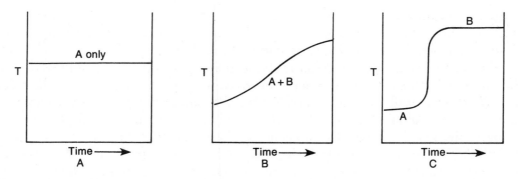

Figure 14.3 Three types of temperature behavior during a simple distillation. **(A)** A single pure component. **(B)** Two components of similar boiling points. **(C)** Two components with widely differing boiling points. Good separations are achieved in **A** and **C**.

For a liquid mixture, the composition of the vapor in equilibrium with the heated solution is different from the composition of the solution itself. This is shown in Figure 14.4, which is a phase diagram of the typical vapor–liquid relation for a two-component system (A + B).

On Figure 14.4, horizontal lines represent constant temperatures. The upper curve represents vapor composition, and the lower curve represents liquid composition. For any horizontal line (constant temperature), like that shown at t, the intersections of the line with the curves give the compositions of the liquid and the vapor that are in equilibrium with each other at that temperature. In the diagram, at temperature t, the intersection of the curve at x indicates that liquid of composition w will be in equilibrium with vapor of composition z, which corresponds to

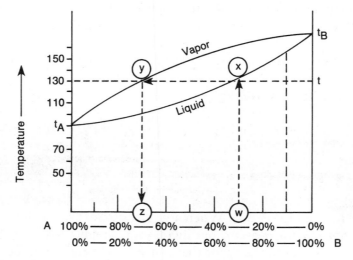

Figure 14.4 Phase diagram for a typical liquid mixture of two components.

the intersection at y. Composition is given as a mole percentage of A and B in the mixture. Pure A, which boils at temperature t_A, is represented at the left. Pure B, which boils at temperature t_B, is represented on the right. For either pure A or pure B, the vapor and liquid curves meet at the boiling point. Thus, either pure A or pure B will distill at a constant temperature (t_A or t_B). Both the vapor and the liquid must have the same composition in either of these cases. This is not the case for mixtures of A and B.

A mixture of A and B of composition w will have the following behavior when heated. The temperature of the liquid mixture will increase until the boiling point of the mixture is reached. This corresponds to following line wx from w to x, the boiling point of the mixture t. At temperature t the liquid begins to vaporize, which corresponds to line xy. The vapor has the composition corresponding to z. In other words, the first vapor obtained in distilling a mixture of A and B does not consist of pure A. It is richer in A than the original mixture but still contains a significant amount of the higher boiling component B, *even from the very beginning of the distillation.* The result is that it is never possible to separate a mixture completely by a simple distillation. However, in two cases it is possible to get an acceptable separation into relatively pure components. In the first case, if the boiling points of A and B differ by a large amount (>100°C), and if the distillation is carried out carefully, it will be possible to get a fair separation of A and B. In the second case, if A contains a fairly small amount of B (<10%), a reasonable separation of A and B can be achieved. When the boiling-point differences are not large, and when highly pure components are desired, it is necessary to perform a **fractional distillation.** Fractional distillation is described in Chapter 15, where the behavior during a simple distillation is also considered in detail. Note only that as vapor distills from the mixture of composition w (Fig. 14.4), it is richer in A than is the solution. Thus, the composition of the material left behind in the distillation becomes richer in B (moves to the right from w toward pure B in the graph). A mixture of 90% B (dotted line on the right side in Fig. 14.4) has a higher boiling point than at w. Hence, the temperature of the liquid in the distillation flask will increase during the distillation, and the composition of the distillate will change (as is shown in Fig. 14.3B).

When two components that have a large boiling-point difference are distilled, the temperature remains constant while the first component distills. If the temperature remains constant, a relatively pure substance is being distilled. After the first substance distills, the temperature of the vapors rises, and the second component distills, again at a constant temperature. This is shown in Figure 14.3C. A typical application of this type of distillation might be an instance of a reaction mixture containing the desired component A (bp 140°C) contaminated with a small amount of undesired component B (bp 250°C) and mixed with a solvent such as diethyl ether (bp 36°C). The ether is removed easily at low temperature. Pure A is removed at a higher temperature and collected in a separate receiver. Component B can then be distilled, but it usually is left as a residue and not distilled. This separation is not difficult and represents a case in which simple distillation might be used to advantage.

Part B. Macroscale Distillation

14.3 SIMPLE DISTILLATION—MACROSCALE METHODS

For a simple distillation, the apparatus shown in Figure 14.5 is used. Six pieces of specialized glassware are used:

1. Distilling flask
2. Distillation head
3. Thermometer adapter
4. Water condenser
5. Vacuum takeoff adapter
6. Receiving flask

The apparatus is usually heated electrically, using a heating mantle. The distilling flask, condenser, and vacuum adapter should be clamped. Two different methods of clamping this apparatus were shown in Chapter 7 (Fig. 7.2, page 80 and Fig. 7.4,

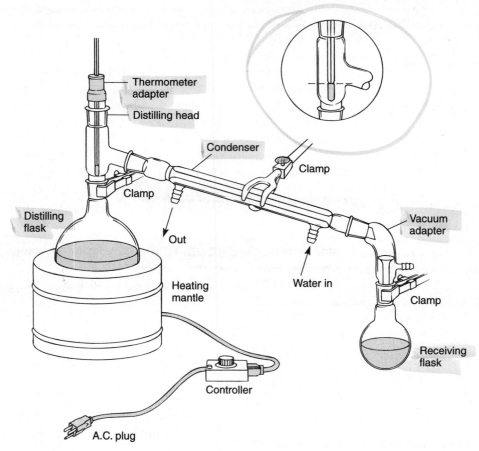

Figure 14.5 Distillation with the standard macroscale laboratory kit.

page 82). The receiving flask should be supported by removable wooden blocks or a wire gauze on an iron ring attached to a ring stand. The various components are each discussed below, along with some other important points.

Distilling Flask.   The distilling flask should be a round-bottom flask. This type of flask is designed to withstand the required input of heat, and to accommodate the boiling action. It gives a maximized heating surface. The size of the distilling flask should be chosen so that it is never filled more than two-thirds full. When the flask is filled beyond this point, the neck constricts and "chokes" the boiling action, resulting in bumping. The surface area of the boiling liquid should be kept as large as possible. However, too large a distilling flask should also be avoided. With too large a flask, the **holdup** is excessive; the holdup is the amount of material that cannot distill since some vapor must fill the empty flask. When you cool the apparatus at the end, this material drops back into the distilling flask.

Boiling Stones.   A boiling stone (Chapter 7, Section 7.4, page 86) should be used during distillation to prevent bumping. As an alternative, the liquid being distilled may be rapidly stirred using a magnetic stirrer and stir bar (Chapter 7, Section 7.3, page 85). If you forget a boiling stone, cool the mixture before adding it. If you add a boiling stone to a hot superheated liquid, it may "erupt" into vigorous boiling, breaking your apparatus and spilling hot solvent everywhere.

Grease.   In most cases, it is unnecessary to grease standard-taper joints for a simple distillation. The grease makes cleanup more difficult, and it may contaminate your product.

Distillation Head.   The distillation head directs the distilling vapors into the condenser and allows the connection of a thermometer via the thermometer adapter. The thermometer should be positioned in the distillation head so that it is placed directly in the stream of vapor that is distilling. This can be accomplished if the entire bulb of the thermometer is positioned *below* the sidearm of the distilling head (see the circular inset in Fig. 14.5). The entire bulb must be immersed in the vapor to achieve an accurate temperature reading. When distilling, you should be able to see a reflux ring (Chapter 7, Section 7.2, page 83) positioned well above both the thermometer bulb and the bottom of the side arm.

Thermometer Adapter.   The thermometer adapter connects to the top of the distillation head (see Fig. 14.5). There are two parts to the thermometer adapter: a glass joint with an open rolled edge on the top, and a rubber adapter that fits over the rolled edge and holds the thermometer. The thermometer fits in a hole in the top of the rubber adapter and can be adjusted upward and downward by sliding it in the hole. Adjust the bulb to a point below the sidearm.

Water Condenser.   The joint between the distillation head and the water condenser is the joint most prone to leak in this entire apparatus. Since the distilling liquid is both hot and vaporized when it reaches this joint, it will leak out of any small opening between the two joint surfaces. The odd angle of the joint, neither vertical or horizontal, also makes a good connection more difficult. Be sure this joint is well sealed. If possible, use one of the plastic joint clips described in Chapter 7, Figure 7.3, page 81. Otherwise, adjust your clamps to be sure that the joint surfaces are pressed together, and not pulled apart.

The condenser will remain full of cooling water only if the water flows *upward*, not downward. The water input hose should be connected to the lower opening in the jacket, and the exit hose should be attached to the upper opening. Place the other end of the exit hose in a sink. A moderate water flow will perform a good deal of cooling. A high rate of water flow may cause the tubing to pop off the joints and cause a flood. If you hold the exit hose horizontally and point the end into a sink, the flow rate is correct if the water stream continues horizontally for about 2 inches before bending downward.

If a distillation apparatus is to be left untended for a period of time, it is a good idea to wrap copper wire around the ends of the tubing and twist it tight. This will help to prevent the hoses from popping off of the connectors if there is an unexpected water pressure change.

Vacuum Adapter.   In a simple distillation the vacuum adapter is not connected to a vacuum, but is left open. It is merely an opening to the outside air so that pressure does not build up in the distillation system. If you plug this opening, you will have a **closed system** (no outlet). It is always dangerous to heat a closed system. Enough pressure can build up in the closed system so that it can explode. The vacuum adapter, in this case, merely directs the distillate into the receiving, or collection, flask.

If the substance you are distilling is water sensitive, you can attach a calcium chloride drying tube to the vacuum connection to protect the freshly distilled liquid from atmospheric water vapor. Air that enters the apparatus will have to pass through the calcium chloride and be dried. Depending on the severity of the problem, drying agents other than calcium chloride may also be used.

The vacuum adapter has a disturbing tendency to obey the laws of Newtonian physics and fall off the slanted condenser onto the desk and break. If they are available, it is a good idea to use plastic joint clips (page 81) on both ends of this piece. The top clip will secure the vacuum adapter to the condenser, and the bottom clip will secure the receiving flask, preventing it from falling.

Rate of Heating.   The rate of heating for the distillation can be adjusted to the proper rate of **takeoff,** the rate at which distillate leaves the condenser, by watching drops of liquid emerge from the bottom of the vacuum adapter. A rate of from one to three drops per second is considered a proper rate of takeoff for most applications. At a greater rate, equilibrium is not established within the distillation apparatus, and the separation may be poor. A slower rate of takeoff is also unsatisfactory since the temperature recorded on the thermometer is not maintained by a constant vapor stream, thus leading to an inaccurately low observation of the boiling point.

Receiving Flask.   The receiving flask collects the distilled liquid, and is usually a round-bottom flask. If the liquid you are distilling is extremely volatile, and there is danger of losing some of it to evaporation, it is sometimes advisable to cool the receiving flask in an ice-water bath.

Fractions.   The material being distilled is called the **distillate.** Frequently a distillate is collected in contiguous portions, called **fractions.** This is accomplished by replacing the collection flask with clean ones at regular intervals. Time can be saved if each flask is tared before use. If a small amount of liquid is collected at the beginning of a distillation and not saved or used further, it is called a **forerun.** Subsequent fractions will have higher boiling ranges, and each fraction should be labeled with its correct boiling range when the fraction is taken. For a simple distillation of a pure material, most of the material will be collected in a single large **midrun** fraction, with only a small forerun. In some small-scale distillations, the volume of the forerun will be so small that you will not be able to collect it separately from the midrun fraction. The material left behind is called the **residue.** It is usually advised that you discontinue a distillation before the distilling flask becomes empty. Typically, the residue becomes increasingly dark in color during distillation, and it frequently contains thermal decomposition products. In addition, a dry residue may explode on overheating, or the flask may melt or crack when it becomes dry. Don't distill until the distilling flask is completely dry!

Part C. Microscale Distillation

14.4   SIMPLE DISTILLATION—MICROSCALE METHODS

Most large-scale distillation equipment requires the distilled liquid to travel a long distance from the distillation flask, through the condenser, to the receiving flask. When working at the microscale level, a long distillation path must be avoided. With small quantities of liquid, there are too many opportunities to lose all the sample. The liquid will adhere to, or *wet*, surfaces and get lost in every little nook and cranny of the system. A system with a long path also has a large volume, and a small amount of liquid may not produce enough vapor to fill it. Small-scale distillation requires a "short path" distillation. To make the distilling path as short as possible, the **Hickman head** has been adopted as the principal receiving device for most microscale distilling operations.

The Hickman Head.   Two types of Hickman head (also called a Hickman "still") are shown in Figure 14.6. One of these variations has a convenient opening, or port, in the side, making removal of liquid that has collected in it easier. In operation, the liquid to be distilled is placed in a flask or vial attached to the bottom joint of the Hickman head and heated. If desired, you can attach a condenser to the top joint. Either a magnetic spin vane or a boiling stone is used to prevent bumping. Some typical

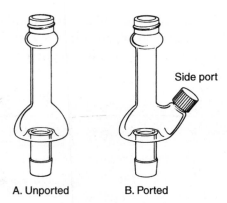

A. Unported   B. Ported

Figure 14.6 The Hickman head.

assemblies are shown in Figures 14.7 and 14.9. The vapors of the heated liquid rise upward and are cooled and condensed on either the walls of the condenser, or, if no condenser is used, on the inside walls of the Hickman head itself. As liquid drains downward, it collects in the circular well at the bottom of the still.

Collecting Fractions.   The liquid that distills is called the **distillate.** Portions of the distillate collected during the course of a distillation are called **fractions.** A small fraction (usually discarded) collected before the distillation is begun in earnest is called a **forerun.** The well in a Hickman head can contain anywhere from 1 to 2 mL of liquid. In the style with the side port, fractions may be removed by opening the port and inserting a Pasteur pipet (Fig. 14.8B). The unported head works equally well, but the head is emptied from the top by using a Pasteur pipet (Fig. 14.8A). If a condenser or an internal thermometer is used, the distilling apparatus must be partially disassembled to remove liquid when the well fills. In some stills the inner diameter of the head is small, and it is difficult to reach in at an angle with the pipet and make contact with the liquid. To remedy this problem, you may be able to use the longer (9-inch) Pasteur pipet instead of the shorter (5¾-inch) one. The longer pipet has a much longer narrow section (tip) and can adapt more effectively to the required angle. The disadvantage of the longer tip is that you are more likely to break it off inside the still. You may prefer to modify a short pipet by bending its tip slightly in a flame (Fig. 14.8C).

Choice of Condenser.   If you are careful (slow heating) or if the liquid to be distilled has a high boiling point, it may not be necessary to use a condenser with the Hickman head (Fig. 14.9). In this case, the liquid being distilled must condense on the cooler sides of the head itself without any being lost through evaporation. If the liquid has a low boiling point or is very volatile, a condenser must be used. With very volatile liquids, a water-cooled condenser must be used; however, an air-cooled condenser may suffice for less demanding cases. When using a water condenser, remember that water should enter the lower opening and exit from the upper one. If the hoses carrying the water in and out are connected in reverse fashion, the water jacket of the condenser will not fill completely.

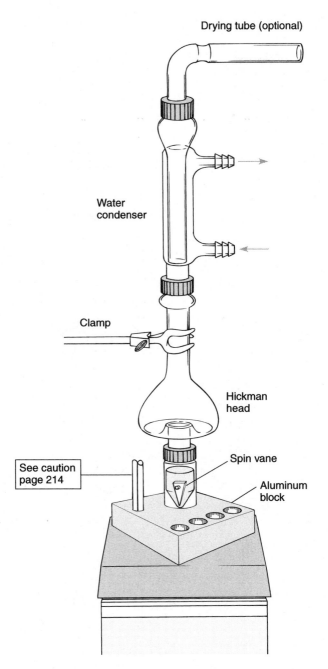

Figure 14.7 Basic microscale distillation (external monitoring of temperature).

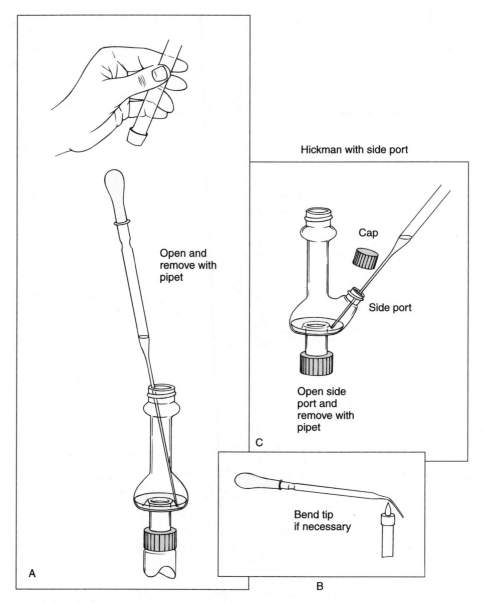

Figure 14.8 Removing fractions.

Sealed Systems.   Whenever you perform a distillation, be sure the system you are heating is not sealed off completely from the outside atmosphere. During a distillation, the air and vapors inside the system will both expand and contract. If pressure builds up inside a sealed system, the apparatus may explode. In performing a distillation, you should leave a small opening at the far end of the system. If water vapor

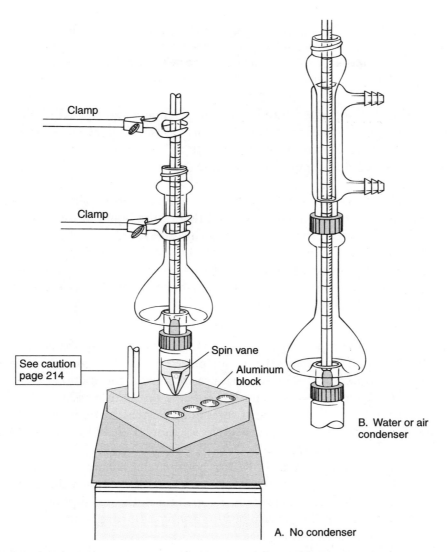

Clamp

Clamp

See caution
page 214

Spin vane

Aluminum
block

B. Water or air
condenser

A. No condenser

Figure 14.9 Basic microscale distillation (internal monitoring of temperature).

could be harmful to the substances being distilled, a calcium chloride drying tube may be used to protect the system from moisture. Carefully examine each system discussed to see how an opening to the outside is provided.

External Monitoring of Temperature.   The simple assembly using the Hickman head shown in Figure 14.7 does not monitor the temperature inside the apparatus. Instead, the temperature is monitored externally with a thermometer placed in an aluminum block.

CAUTION You should not use a mercury thermometer with an aluminum block. If it breaks, the mercury will vaporize on the hot surface. Instead, use a nonmercury glass thermometer, a metal dial thermometer, or a digital electronic temperature-measuring device.

External monitoring of the temperature has the disadvantage that the exact temperature at which liquid distills is never known. In many cases, this does not matter or is unavoidable, and the boiling point of the distilled liquid can be checked later by performing a microscale boiling-point determination (Chapter 13, Section 13.2, page 192).

As a rule, there is at least a 15-degree difference in temperature between the temperature of the aluminum block or sand bath and that of the liquid in the heated distillation vial or flask. However, the magnitude of this difference cannot be relied on. Keep in mind that the liquid in the vial or flask may be at a different temperature than the vapor that is distilling. Because this method of monitoring the temperature is rather approximate, you will need to make the actual heater setting based on what is supposed to be occurring in the vial or flask.

Internal Monitoring of Temperature.   When you wish to monitor the actual temperature of a distillation, a thermometer must be placed inside the apparatus. Figures 14.9 and 14.10 show distillation assemblies that use an internal thermometer. The apparatus in Figure 14.9A represents the simplest possible distillation assembly. It does not use a condenser and the thermometer is suspended from a clamp. It is possible to add either an air or a water condenser to this basic assembly (Fig. 14.9B) and maintain internal monitoring of the temperature.

In the arrangement shown in Figure 14.10, a thermometer adapter is used. A thermometer adapter (Fig. 14.11A) provides a convenient way of holding a thermometer in place. The Claisen head is used to provide an opening to the atmosphere, thereby avoiding a sealed system. With the Claisen head, a drying tube may be used to protect the system from atmospheric moisture.

If protection from atmospheric moisture is not required, the multipurpose adapter may be used. The multipurpose adapter (Fig. 14.11B) replaces both the thermometer adapter and the Claisen head. With this adapter, the necessary opening to the atmosphere is provided by the sidearm. The threaded joint holds the thermometer in place.

Carefully notice the position of the thermometer in Figures 14.9 and 14.10. The bulb of the thermometer must be placed in the stem of the Hickman head, *just below the well*, or it will not read the temperature correctly.

NOTE: It is good practice to monitor the temperature internally whenever possible.

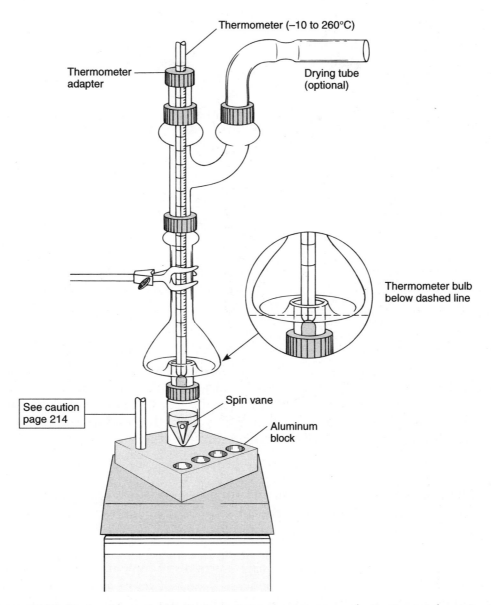

Figure 14.10 Basic microscale distillation using a thermometer adapter (internal monitoring of temperature).

Boiling Stones or Stirring.   A boiling stone should be used during distillation in order to prevent bumping. As an alternative, the liquid being distilled may be rapidly stirred. A triangular spin vane of the correct size should be used when distilling from a conical vial, whereas a stirring bar should be used when distilling from a round-bottom flask.

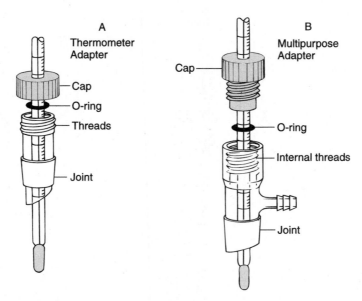

Figure 14.11 Two adapters.

Size of Distillation Flask.   As a rule, the distillation flask or vial should not be filled to more than two-thirds of its total capacity. This allows room for boiling and stirring action, and it prevents contamination of the distillate by bumping. A flask that is too large should also be avoided. With too large a flask, the **holdup** is excessive; the holdup is the amount of material that cannot distill because some vapor must fill the empty flask.

Assembling the Apparatus.   You should not grease the joints when assembling the apparatus. Ungreased joints seal well enough to allow you to perform a simple distillation. Stopcock grease can introduce a serious contaminant into your product.

Rate and Degree of Heating.   Take care not to distill too quickly. If you vaporize liquid at a rate faster than it can be recondensed, some of your product may be lost by evaporation. On the other hand, you should not distill too slowly. This may also lead to loss of product, because there is a longer time period during which vapors can escape. Carefully examine your apparatus during distillation to monitor the position of either a reflux ring or a wet appearance on the surface of the glass. Either of these indicates the place at which condensation is occurring. The position at which condensation occurs should be well inside the Hickman head. Be sure that liquid is collecting in the well. If all the surfaces are shiny (wet) and there is no distillate, you are losing material.

NOTE: A slower rate of heating also helps avoid bumping.

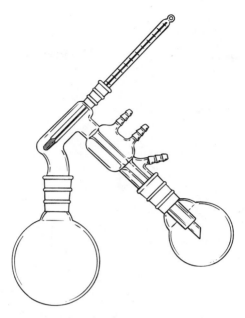

Figure 14.12 A research-style short-path distillation apparatus.

If you are using a sand bath, material may be lost because the hot sand bath radiates too much heat upward and warms the Hickman still. If you believe this to be the case, it can often be remedied by placing a small square of aluminum foil over the top of the sand bath. Make a tear from one edge to the center of the foil to wrap it around the apparatus.

Microscale—Research Equipment. Figure 14.12 shows a very-well designed research-style, short-path distillation head. Note how the equipment has been "unitized," eliminating several joints, and decreasing the holdup. Compare this apparatus to the one shown in Figure 14.1.

Part D. Semimicroscale Distillation

14.5 SIMPLE DISTILLATION—SEMIMICROSCALE METHODS

There are times when the amount of material to be distilled is too large (greater than 2–3 mL) to use a microscale apparatus with a Hickman head, yet too small to use macroscale equipment. Most manufacturers of microscale equipment make two pieces of conventional distillation equipment sized to work with the ℥ 14/10 microscale kit components. These two pieces, the **distillation head** and the **bent vacuum adapter,** are not provided in student microscale kits, but must be purchased

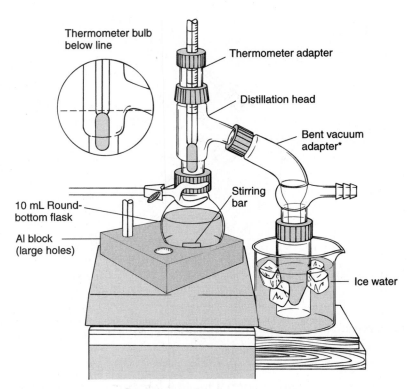

Figure 14.13 Semimicroscale distillation (*requires special pieces).

separately. Figure 14.13 shows a semimicroscale assembly using these components. Note, that the bulb of the thermometer must be placed *below the sidearm* if it is to be bathed in vapor and give a correct temperature reading. This apparatus assumes that a condenser is not necessary; however, you could easily insert one between the distilling head and the bent vacuum adapter. This insertion would produce a completely traditional distillation apparatus but would use microscale equipment. (See Figure 14.1 on page 201.)

PROBLEMS

1. Using Figure 14.3, answer the following questions.
   (a) What is the molar composition of the vapor in equilibrium with a boiling liquid that has a composition of 60% A and 40% B?
   (b) A sample of vapor has the composition 50% A and 50% B. What is the composition of the boiling liquid that produced this vapor?
2. Use an apparatus similar to that shown in Figure 14.5, and assume that the round-bottom flask holds 100 mL and the distilling head has an internal volume of 12 mL in the vertical section. At the end of a distillation, vapor would fill this volume, but it could not be forced through the system. No liquid would remain in

the distillation flask. Assuming this holdup volume of 112 mL, use the ideal gas law and assume a boiling point of 100°C (760 mm Hg) to calculate the number of milliliters of liquid ($d = 0.9$ g/mL, $MW = 200$) that would recondense into the distillation flask upon cooling.

3. Use an apparatus similar to that shown in Figure 14.13, and assume that the round-bottom flask holds 10 mL and that the distillation head has an internal volume of about 2 mL in the vertical section. At the end of a distillation, vapor would fill this volume, but it could not be forced through the system. No liquid would remain in the distillation flask. Assuming this *holdup volume* of 12 mL, use the ideal gas law and assume a boiling point of 100°C (760 mm Hg) to calculate the number of microliters of liquid ($d = 0.9$ g/mL, $MW = 200$) that would recondense into the distillation flask on cooling.

4. What is the approximate difference between the temperature of a boiling liquid in a conical vial and the temperature read on an *external* thermometer when both are placed on an aluminum block?

5. Explain the significance of a horizontal line connecting a point on the lower curve with a point on the upper curve (such as line *xy*) in Figure 14.4.

6. Using Figure 14.4, determine the boiling point of a liquid having a molar composition of 50% A and 50% B.

7. Where should the thermometer bulb be located in
   (a) a microscale distillation apparatus using a Hickman head?
   (b) a macroscale distillation apparatus using a distilling head, condenser, and vacuum takeoff adapter?

8. Under what conditions can a good separation be achieved with a simple distillation?

CHAPTER 15

Fractional Distillation; Azeotropes

Simple distillation, described in Chapter 14, works well for most routine separation and purification procedures for organic compounds. When boiling-point differences of components to be separated are small, however, **fractional distillation** must be used to achieve a good separation.

A typical fractional distillation apparatus (semimicroscale) is shown in Figure 15.1. The major difference between this apparatus and one used for simple distillation

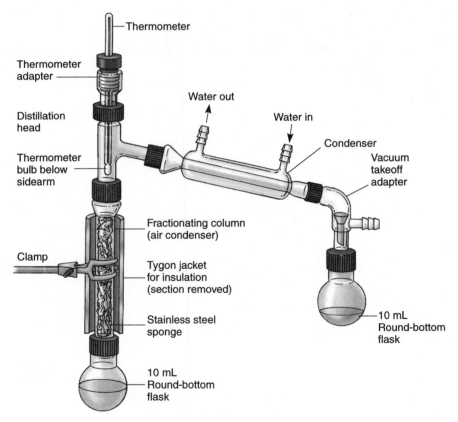

Thermometer

Thermometer adapter

Distillation head

Thermometer bulb below sidearm

Water out

Water in

Condenser

Vacuum takeoff adapter

Clamp

Fractionating column (air condenser)

Tygon jacket for insulation (section removed)

Stainless steel sponge

10 mL Round-bottom flask

10 mL Round-bottom flask

Figure 15.1 A typical apparatus for fractional distillation.

(Figure 14.1) is the insertion of a **fractionating column** between the distilling flask and the distillation head. The fractionating column is filled with a **packing,** a material that causes the liquid to condense and revaporize repeatedly as it passes through the column. With a good fractionating column better separations are possible, and liquids with small boiling-point differences may be separated by using this technique.

Part A. Theory

15.1 Differences between Simple and Fractional Distillation

When an ideal solution of two liquids, such as benzene (bp 80°C) and toluene (bp 110°C), is distilled by simple distillation, the first vapor produced will be enriched in the lower-boiling component (benzene). However, when that initial vapor is condensed and analyzed, the distillate will not be pure benzene. The boiling point difference of benzene and toluene (30°C) is too small to achieve a complete separation by simple distillation. Following the principles outlined in Chapter 14, Section 14.2 (pages 203–205), and using the vapor–liquid composition curve given in Figure 15.2, you can see what would happen if you started with an equimolar mixture of benzene and toluene.

Following the dashed lines shows that an equimolar mixture (50 mole percent benzene) would begin to boil at about 91°C and, far from being 100% benzene, the distillate would contain about 74 mole percent benzene and 26 mole percent toluene. As the distillation continued, the composition of the undistilled liquid would move in the direction of A′ (there would be increased toluene, due to removal of more benzene than toluene), and the corresponding vapor would contain a progressively smaller amount of benzene. In effect, the temperature of the distillation would continue to increase throughout the distillation (as in Figure 14.3B, page 204), and it would be impossible to obtain any fraction that consisted of pure benzene.

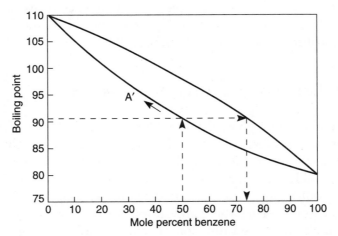

Figure 15.2 The vapor–liquid composition curve for mixtures of benzene and toluene.

Suppose, however, that we are able to collect a small quantity of the first distillate that was 74 mole percent benzene and redistill it. Using Figure 15.2, we can see that this liquid would begin to boil at about 84°C and would give an initial distillate containing 90 mole percent of benzene. If we were experimentally able to continue taking small fractions at the beginning of each distillation, and redistill them, we would eventually reach a liquid with a composition of nearly 100 mole percent benzene. However, since we took only a small amount of material at the beginning of each distillation, we would have lost most of the material with which we started. To recapture a reasonable amount of benzene, we would have to process each of the fractions left behind in the same way as our early fractions. As each of them was partially distilled, the material advanced would become progressively richer in benzene, while that left behind would become progressively richer in toluene. It would require thousands (maybe millions) of such microdistillations to separate benzene from toluene.

Obviously, the procedure just described would be very tedious; fortunately, it need not be performed in usual laboratory practice. **Fractional distillation** accomplishes the same result. You simply have to use a column inserted between the distillation flask and the distilling head, as shown in Figure 15.1. This **fractionating column** is filled, or **packed,** with a suitable material such as a stainless steel sponge. This packing allows a mixture of benzene and toluene to be subjected continuously to many vaporization–condensation cycles as the material moves up the column. With each cycle within the column, the composition of the vapor is progressively enriched in the lower-boiling component (benzene). Nearly pure benzene (bp 80°C) finally emerges from the top of the column, condenses, and passes into the receiving head or flask. This process continues until all the benzene is removed. The distillation must be carried out slowly to ensure that numerous vaporization–condensation cycles occur. When nearly all the benzene has been removed, the temperature begins to rise and a small amount of a second fraction, which contains some benzene and toluene, may be collected. When the temperature reaches 110°C, the boiling point of pure toluene, the vapor is condensed and collected as the third fraction. A plot of boiling-point versus volume of condensate (distillate) would resemble Figure 14.3C (page 204). This separation would be much better than that achieved by simple distillation (Figure 14.3B, page 204).

15.2 VAPOR–LIQUID COMPOSITION DIAGRAMS.

A vapor–liquid composition phase diagram like the one in Figure 15.3 can be used to explain the operation of a fractionating column with an **ideal solution** of two liquids, A and B. An ideal solution is one in which the two liquids are chemically similar, miscible (mutually soluble) in all proportions, and do not interact. Ideal solutions obey **Raoult's Law.** Raoult's Law is explained in detail in Section 15.3.

The phase diagram relates the compositions of the boiling liquid (lower curve) and its vapor (upper curve) as a function of temperature. Any horizontal line drawn across the diagram (a constant-temperature line) intersects the diagram in two places.

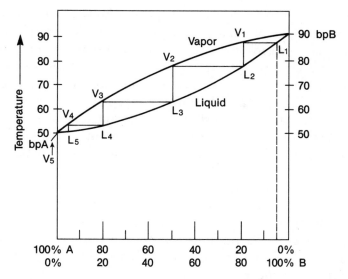

Figure 15.3 Phase diagram for a fractional distillation of an ideal two-component system.

These intersections relate the vapor composition to the composition of the boiling liquid that produces that vapor. By convention, composition is expressed either in *mole fraction* or in *mole percentage*. The mole fraction is defined as follows:

$$\text{Mole fraction A} = N_A = \frac{\text{Moles A}}{\text{Moles A} + \text{Moles B}}$$

$$\text{Mole fraction B} = N_B = \frac{\text{Moles B}}{\text{Moles A} + \text{Moles B}}$$

$$N_A + N_B = 1$$

$$\text{Mole percentage A} = N_A \times 100$$

$$\text{Mole percentage B} = N_B \times 100$$

The horizontal and vertical lines shown in Figure 15.3 represent the processes that occur during a fractional distillation. Each of the **horizontal lines** (L_1V_1, L_2V_2, etc.) represents the **vaporization** step of a given vaporization–condensation cycle and represents the composition of the vapor in equilibrium with liquid at a given temperature. For example, at 63°C a liquid with a composition of 50% A (L_3 on the diagram) would yield vapor of composition 80% A (V_3 on the diagram) at equilibrium. The vapor is richer in the lower-boiling component A than the original liquid was.

Each of the **vertical lines** (V_1L_2, V_2L_3, etc.) represents the **condensation** step of a given vaporization–condensation cycle. The composition does not change as the temperature drops on condensation. The vapor at V_3, for example, condenses to give a liquid (L_4 on the diagram) of composition 80% A with a drop in temperature from 63 to 53°C.

In the example shown in Figure 15.3, pure A boils at 50°C and pure B boils at 90°C. These two boiling points are represented at the left- and right-hand edges of the diagram, respectively. Now consider a solution that contains only 5% of A but 95% of B. (Remember that these are *mole* percentages.) This solution is heated (following the dashed line) until it is observed to boil at L_1 (87°C). The resulting vapor has composition V_1 (20% A, 80% B). The vapor is richer in A than the original liquid, but it is by no means pure A. In a simple distillation apparatus, this vapor would be condensed and passed into the receiver in a very impure state. However, with a fractionating column in place, the vapor is condensed in the **column** to give liquid L_2 (20% A, 80% B). Liquid L_2 is immediately revaporized (bp 78°C) to give a vapor of composition V_2 (50% A, 50% B), which is condensed to give liquid L_3. Liquid L_3 is revaporized (bp 63°C) to give vapor of composition V_3 (80% A, 20% B), which is condensed to give liquid L_4. Liquid L_4 is revaporized (bp 53°C) to give vapor of composition V_4 (95% A, 5% B). This process continues to V_5, which condenses to give nearly pure liquid A. The fractionating process follows the stepped lines in the figure downward and to the left.

As this process continues, all of liquid A is removed from the distillation flask or vial, leaving nearly pure B behind. If the temperature is raised, liquid B may be distilled as a nearly pure fraction. Fractional distillation will have achieved a separation of A and B, a separation that would have been nearly impossible with simple distillation. Notice that the boiling point of the liquid becomes lower each time it vaporizes. Because the temperature at the bottom of a column is normally higher than the temperature at the top, successive vaporizations occur higher and higher in the column as the composition of the distillate approaches that of pure A. This process is illustrated in Figure 15.4, where the composition of the liquids, their boiling points, and the composition of the vapors present are shown alongside the fractionating column.

15.3 RAOULT'S LAW

Two liquids (A and B) that are miscible and that do not interact form an **ideal solution** and follow Raoult's Law. The law states that the partial vapor pressure of component A in the solution (P_A) equals the vapor pressure of pure A (P_A^0) times its mole fraction (N_A) (Eq. 1). A similar expression can be written for component B (Eq. 2). The mole fractions N_A and N_B were defined in Section 15.2.

$$\text{Partial vapor pressure of A in solution} = P_A = (P_A^0)(N_A) \qquad [1]$$

$$\text{Partial vapor pressure of B in solution} = P_B = (P_B^0)(N_B) \qquad [2]$$

P_A^0 is the vapor pressure of pure A, independent of B. P_B^0 is the vapor pressure of B, independent of A. In a mixture of A and B, the partial vapor pressures are added to give the total vapor pressure above the solution (Eq. 3). When the total pressure (sum of the partial pressures) equals the applied pressure, the solution boils.

$$P_{\text{total}} = P_A + P_B = P_A^0 N_A + P_B^0 N_B \qquad [3]$$

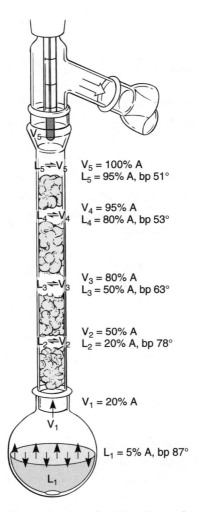

V_5 = 100% A
L_5 = 95% A, bp 51°

V_4 = 95% A
L_4 = 80% A, bp 53°

V_3 = 80% A
L_3 = 50% A, bp 63°

V_2 = 50% A
L_2 = 20% A, bp 78°

V_1 = 20% A

L_1 = 5% A, bp 87°

Figure 15.4 Vaporization–condensation in a fractionation column.

The composition of A and B in the vapor produced is given by Equations 4 and 5.

$$N_A \text{ (vapor)} = \frac{P_A}{P_{total}} \qquad [4]$$

$$N_B \text{ (vapor)} = \frac{P_B}{P_{total}} \qquad [5]$$

Several exercises involving applications of Raoult's Law are illustrated in Figure 15.5. Note, particularly in the result from Equation 4, that the vapor is richer ($N_A = 0.67$) in the lower-boiling (higher vapor pressure) component A than it was before vaporization ($N_A = 0.50$). This proves mathematically what was described in Section 15.2.

Consider a solution at 100°C where $N_A = 0.5$ and $N_B = 0.5$.

1. What is the partial vapor pressure of A in the solution if the vapor pressure of pure A at 100°C is 1020 mmHg?

   Answer: $P_A = P°_A N_A = (1020)(0.5) = 510$ mmHg

2. What is the partial vapor pressure of B in the solution if the vapor pressure of pure B at 100°C is 500 mmHg?

   Answer: $P_B = P°_B N_B = (500)(0.5) = 250$ mmHg

3. Would the solution boil at 100°C if the applied pressure were 760 mmHg?

   Answer: Yes. $P_{total} = P_A + P_B = (510 + 250) = 760$ mmHg

4. What is the composition of the vapor at the boiling point?

   Answer: The boiling point is 100°C.

$$N_A \text{ (vapor)} = \frac{P_A}{P_{total}} = 510/760 = 0.67$$

$$N_B \text{ (vapor)} = \frac{P_B}{P_{total}} = 250/760 = 0.33$$

Figure 15.5 Sample calculations with Raoult's Law.

The consequences of Raoult's Law for distillations are shown schematically in Figure 15.6. In Part A the boiling points are identical (vapor pressures the same), and no separation is attained regardless of how the distillation is conducted. In Part B a fractional distillation is required, while in Part C a simple distillation provides an adequate separation.

When a solid B (rather than another liquid) is dissolved in a liquid A, the boiling point is increased. In this extreme case, the vapor pressure of B is negligible, and the vapor will be pure A no matter how much solid B is added. Consider a solution of salt in water.

$$P_{total} = P^0_{water} N_{water} + P^0_{salt} N_{salt}$$

$$P^0_{salt} = 0$$

$$P_{total} = P^0_{water} N_{water}$$

A solution whose mole fraction of water is 0.7 will not boil at 100°C, because $P_{total} = (760)(0.7) = 532$ mm Hg and is less than atmospheric pressure. If the solution is heated to 110°C, it will boil because $P_{total} = (1085)(0.7) = 760$ mm Hg. Although the solution must be heated at 110°C to boil it, the vapor is pure water and has a boiling-point temperature of 100°C. (The vapor pressure of water at 110°C can be looked up in a handbook; it is 1085 mm Hg.)

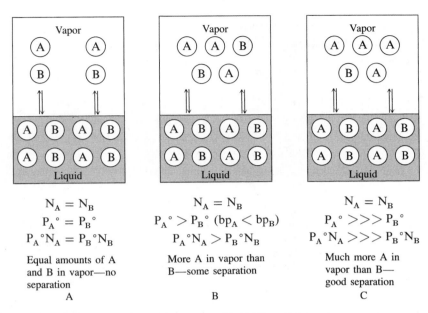

$$N_A = N_B$$
$$P_A^\circ = P_B^\circ$$
$$P_A^\circ N_A = P_B^\circ N_B$$

Equal amounts of A
and B in vapor—no
separation

A

$$N_A = N_B$$
$$P_A^\circ > P_B^\circ \; (bp_A < bp_B)$$
$$P_A^\circ N_A > P_B^\circ N_B$$

More A in vapor than
B—some separation

B

$$N_A = N_B$$
$$P_A^\circ >>> P_B^\circ$$
$$P_A^\circ N_A >>> P_B^\circ N_B$$

Much more A in
vapor than B—
good separation

C

Figure 15.6 Consequences of Raoult's Law. **(A)** Boiling points (vapor pressures) are identical—no separation. **(B)** Boiling points are somewhat less for A than for B—requires fractional distillation. **(C)** Boiling points are much less for A than for B—simple distillation will suffice.

15.4 COLUMN EFFICIENCY

A common measure of the efficiency of a column is given by its number of **theoretical plates.** The number of theoretical plates in a column is related to the number of vaporization–condensation cycles that occur as a liquid mixture travels through it. Using the example mixture in Figure 15.3, if the first distillate (condensed vapor) had the composition at L_2 when starting with liquid of composition L_1, the column would be said to have *one theoretical plate.* This would correspond to a simple distillation, or one vaporization–condensation cycle. A column would have two theoretical plates if the first distillate had the composition at L_3. The two-theoretical-plate column essentially carries out "two simple distillations." According to Figure 15.3, *five theoretical plates* would be required to separate the mixture that started with composition L_1. Notice that this corresponds to the number of "steps" that need to be drawn in the figure to arrive at a composition of 100% A.

Most columns do not allow distillation in discrete steps, as indicated in Figure 15.3. Instead, the process is *continuous,* allowing the vapors to be continuously in contact with liquid of changing composition as they pass through the column. Any material can be used to pack the column as long as it can be wetted by the liquid and as long as it does not pack so tightly that vapor cannot pass.

Table 15.1 Theoretical Plates Required to Separate Mixtures, Based on Boiling-Point Differences of Components

Boiling-Point Difference	Number of Theoretical Plates
108	1
72	2
54	3
43	4
36	5
20	10
10	20
7	30
4	50
2	100

The approximate relationship between the number of theoretical plates needed to separate an ideal two-component mixture and the difference in boiling points is given in Table 15.1. Notice that more theoretical plates are required as the boiling-point differences between the components decrease. For instance, a mixture of A (bp 130°C) and B (bp 166°C) with a boiling-point difference of 36°C would be expected to require a column with a minimum of five theoretical plates.

The **reflux ratio** is defined as the ratio of the number of drops of distillate that returns to the distillation flask compared to the number of drops of distillate collected. In an efficient column, the reflux ratio should equal or exceed the number of theoretical plates. A high reflux ratio ensures that the column will achieve temperature equilibrium and achieve its maximum efficiency. This ratio is not easy to determine; in fact, it is impossible to determine using a Hickman head, and it should not concern a beginning student. In some cases, the **throughput,** or **rate of takeoff,** of a column may be specified. This is expressed as the number of milliliters of distillate that can be collected per unit of time, usually as mL/minute.

15.5 TYPES OF FRACTIONATING COLUMNS AND PACKINGS

Several types of fractionating columns are shown in Figure 15.7. The Vigreux column, shown in Part A, has indentations that incline downward at angles of 45° and are in pairs on opposite sides of the column. The projections into the column provide increased possibilities for condensation and for the vapor to equilibrate with liquid. Vigreux columns are popular in cases in which only a small number of theoretical plates is required. They are not very efficient (a 20-cm column might have only 2.5 theoretical plates), but they allow for rapid distillation and have a small **holdup** (the amount of liquid retained by the column). A column packed with a stainless steel sponge is a more effective fractionating column than a Vigreux column, but not by a large margin. Glass

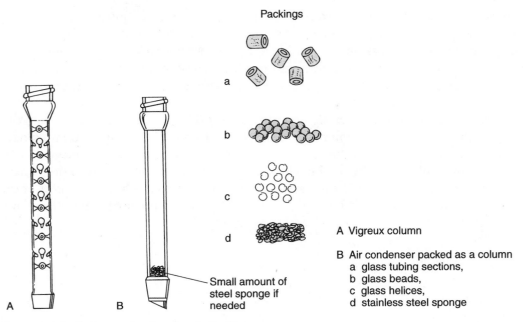

Figure 15.7 Columns for fractional distillation.

beads, or glass helices, can also be used as a packing material, and they have a slightly greater efficiency. The air condenser or the water condenser can be used as an improvised column if an actual fractionating column is unavailable. If a condenser is packed with glass beads, glass helices, or sections of glass tubing, the packing must be held in place by inserting a small plug of stainless steel sponge into the bottom of the condenser.

The most effective type of column is the **spinning-band column.** In the most elegant form of this device, a tightly fitting, twisted platinum screen or a Teflon rod with helical threads is rotated rapidly inside the bore of the column (Fig. 15.8). A

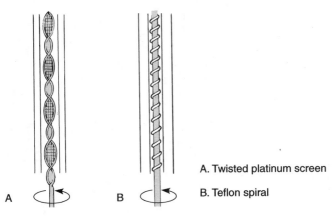

Figure 15.8 Bands for spinning-band columns.

spinning-band column that is available for microscale work is shown in Figure 15.9. This spinning-band column has a band about 2–3 cm in length and provides 4–5 theoretical plates. It can separate 1–2 mL of a mixture with a 30°C boiling-point difference. Larger research models of this spinning-band column can provide as many as 20 or 30 theoretical plates and can separate mixtures with a boiling-point difference of as little as 5–10°C.

Manufacturers of fractionating columns often offer them in a variety of lengths. Because the efficiency of a column is a function of its length, longer columns have more theoretical plates than shorter ones. It is common to express efficiency of a column in a unit called **HETP,** the **H**eight of a column that is **E**quivalent to one **T**heoretical **P**late. HETP is usually expressed in units of cm/plate. When the height of the column (in centimeters) is divided by this value, the total number of theoretical plates is specified.

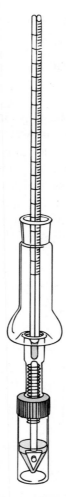

Figure 15.9 A commercially available microscale spinning-band column.

Fractionating columns must be insulated so that temperature equilibrium is maintained at all times. Additional insulation will not be required for columns that have an evacuated outer jacket, but those that do not can benefit from being wrapped in insulation.

Glass wool and aluminum foil (shiny side in) are often used for insulation. You can wrap the column with glass wool and then use a wrapping of the aluminum foil to keep it in place. An especially effective method is to make an insulation blanket by placing a layer of glass wool or cotton between two rectangles of aluminum foil, placed shiny side in. The sandwich is bound together with duct tape. This blanket, which is reusable, can be wrapped around the column and held in place with twist ties or tape.

Part B. Macroscale Distillation

15.6 FRACTIONAL DISTILLATION—MACROSCALE METHODS

Figure 15.10 illustrates a typical fractional distillation assembly that is used for larger-scale distillations. It has a glass-jacketed column that is packed with a stainless steel sponge. This apparatus would be common in situations in which quantities of liquid in excess of 10 mL are to be distilled.

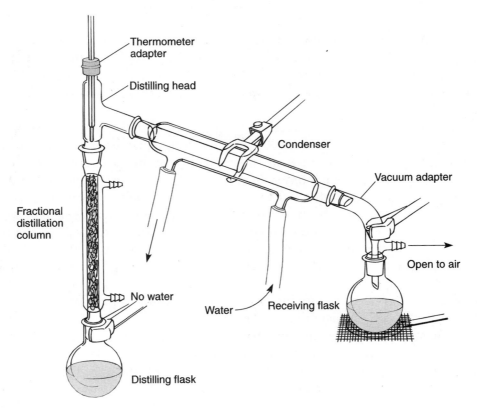

Figure 15.10 A macroscale fractional distillation apparatus.

In performing a fractional distillation, the column should be clamped in a vertical position. The distilling flask would normally be heated by a heating mantle, since it allows a precise adjustment of the temperature. A proper rate of distillation is extremely important. The distillation should be conducted as slowly as possible to allow as many vaporization–condensation cycles as possible to occur as the vapor passes through the column. However, the rate of distillation must be steady enough to produce a constant temperature reading at the thermometer. Conversely, a rate that is too fast will cause the column to "flood" or "choke." In this instance, there is so much condensing liquid flowing downward in the column that the vapor cannot rise upward, and the column fills with liquid. Flooding can also occur if the column is not well insulated and has a large temperature difference from bottom to top. This situation can be remedied by employing one of the insulation methods that uses glass wool or aluminum foil, as described in Section 15.5. It may also be necessary to insulate the distilling head at the top of the column. If the distilling head is cold, it will stop the progress of the distilling vapor.

Part C. Microscale Distillation

15.7 FRACTIONAL DISTILLATION—MICROSCALE METHODS

Figure 15.11 illustrates a fractional distillation assembly that is used for microscale distillations. It has a glass column (air condenser) that is packed with a stainless steel sponge (about 0.9 g of stainless steel cleaning pad material). This apparatus would be common in situations in which quantities of liquid of less than 10 mL are to be distilled.

The column shown is insulated with a piece of large-diameter Tygon tubing (½-inch by ⅝-inch outer diameter by 3½-inch long) that has been slit along its length with a pair of scissors and then snapped over the glass column. A second layer of Tygon tubing (⅝-inch inner diameter by ⅞-inch outer diameter by 3½-inch long) may be added for additional insulation. Alternatively, cotton or glass wool, sandwiched between two sheets of aluminum foil (3½-inch square, shiny side in), may be used. The edges of the sandwich are sealed with duct tape, and the pad is wrapped around the column and held in place with twist ties or copper wire. When you distill, pay attention to the comments on rate of heating that are presented at the end of Section 15.6.

If you have access to a spinning-band column like the one illustrated in Figure 15.9, you may have to insulate the small shaft that carries the Teflon band. A particular problem with this type of spinning band occurs if the distilling flask is not exactly centered on the hot plate–magnetic stirrer. In that case the band will rotate backward, forcing the liquid down the column! It is also necessary to rotate the band very rapidly in order for it to function correctly.

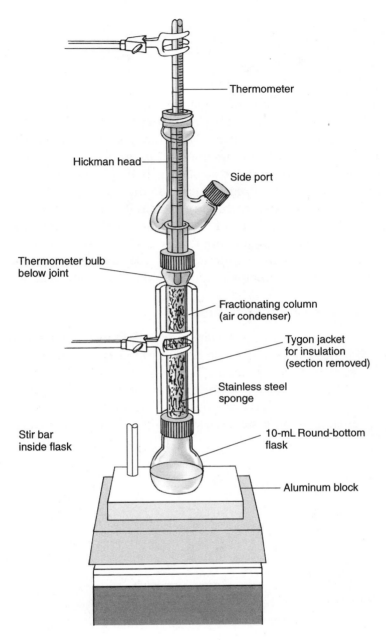

Figure 15.11 A microscale fractional distillation apparatus.

Part D. Azeotropes

15.8 NONIDEAL SOLUTIONS: AZEOTROPES

Some mixtures of liquids, because of attractions or repulsions between the molecules, do not behave ideally; they do not follow Raoult's Law. There are two types of vapor–liquid composition diagrams that result from this nonideal behavior: **minimum-boiling-point** and **maximum-boiling-point** diagrams. The minimum or maximum points in these diagrams correspond to a constant-boiling mixture called an **azeotrope.** An azeotrope is a mixture with a fixed composition that cannot be altered by either simple or fractional distillation. An azeotrope behaves as if it were a pure compound, and it distills from the beginning to the end of its distillation at a constant temperature, giving a distillate of constant (azeotropic) composition. The vapor in equilibrium with an azeotropic liquid has the same composition as the azeotrope. Because of this, an azeotrope is represented as a *point* on a vapor–liquid composition diagram.

A. Minimum-Boiling-Point Diagrams

A minimum-boiling-point azeotrope results from a slight incompatibility (repulsion) between the liquids being mixed. This incompatibility leads to a higher than expected combined vapor pressure from the solution. This higher combined vapor pressure brings about a lower boiling point for the mixture than is observed for the pure components. The most common two-component mixture that gives a minimum-boiling-point azeotrope is the ethanol–water system shown in Figure 15.12. The azeotrope at V_3 has a

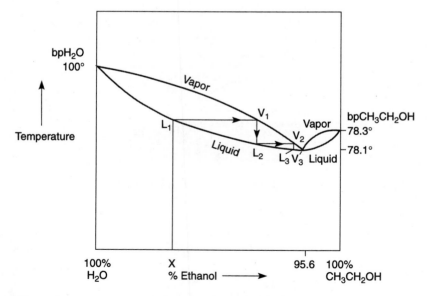

Figure 15.12 An ethanol–water minimum-boiling-point phase diagram.

Table 15.2 Common Minimum-Boiling Azeotropes

Azeotrope	Composition (Weight Percentage)	Boiling Point (°C)
Ethanol–water	95.6% C_2H_5OH, 4.4% H_2O	78.17
Benzene–water	91.1% C_6H_6, 8.9% H_2O	69.4
Benzene–water–ethanol	74.1% C_6H_6, 7.4% H_2O, 18.5% C_2H_5OH	64.9
Methanol–carbon tetrachloride	20.6% CH_3OH, 79.4% CCl_4	55.7
Ethanol–benzene	32.4% C_2H_5OH, 67.6% C_6H_6	67.8
Methanol–toluene	72.4% CH_3OH, 27.6% $C_6H_5CH_3$	63.7
Methanol–benzene	39.5% CH_3OH, 60.5% C_6H_6	58.3
Cyclohexane–ethanol	69.5% C_6H_{12}, 30.5% C_2H_5OH	64.9
2-Propanol–water	87.8% $(CH_3)_2CHOH$, 12.2% H_2O	80.4
Butyl acetate–water	72.9% $CH_3COOC_4H_9$, 27.1% H_2O	90.7
Phenol–water	9.2% C_6H_5OH, 90.8% H_2O	99.5

composition of 96% ethanol–4% water and a boiling point of 78.1°C. This boiling point is not much lower than that of pure ethanol (78.3°C), but it means that it is impossible to obtain pure ethanol from the distillation of any ethanol–water mixture that contains more than 4% water. Even with the best fractionating column, you cannot obtain 100% ethanol. The remaining 4% of water can be removed by adding benzene and removing a different azeotrope, the ternary benzene–water–ethanol azeotrope (bp 65°C). Once the water is removed, the excess benzene is removed as an ethanol–benzene azeotrope (bp 68°C). The resulting material is free of water and is called "absolute" ethanol.

The fractional distillation of an ethanol–water mixture of composition X can be described as follows. The mixture is heated (follow line XL_1) until it is observed to boil at L_1. The resulting vapor at V_1 will be richer in the lower-boiling component, ethanol, than the original mixture.[1] The condensate at L_2 is vaporized to give V_2. The process continues, following the lines to the right, until the azeotrope is obtained at V_3. The liquid that distills is not pure ethanol, but it has the azeotropic composition of 96% ethanol and 4% water, and it distills at 78.1°C. The azeotrope, which is richer in ethanol than the original mixture, continues to distill. As it distills, the percentage of water left behind in the distillation flask continues to increase. When all the ethanol has been distilled (as the azeotrope), pure water remains behind in the distillation flask, and it distills at 100°C.

If the azeotrope obtained by the preceding procedure is redistilled, it distills from the beginning to the end of the distillation at a constant temperature of 78.1°C as if it were a pure substance. There is no change in the composition of the vapor during the distillation.

Some common minimum-boiling azeotropes are given in Table 15.2. Numerous other azeotropes are formed in two- and three-component systems; such azeotropes are common. Water forms azeotropes with many substances; therefore, water must be

---

[1] Keep in mind that this distillate is not pure ethanol but is an ethanol–water mixture.

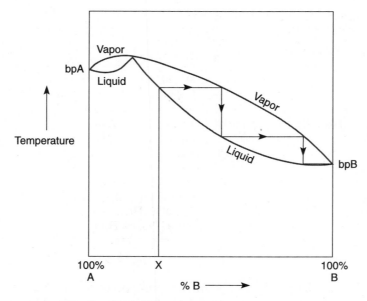

Figure 15.13 A maximum-boiling-point phase diagram.

carefully removed with **drying agents** whenever possible before compounds are distilled. Extensive azeotropic data are available in references such as the *Handbook of Chemistry and Physics*.[2]

B. Maximum-Boiling-Point Diagrams

A maximum-boiling-point azeotrope results from a slight attraction between the component molecules. This attraction leads to lower combined vapor pressure than expected in the solution. The lower combined vapor pressures cause a higher boiling point than what would be characteristic for the components. A two-component maximum-boiling-point azeotrope is illustrated in Figure 15.13. Because the azeotrope has a higher boiling point than any of the components, it will be concentrated in the distillation flask as the distillate (pure B) is removed. The distillation of a solution of composition X would follow to the right along the lines in Figure 15.13. Once the composition of the material remaining in the flask has reached that of the azeotrope, the temperature will rise, and the azeotrope will begin to distill. The azeotrope will continue to distill until all the material in the distillation flask has been exhausted.

Some maximum-boiling-point azeotropes are listed in Table 15.3. They are not nearly as common as minimum-boiling-point azeotropes.[3]

---

[2]More examples of azeotropes, with their compositions and boiling points, can be found in the *CRC Handbook of Chemistry and Physics*; also in L. H. Horsley, ed., *Advances in Chemistry Series*, no. 116. Azeotropic Data, III (Washington, DC: American Chemical Society, 1973).

[3]See Footnote 2.

Table 15.3 Maximum-Boiling Azeotropes

Azeotrope	Composition (Weight Percentage)	Boiling Point (°C)
Acetone–chloroform	20.0% CH_3COCH_3, 80.0% $CHCl_3$	64.7
Chloroform–methyl ethyl ketone	17.0% $CHCl_3$, 83.0% $CH_3COCH_2CH_3$	79.9
Hydrochloric acid	20.2% HCl, 79.8% H_2O	108.6
Acetic acid–dioxane	77.0% CH_3COOH, 23.0% $C_4H_8O_2$	119.5
Benzaldehyde–phenol	49.0% C_6H_5CHO, 51.0% C_6H_5OH	185.6

C. Generalizations

There are some generalizations that can be made about azeotropic behavior. They are presented here without explanation, but you should be able to verify them by thinking through each case using the phase diagrams given. (Note that pure A is always to the left of the azeotrope in these diagrams, while pure B is to the right of the azeotrope.)

Minimum-Boiling-Point Azeotropes

Initial Composition	Experimental Result
To left of azeotrope	Azeotrope distills first, pure A second
Azeotrope	Unseparable
To right of azeotrope	Azeotrope distills first, pure B second

Maximum-Boiling-Point Azeotropes

Initial Composition	Experimental Result
To left of azeotrope	Pure A distills first, azeotrope second
Azeotrope	Unseparable
To right of azeotrope	Pure B distills first, azeotrope second

15.9 AZEOTROPIC DISTILLATION: APPLICATIONS

There are numerous examples of chemical reactions in which the amount of product is low because of an unfavorable equilibrium. An example is the direct acid-catalyzed esterification of a carboxylic acid with an alcohol:

$$R-\overset{\displaystyle O}{\overset{\|}{C}}-OH + R-O-H \overset{H^+}{\rightleftharpoons} R-\overset{\displaystyle O}{\overset{\|}{C}}-OR + H_2O$$

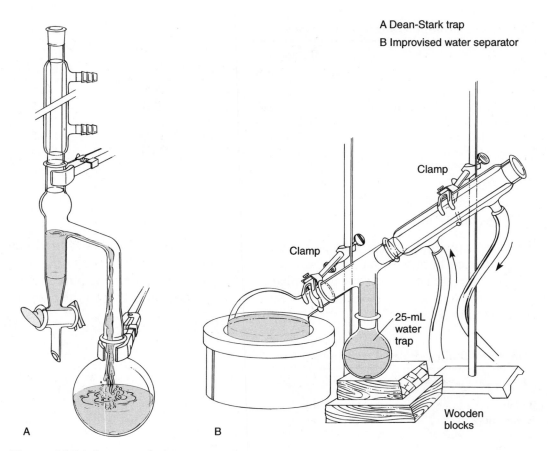

A Dean-Stark trap

B Improvised water separator

Clamp

Clamp

25-mL
water
trap

Wooden
blocks

A   B

Figure 15.14 Large-scale water separators.

Because the equilibrium does not favor formation of the ester, it must be shifted to the right, in favor of the product, by using an excess of one of the starting materials. In most cases, the alcohol is the least expensive reagent and is the material used in excess.

Another way of shifting the equilibrium to the right is to remove one of the products from the reaction mixture as it is formed. In the above example, water can be removed as it is formed by **azeotropic distillation.** A common large-scale method is to use the Dean–Stark water separator shown in Figure 15.14A. In this technique, an inert solvent, commonly benzene or toluene, is added to the reaction mixture contained in the round-bottom flask. The side-arm of the water separator is also filled with this solvent. If benzene is used, as the mixture is heated under reflux, the benzene–water azeotrope (bp 69.4°C, Table 15.2) distills out of the flask.[4] When

---

[4]Actually, with ethanol, a lower-boiling-point, three-component azeotrope distills at 64.9°C (see Table 15.2). It consists of benzene–water–ethanol. Because some ethanol is lost in the azeotropic distillation, a large excess of ethanol is used in esterification reactions. The excess also helps to shift the equilibrium to the right.

the vapor condenses, it enters the side-arm directly below the condenser, and water separates from the benzene–water condensate; benzene and water mix as vapors, but they are not miscible as cooled liquids. Once the water (lower phase) separates from the benzene (upper phase), liquid benzene overflows from the side-arm back into the flask. The cycle is repeated continuously until no more water forms in the side-arm. You may calculate the weight of water that should theoretically be produced and compare this value with the amount of water collected in the side-arm. Because the density of water is 1.0, the volume of water collected can be compared directly with the calculated amount, assuming 100% yield.

An improvised water separator, constructed from the components found in the macroscale organic kit, is shown in Figure 15.14B. Although this requires the condenser to be placed in a nonvertical position, it works quite well.

At the microscale level, water separation can be achieved using a standard distillation assembly with a water condenser and a Hickman head (Fig. 15.15). The side-ported variation of the Hickman head is the most convenient one to use for

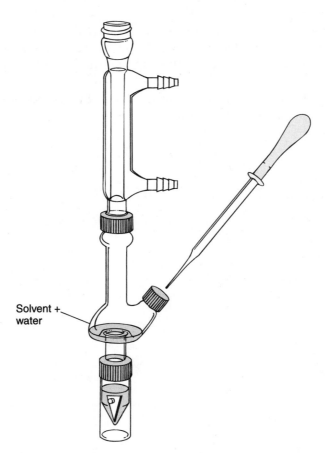

Solvent + water

Figure 15.15 Microscale water separator (both layers are removed).

this purpose, but it is not essential. In this variation, you simply remove all the distillate (both solvent and water) several times during the course of the reaction. Use a Pasteur pipet to remove the distillate, as shown in Chapter 14 (Fig. 14.8, page 212). Because both the solvent and water are removed in this procedure, it may be desirable to add more solvent from time to time, adding it through the condenser with a Pasteur pipet.

The most important consideration in using azeotropic distillation to prepare an ester (described on page 237) is that the azeotrope containing water must have a **lower boiling point** than the alcohol used. With ethanol, the benzene–water azeotrope boils at a much lower temperature (69.4°C) than ethanol (78.3°C), and the technique previously described works well. With higher-boiling-point alcohols, azeotropic distillation works well because of the large boiling-point difference between the azeotrope and the alcohol.

With methanol (bp 65°C), however, the boiling point of the benzene–water azeotrope is actually *higher* by about 5°C, and methanol distills first. Thus, in esterifications involving methanol, a totally different approach must be taken. For example, you can mix the carboxylic acid, methanol, the acid catalyst, and *1,2-dichloroethane* in a conventional reflux apparatus (Chapter 7, Fig. 7.6, page 83) without a water separator. During the reaction, water separates from the 1,2-dichloroethane because it is not miscible; however, the remainder of the components are soluble, so the reaction can continue. The equilibrium is shifted to the right by the "removal" of water from the reaction mixture.

Azeotropic distillation is also used in other types of reactions, such as ketal or acetal formation, and in enamine formation.

PROBLEMS

1. In the accompanying table are approximate vapor pressures for benzene and toluene at various temperatures.

Temp (°C)	mm Hg	Temp (°C)	mm Hg
Benzene 30	120	Toluene 30	37
40	180	40	60
50	270	50	95
60	390	60	140
70	550	70	200
80	760	80	290
90	1010	90	405
100	1340	100	560
		110	760

   (a) What is the mole fraction of each component if 3.9 g of benzene C_6H_6 is dissolved in 4.6 g of toluene C_7H_8?

   (b) Assuming that this mixture is ideal, that is, it follows Raoult's Law, what is the partial vapor pressure of benzene in this mixture at 50°C?

   (c) Estimate to the nearest degree the temperature at which the vapor pressure of the solution equals 1 atm (bp of the solution). Use a linear interpolation method.

   (d) Calculate the composition of the vapor (mole fraction of each component) that is in equilibrium in the solution at the boiling point of this solution.

   (e) Calculate the composition in weight percentage of the vapor that is in equilibrium with the solution.

2. Estimate how many theoretical plates are needed to separate a mixture that has a mole fraction of B equal to 0.70 (70% B) in Figure 15.3.

3. Two moles of sucrose are dissolved in 8 moles of water. Assume that the solution follows Raoult's Law and that the vapor pressure of sucrose is negligible. The boiling point of water is 100°C. The distillation is carried out at 1 atm (760 mm Hg).

   (a) Calculate the vapor pressure of the solution when the temperature reaches 100°C.

   (b) What temperature would be observed during the entire distillation?

   (c) What would be the composition of the distillate?

   (d) If a thermometer were immersed below the surface of the liquid of the boiling flask, what temperature would be observed?

4. Explain why the boiling point of a two-component mixture rises slowly throughout a simple distillation when the boiling-point differences are not large.

5. Given the boiling points of several known mixtures of A and B (mole fractions are known) and the vapor pressures of A and B in the pure state (P_A^0 and P_B^0) at these same temperatures, how would you construct a boiling-point-composition phase diagram for A and B? Give a stepwise explanation.

6. Describe the behavior on distillation of a 98% ethanol solution through an efficient column. Refer to Figure 15.12.

7. Construct an approximate boiling-point-composition diagram for a benzene–methanol system. The mixture shows azeotropic behavior (see Table 15.2). Include on the graph the boiling points of pure benzene and pure methanol and the boiling point of the azeotrope. Describe the behavior for a mixture that is initially rich in benzene (90%) and then for a mixture that is initially rich in methanol (90%).

8. Construct an approximate boiling-point-composition diagram for an acetone–chloroform system, which forms a maximum boiling azeotrope (Table 15.3). Describe the behavior on distillation of a mixture that is initially rich in acetone (90%), then describe the behavior of a mixture that is initially rich in chloroform (90%).

9. Two components have boiling points of 130 and 150°C. Estimate the number of theoretical plates needed to separate these substances in a fractional distillation.

10. A spinning-band column has an HETP of 0.25 inch/plate. If the column has 12 theoretical plates, how long is it?

CHAPTER 16

Vacuum Distillation; Manometers

Vacuum distillation (distillation at reduced pressure) is used for compounds that have high boiling points (above 200°C). Such compounds often undergo thermal decomposition at the temperatures required for their distillation at atmospheric pressure. The boiling point of a compound is lowered substantially by reducing the applied pressure. Vacuum distillation is also used for compounds that, when heated, might react with the oxygen present in air. It is also used when it is more convenient to distill at a lower temperature because of experimental limitations. For instance, a heating device may have difficulty heating to a temperature in excess of 250°C.

Part A. Theory

16.1 THE EFFECT OF PRESSURE ON BOILING POINT

As a rule of thumb, you can expect the boiling point of many liquids to drop about 0.5°C for each 10-mm decrease in pressure when in the vicinity of 760 mm Hg. At lower pressures, a 10°C drop in boiling point is expected for each halving of the pressure. The effect of pressure on the boiling point is discussed more thoroughly in Chapter 13 (Section 13.1, page 190). A very useful nomograph is given (Fig. 13.2, page 191) that allows you to estimate the boiling point of a liquid at a pressure different from the one at which it is reported. For instance, using this nomograph, a liquid reported to boil at 200°C at 760 mm Hg would be expected to boil at 90°C at 20 mm Hg. This is a significant decrease in temperature, and it would be advantageous to use a vacuum distillation for this substance if any decomposition problems were to be expected at 200°C.

Counterbalancing the effect of a decrease in temperature in a vacuum distillation is the fact that separations of liquids of different boiling points may not be as effective with a vacuum distillation as with a simple distillation. In addition, it is not always possible to keep the vacuum pressure constant during a vacuum distillation. It is important to watch temperature and pressure variations carefully during the course of a vacuum distillation.

Part B. Macroscale Distillation

16.2 VACUUM DISTILLATION—MACROSCALE METHODS

A basic apparatus similar to the one shown in Figure 16.1 may be used for vacuum distillations. The major differences to be found when comparing this assembly to one for simple distillation (Fig. 14.5, page 206) are that a Claisen head has been inserted between the distillation flask and the distilling head, and that the opening to the atmosphere has been replaced by a connection **A** to a vacuum source. In addition, an air inlet tube **B** has been added to the top of the Claisen head. When connecting to a vacuum

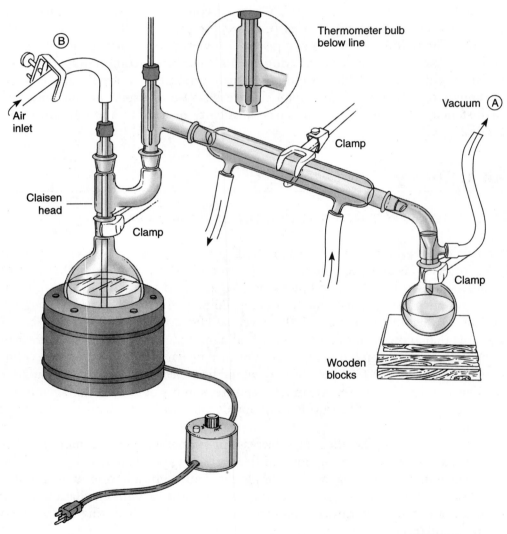

Figure 16.1 Macroscale vacuum distillation using the standard organic laboratory kit.

source, an aspirator (Chapter 8, Section 8.3, page 107), a mechanical vacuum pump (Section 16.10, page 263), or a "house" vacuum system (one piped directly to the laboratory bench) may be used. The aspirator is probably the simplest of these sources and the vacuum source most likely to be available. However, if pressures below 10–20 mm Hg are required, a mechanical vacuum pump must be used.

When working with glassware that is to be evacuated, you should wear safety glasses at all times. There is always danger of an implosion.

⚠️ CAUTION Safety glasses must be worn at all times during vacuum distillation.

It is a good idea to work in a hood when performing a vacuum distillation. If the experiment will involve high temperatures (>220°C) for distillation, or an extremely low pressure (<0.1 mm Hg), for your own safety you should definitely work in a hood, behind a shield.

Assembling the Apparatus

When assembling an apparatus for vacuum distillation, it is important to check the following points.

Glassware.   Before assembly, check all glassware to be sure there are no cracks and that there are no chips in the standard-taper joints. Cracked glassware may break when evacuated. Joints that have chips may not be airtight and they will leak.

Greasing Joints.   With macroscale equipment it is necessary to grease all standard-taper joints lightly. Take care not to use too much grease. Grease can become a very serious contaminant if it oozes out the bottom of the joints into your system. Apply a small amount of grease (thin film) completely around the top of the *inner* joint; then mate the joints and press or turn them slightly to spread the grease evenly. If you have used the correct amount of grease, it will not ooze out the bottom of the joint; rather the entire joint will appear clear and without striations or uncovered areas.

Claisen Head.   The Claisen head is placed between the distilling flask and the distilling head to help prevent material from "bumping over" into the condenser.

Ebulliator Tube.   The air inlet tube on top of the Claisen head is called an **ebulliator** *(ebb-u-lay-tor)* tube. Using the screw clamp **B** on the attached heavy-wall tubing (see pressure tubing below), the ebulliator is adjusted to admit a slow continuous stream of air bubbles into the distillation flask while you are distilling. Since boiling stones will not work in a vacuum, these bubbles keep the solution stirred and help to prevent bumping. The ebulliator tube is drawn to a point at its lower end. The end of the tube should be adjusted so that it is just above the bottom of the distilling flask.

Most macroscale ground-glass kits contain an ebulliator tube. If one is not available, an ebulliator can be prepared easily by heating a section of glass tubing and drawing it out about 3 cm. The glass is then scored in the middle of this drawn-out section and broken, making two tubes at once. In Figure 16.1, the ebulliator is inserted into a thermometer adapter. If you do not have a second thermometer adapter, a one-hole rubber stopper may be used, placing the stopper directly into the joint on top of the Claisen head.

Wooden Applicator Sticks.   An alternative to an ebulliator tube that is sometimes used is the wooden pine splint or wooden applicator stick. Air is trapped in the pores of the wood. Under vacuum, the stick will emit a slow stream of bubbles to stir the solution. The disadvantage is that each time you open the system, you must use a new stick.

Thermometer Placement.   Be sure that the thermometer is positioned so that the entire mercury bulb is below the sidearm in the distilling head (see the circular insert in Fig. 16.1). If it is placed higher, it may not be surrounded by a constant stream of vapor from the material being distilled. If the thermometer is not exposed to a continuous stream of vapor, it may not reach temperature equilibrium. As a result, the temperature reading would be incorrect (low).

Joint Clips.   If plastic joint clips are available (Chapter 7, Figure 7.3, page 81), they should be used to secure the greased joints, particularly those on either side of the condenser, and the one at the bottom of the vacuum adapter where the receiving flask is attached.

Pressure Tubing.   The connection to the vacuum source **A** is made using pressure tubing. Pressure tubing (also called vacuum tubing), unlike the more common thin-walled tubing used to carry water or gas, has heavy walls and will not collapse inward when it is evacuated. A comparison of the two types of tubing is shown in Figure 16.2.

Make doubly sure that any connections to pressure tubing are tight. If a tight connection cannot be made, you may have the wrong size of tubing (either the rubber tubing or the glass tubing to which it is attached). Keep the lengths of pressure tubing relatively short. The pressure tubing should be relatively new and without cracks. If the tubing shows cracks when you stretch it or bend it, it may be old and leak air into the system. Replace any tubing that shows its age.

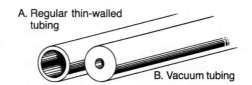

Figure 16.2 Comparison of tubing.

Rubber Stoppers.   Always use soft rubber stoppers in a vacuum apparatus; corks will not give an airtight seal. Rubber stoppers harden with age and use. If a rubber stopper is not soft (will not squeeze), discard it. Glass tubing should fit securely into any rubber stoppers. If you can move the tubing up and down with only gentle force, it is too loose and you should obtain a larger size.

Receiving Flask.   When more than one fraction is expected from a vacuum distillation, it is considered good practice to have several preweighed receiver flasks, including the original, available before the distillation begins. Such preparation permits the rapid changing of receiving flasks during the distillation. The preweighing allows easy calculation of the weight of distillate in each fraction without the need to transfer the distillate to yet another flask.

To change receiving flasks, heating must be stopped, and the system vented at both ends, before replacing a flask. Complete directions for this procedure are given in the next section.

Vacuum Traps.   When performing a vacuum distillation, it is customary to place a "trap" in the line that connects to the vacuum source. Two common trap arrangements are shown in Figures 16.3 and 16.4. This type of trap is essential if an aspirator or a house vacuum is used as the source of vacuum. A mechanical vacuum pump requires a different type of trap (see Fig. 16.12, page 263). Variations in pressure are to be expected when using an aspirator or a house vacuum. With an aspirator, if the pressure drops low enough, the vacuum in the system will draw water from the aspirator into the connecting line. The trap allows you to see this happening and take corrective

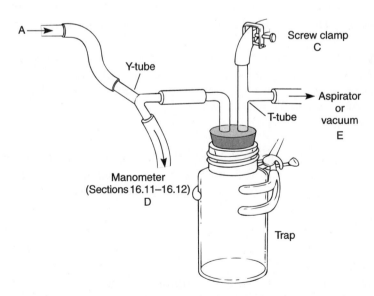

Figure 16.3 A vacuum trap using a gas bottle. The assembly connects to Figure 16.1 by joining the tubing at the point marked **A.** (The Y-tube connection to a manometer is optional.)

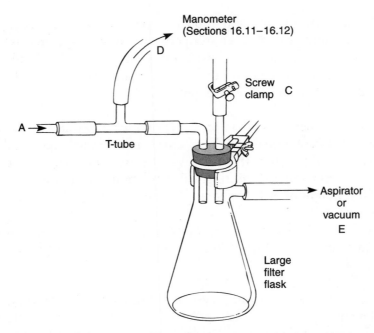

Figure 16.4 Vacuum trap using a heavy-wall filter flask. The assembly connects to Figure 16.1 by joining the tubing at the point marked **A**. (The T-tube connection to the manometer is optional.)

action (i.e., prevent water from entering the distillation apparatus). The correct action for anything but a small amount of water is to "vent" the system. This can be accomplished by opening the screw clamp **C** at the top of the trap to let air into the system. This is also the way air is admitted into the system at the end of the distillation.

> ▰▰▰ **CAUTION** Note also that it is always necessary to vent the system *before* the aspirator is stopped. If you fail to vent the system, water may be drawn into it, contaminating your product. Be sure, however, that you vent *both* ends of the system. After venting the vacuum trap, you should immediately open the screw clamp on top of the ebulliator tube.

The trap, which contains a large volume, also acts as a buffer to pressure changes, evening out small variations in the line. In the house vacuum system, it prevents oil and water (often present in house lines) from entering your system.

Manometer Connection. A manometer allows measurement of pressure. A Y-tube (or T-tube) connection **D** is shown in the line from the apparatus to the trap. This branching connection is optional, but is required if you wish to monitor the actual pressure of your system when using a manometer. The operation of manometers is discussed in Sections 16.11 and 16.12. A suitable manometer should be included in the system at least part of the time during the distillation to measure the pressure at which the distillation is being conducted. A boiling point is of little value if the pressure is not known! After use, the manometer can be removed if a screw clamp is used to close the connection.

CAUTION The manometer must be vented very slowly to prevent a rush of mercury from breaking out the end of the tubing.

A manometer is also very useful in troubleshooting your system. It can be attached to the aspirator or house vacuum to determine the working pressure. In this way a defective aspirator (not uncommon) can be spotted and replaced. When you connect your apparatus, you can adjust all the joints and connections to obtain the best working pressure *before* you begin to distill. Generally a working pressure of from 25 to 50 mm Hg is adequate for the typical student laboratory experiment.

Aspirators. In many laboratories the most convenient source of vacuum for a reduced-pressure distillation is the aspirator. The aspirator, or other vacuum source, is attached to the trap. The aspirator can theoretically pull a vacuum equal to the vapor pressure of the water flowing through it. The vapor pressure of flowing water depends on its temperature (24 mm Hg at 25°C; 18 mm Hg at 20°C; 9 mm Hg at 10°C). However, in the typical instructional laboratory, the pressures attained are higher than expected due to reduced water pressure when many students are using their aspirators simultaneously. Good laboratory practice requires that only a few students on a given bench use the aspirator at the same time. It may be necessary to establish a schedule for aspirator use, or at least to have some students wait until others are finished.

House Vacuum. As stated for aspirators, depending on the capacity of the system, it may not be possible for everyone to use the vacuum system at once. Students may have to take turns or work in rotation. A typical house vacuum system will have a base pressure of about 35–100 mm Hg when it is not overloaded.

16.3 VACUUM DISTILLATION: STEPWISE MACROSCALE DIRECTIONS

The procedures in applying vacuum distillation are described in this section.

CAUTION Safety glasses must be worn at all times during vacuum distillation.

A. Evacuating the Apparatus

1. Assemble the apparatus shown in Figure 16.1 as discussed in Section 16.2, and attach a trap (either Fig. 16.3 or 16.4). The connection is made at the points labeled **A**. Next, attach the trap to either an aspirator or a house vacuum system at point **E**. Do not close any clamps at this time.
2. Weigh each empty receiving flask to be used in collecting the various fractions during the distillation.

3. Concentrate the material to be distilled in an Erlenmeyer flask or beaker by removing all volatile solvents, such as ether, using a steam bath or a water bath in the hood. Use boiling stones and a stream of air to help the solvent removal.

4. Remove the distilling flask from the vacuum distillation apparatus, remove the grease by wiping with a towel, and transfer the concentrate to the flask, using a funnel. Complete the transfer by rinsing with a *small* amount of solvent. Again, concentrate the material until no additional volatile solvent can be removed (boiling will cease). The flask should be no more than half-full after concentration. Regrease the joint and reattach the flask to the distilling apparatus. Make sure all joints are tight.

5. On the trap assembly (Fig. 16.3 or 16.4) open the clamp at **C** and attach a manometer at point **D**.

6. Turn on the aspirator (or house vacuum) (Fig. 16.3) to the maximum extent.

7. Tighten the screw clamp at **B** (Fig. 16.1) until the tubing is nearly closed.

8. Going back to the trap (Fig. 16.3), slowly tighten the screw clamp at point **C**. Watch the bubbling action of the ebulliator tube to see that it is not too vigorous or too slow. Any volatile solvents you could not remove during concentration will be removed now. Once the loss of volatiles slows down, close screw clamp **C** to the fullest extent.

9. Adjust the ebulliator tube at **B** until a fine steady stream of bubbles is formed.

10. Wait a few minutes and then record the pressure obtained.

11. If the pressure is not satisfactory, check all connections to see that they are tight. Gently twist any hoses to snug them down. Press down on any rubber stoppers. Check the fit of all glass tubing. Press any joints together until they appear evenly greased and well joined. If you crimp the rubber tubing between the apparatus and the trap with your hand and the pressure decreases, you will know that there is a leak in the glassware assembly. If there is no change, the problem may be with the aspirator or the trap. Readjust the ebulliator screw clamp at **B** if necessary.

> **NOTE:** Do not proceed until you have a good vacuum. Ask your instructor for help if necessary.

12. Once your vacuum has been established, record the pressure. The manometer may now be removed for use by another student if necessary. Place a screw clamp ahead of the manometer at **D** and tighten it. With careful venting, the manometer may now be removed.

B. Beginning Distillation

13. Raise the heat source into position with wooden blocks, or other means, and begin to heat.

14. Increase the temperature. Eventually a reflux ring will contact the thermometer bulb, and distillation will begin.

15. Record the temperature range and the pressure range (if the manometer is still connected) during the distillation. The distillate should be collected at a rate of about 1 drop per second.

16. If the reflux ring is in the Claisen head but will not rise into the distilling head, it may be necessary to insulate these pieces by wrapping them with glass wool and aluminum foil (shiny side in). The insulation should aid the distillate to pass into the condenser.

17. The boiling point should be relatively constant so long as the pressure is constant. A rapid increase in pressure may be due to increased use of the aspirators in the laboratory (or additional connections to the house vacuum). It could also be due to decomposition of the material being distilled. Decomposition will produce a dense white fog in the distilling flask. If this happens, reduce the temperature of the heat source, or remove it, and *stand back* until the system cools. When the fog subsides you can investigate the cause.

C. Changing Receiving Flasks

18. To change receiving flasks during distillation when a new component begins to distill (higher boiling point at the same pressure), carefully open the clamp on top of the trap assembly at **C** and immediately lower the heat source.

> **NOTE:** Watch the ebulliator for excessive backup! It may also be necessary to open the clamp at **B**.

19. Remove the wooden blocks, or other support under the receiving flask, release the clamp, and replace the flask with a clean, preweighed receiver. Use a small amount of grease, if necessary, to reestablish a good seal.

20. Reclose the clamp at **C** and allow several minutes for the system to reestablish the reduced pressure. If you opened the ebulliator screw clamp at **B** you will have to close and readjust it. Bubbling will not recommence until any liquid is drawn back out of the ebulliator. This liquid may have been forced into the ebulliator when the vacuum was interrupted.

21. Raise the heating source back into position under the distilling flask and continue with the distillation.

22. When the temperature falls at the thermometer, this usually indicates that distillation is complete. If a significant amount of liquid remains, however, the bubbling may have stopped, the pressure may have risen, the heating source may not be hot enough, or perhaps insulation of the distillation head is required. Adjust accordingly.

D. Shutdown

23. At the end of the distillation, remove the heat source, and slowly open the screw clamps at **C** and **B**. When the system is vented you may shut off the aspirator or house vacuum and disconnect the tubing.

24. Remove the receiving flask and clean all glassware as soon as possible after disassembly (let it cool a bit) to keep the ground-glass joints from sticking.

> **NOTE:** If you used grease, thoroughly clean all grease off the joints, or it will contaminate your samples in other procedures.

Part C. Microscale Distillation

16.4 VACUUM DISTILLATION—MICROSCALE METHODS

A basic apparatus similar to the one shown in Figure 16.5 (or Fig. 16.8) may be used for microscale vacuum distillations. As is the case for simple distillation, this apparatus uses the Hickman head as a means to reduce the length of the vapor path. The major difference to be found when comparing this assembly to one for simple distillation (Fig. 14.10, page 215) is that the opening to the atmosphere has been replaced by a connection to a vacuum source (top right-hand side). The usual sources of vacuum are the aspirator (Chapter 8, Section 8.5, page 110), a mechanical vacuum pump, or a "house" vacuum line (one piped directly to the laboratory bench). The aspirator is probably the simplest of these sources and the vacuum source most likely to be available. However, if pressures below 10–20 mm Hg are required, a vacuum pump must be used.

When working with glassware that is to be evacuated, you should wear safety glasses at all times. There is always danger of an implosion.

▨▨▨ CAUTION Safety glasses must be worn at all times during vacuum distillation.

It is a good idea to work in a hood when performing a vacuum distillation. If the experiment will involve high temperatures (>220°C) for distillation, or an extremely low pressure (<0.1 mm Hg), for your own safety you should definitely work in a hood, behind a shield.

Assembling the Apparatus. When assembling an apparatus for vacuum distillation, it is important that all joints and connections be airtight. The joints in the newest microscale kits are standard-taper ground-glass joints, with a compression cap that contains an O-ring seal. Glassware that contains this type of compression joint will hold a vacuum quite easily. Under normal conditions, it is not necessary to grease these joints.

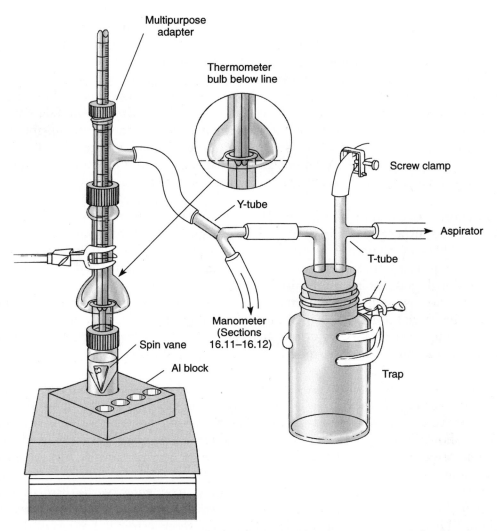

Figure 16.5 Reduced-pressure microscale distillation (internal monitoring of temperature).

NOTE: Normally, you should not grease joints. It is necessary to grease the joints in a vacuum distillation only if you cannot achieve the desired pressure without using grease.

If you must grease joints, take care not to use too much grease. You are working with small quantities of liquid in a microscale distillation, and the grease can become a very serious contaminant if it oozes out the bottom of the joints into your system. Apply a small amount of grease (thin film) completely around the top of the *inner* joint; then mate the joints and turn them slightly to spread the grease evenly. If you have used the correct amount of grease, it will not ooze out the bottom; rather, the entire joint will appear clear and without striations or uncovered areas.

Make doubly sure that any connections to pressure tubing are tight. The pressure tubing itself should be relatively new and without cracks. If the tubing shows cracks when you stretch or bend it, it may be old and leak air into the system. Glass tubing should fit securely into any rubber stoppers. If you can move the tubing up and down with only gentle force, it is too loose, and you should obtain a larger size. Check all glassware to be sure there are no cracks and that there are no chips in the standard-taper joints. Cracked glassware may break when evacuated.

Connecting to Vacuum. In Figure 16.5, the connection to vacuum has been made using a multipurpose adapter (see Fig. 14.11B, page 216). If a multipurpose adapter is not available, an alternative method uses a Claisen head and two thermometer adapters (Fig. 16.6). If two thermometer adapters are not available, a # 0 rubber stopper fitted with glass tubing can be used.

Whichever is used, the connection to the vacuum source is made using **pressure tubing.** Pressure tubing (also called vacuum tubing), unlike the more common thin-walled tubing used to carry water or gas, has heavy walls that will not collapse inward when it is evacuated. Compare the two types of tubing shown in Figure 16.2.

Water Trap. If an aspirator is used as a source of vacuum, a water trap must be placed between it and the distillation assembly. A commonly used type of water trap is shown at the bottom right of Figure 16.5. A different type of trap, which is also common, is shown in Figure 16.8. Variations in water pressure are to be expected when using an aspirator. If the pressure drops low enough, the vacuum in the system will draw water from the aspirator into the connecting line. The trap allows you to see this happening and take corrective action (prevent water from entering the distillation apparatus). The correct action for anything but a small amount of water is to "vent the system." This can be accomplished by opening the screw clamp at the top of the trap to let air into the system. When performing a vacuum distillation, you should also realize that the system should always be vented before stopping the aspirator. If you turn off the aspirator while the system is still under vacuum, water will be drawn into the connecting line and trap.

Manometer Connection. A Y-tube is shown in the line from the apparatus to the trap. This branching connection is optional but is required if you wish to monitor the actual pressure of the system using a manometer. The operation of manometers is discussed in Sections 16.11 and 16.12.

Thermometer Placement. If a thermometer is used, be sure that the bulb is placed in the stem of the Hickman head just below the well. If it is placed higher, it may not be surrounded by a constant stream of vapor from the material being

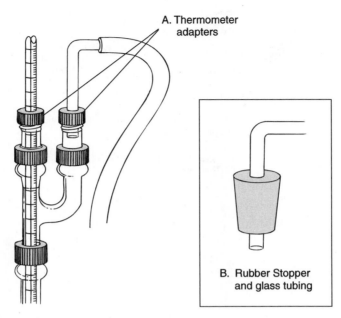

Figure 16.6 Alternative vacuum connections.

distilled. If the thermometer is not exposed to a continuous stream of vapor, it may not reach temperature equilibrium. As a result, the temperature reading would be incorrect (low).

Preventing Bump-Over. When heating a distillation flask, there is always the possibility that the boiling action will become too vigorous (mainly due to superheating) and "bump" some of the undistilled liquid up into the Hickman head. The simplest way to prevent bumping is to stir the boiling liquid with a magnetic spin vane. Stirring rapidly will distribute the heat evenly, keep the boiling action smooth, and prevent bumping. Boiling stones cannot be used for this purpose in a vacuum distillation; they do not work in vacuum. In a conventional vacuum distillation (macroscale), it is customary to maintain smooth boiling action by using an **ebulliator tube.** The ebulliator tube agitates the boiling solution by providing a small, continuous stream of air bubbles. Figure 16.7 shows how a microscale vacuum distillation may be modified to use an ebulliator tube. The amount of air (rate of bubbles) provided by the ebulliator is adjusted by either tightening or loosening the screw clamp at the top. A Pasteur pipet makes an excellent ebulliator tube. As Figure 16.7 shows, the ebulliator tube replaces the thermometer. Hence, the ebulliator should be used only when internal monitoring of temperature is not required. In practice, although this method works satisfactorily, better results are obtained by stirring and distilling slowly.

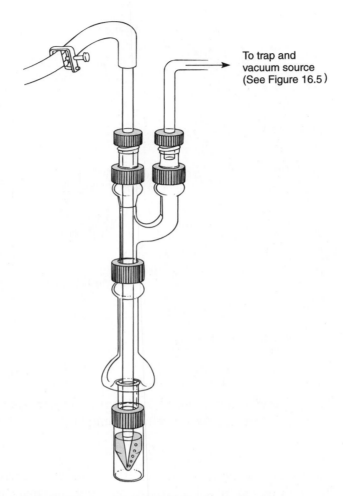

To trap and
vacuum source
(See Figure 16.5)

Figure 16.7 Use of an ebulliator tube instead of a thermometer.

16.5 SIMPLIFIED MICROSCALE APPARATUS

The apparatus shown in Figure 16.8 will often produce very satisfactory results when internal temperature monitoring is not required. It is the apparatus we prefer for microscale experiments that require vacuum distillation. Distillation (heating) should be performed slowly while stirring briskly. Just before the well begins to fill, you will see reflux action (condensation) in the stem. In many cases, this will occur even before there is any evidence of boiling in the heated liquid. In Figure 16.8 an aluminum heating block is shown. The aluminum block is an effective heat source whenever you want a fast heating response or a high temperature.

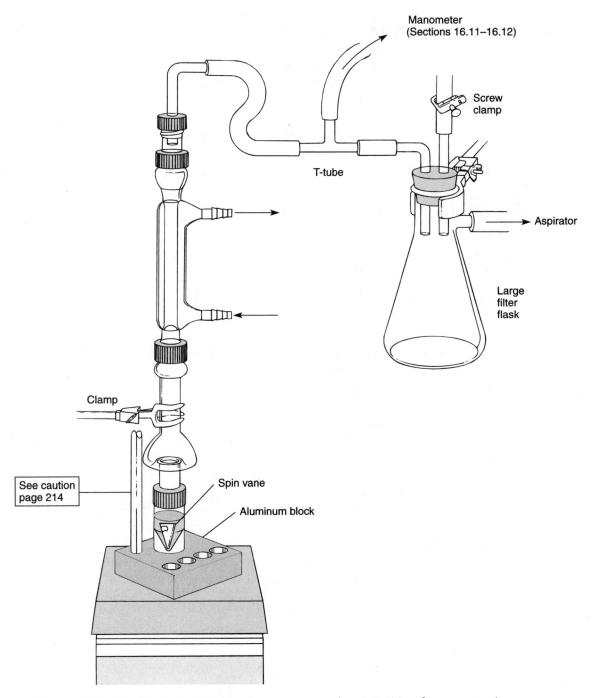

Figure 16.8 Simplified distillation apparatus (external monitoring of temperature).

16.6 Vacuum Distillation: Stepwise Microscale Directions

The following set of instructions is a step-by-step account of how to carry out a vacuum distillation. The microscale apparatus illustrated in Figure 16.5 will be used; however, the procedures apply to any vacuum distillation.

CAUTION Safety glasses must be worn at all times during vacuum distillation.

A. Evacuating the Apparatus

1. Assemble the apparatus as shown in Figure 16.5. It should be held with a clamp attached to the top of the Hickman head and placed *above* the aluminum block.

   NOTE: If you expect the temperature of the distillation to rise above 150°C, omit the threaded cap and O-ring between the conical vial and Hickman head. They will melt at high temperature.

2. If the sample contains solvent, concentrate the sample to be distilled in the conical vial (or round-bottom flask) that you are using. Use one of the solvent-removal methods discussed in Chapter 7. If you have a large volume of solvent to evaporate and the sample does not fit in the conical vial, you must use an Erlenmeyer flask first and then transfer the sample to the conical vial. (Be sure to rinse the Erlenmeyer flask with a little solvent and then reevaporate in the conical vial.) As a rule, the distillation vial or flask should be no more than two-thirds full.
3. Attach the conical vial (or flask) to the apparatus and make sure all joints are sealed.
4. Turn the aspirator on to the maximum extent.
5. Close the screw clamp on the water trap very tightly. (If you are using an ebulliator tube as in Figure 16.7, you next regulate the rate of bubbling by adjusting the tightness of the screw clamp at the top of the tube.)
6. Using the manometer, observe the pressure. It may take a few minutes to remove any residual solvent and evacuate the system. If the pressure is not satisfactory, check all connections to see whether they are tight. (Readjust the ebulliator tube if necessary.)

   NOTE: Do not proceed until you have a good vacuum.

B. Beginning Distillation

7. Lower the apparatus into the aluminum block and begin to heat. Place the external thermometer in the block now if you wish (see caution, page 214).

8. Increase the temperature of the heat source until you begin to see distillate collect in the well of the Hickman head. (Observe very carefully; liquid may appear almost "magically" without any sign of boiling or any obvious reflux ring.)
9. If you are using a thermometer, record the temperature and pressure when distillate begins to appear. (If you are not using an internal thermometer, record the external temperature. If you have two thermometers, record both temperatures.)

C. Collecting a Fraction

10. To collect a fraction, raise the apparatus above the aluminum block and allow it to cool a bit before opening it.
11. Open the screw clamp on the water trap to allow air to enter the system. (If you are using an ebulliator tube, you also need to open the screw clamp at its top *immediately*, or the liquid in the distillation flask will be forced upward into it.)
12. Partially disassemble the apparatus and remove the fraction with a Pasteur pipet, as shown in Figure 14.8A. (If you have a Hickman head with a side port, you may simply open the side port to remove the fraction. This is shown in Figure 14.8C.)

> **NOTE:** If you do not intend to collect a second fraction, go directly to steps 18–20.

13. Reassemble the apparatus (or close the side port) and tighten the clamp at the top of the ebulliator tube.
14. Tighten the screw clamp on the water trap, and reestablish the desired pressure. If the pressure is not satisfactory, check all connections to make sure they are sealed.
15. Lower the apparatus back into the aluminum block and continue the distillation.

D. Shutdown

16. At the end of the distillation, raise the apparatus from the aluminum block and allow it to cool. Also let the aluminum block cool.
17. Open the screw clamp on the water trap first, then immediately open the one at the top of the ebulliator tube.
18. Turn off the water at the aspirator. (Do not do this before step 17!)
19. Remove any distilled material by one of the methods shown in Figure 14.8.
20. Disassemble the apparatus and clean all glassware as soon as possible to prevent the joints from sticking.

> **NOTE:** If you used grease, thoroughly clean all grease off the joints, or it will contaminate your samples in other procedures.

16.7 ROTARY FRACTION COLLECTORS

With the types of apparatus we have discussed previously, the vacuum must be stopped to remove fractions when a new substance (fraction) begins to distill. Quite a few steps are required to perform this change, and it is quite inconvenient when there are several fractions to be collected. Two pieces of semimicroscale apparatus that are designed to alleviate the difficulty of collecting fractions while working under vacuum are shown in Figure 16.9. The collector, which is shown to the right, is sometimes called a "cow" because of its appearance. With these rotary fraction collecting devices, all you need to do is rotate the device to collect fractions.

16.8 BULB-TO-BULB DISTILLATION

The ultimate in microscale methods is to use a bulb-to-bulb distillation apparatus. This apparatus is shown in Figure 16.10. The sample to be distilled is placed in the glass container attached to one of the arms of the apparatus. The sample is frozen solid, usually by using liquid nitrogen, but dry ice in 2-propanol or an ice–salt water

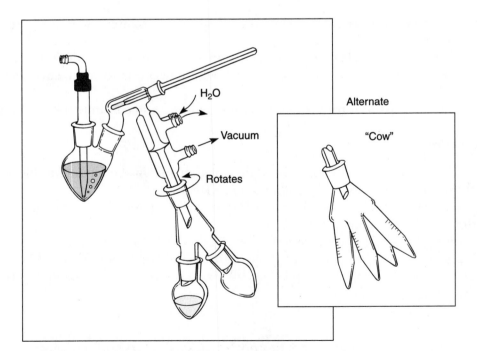

Figure 16.9 A rotary fraction collector.

Vacuum

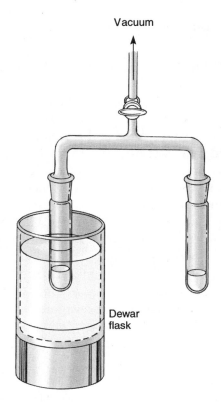

Dewar
flask

Figure 16.10 Bulb-to-bulb distillation.

mixture may also be used. The coolant container shown in the figure is a **Dewar flask.** The Dewar flask has a double wall with the space between the walls evacuated and sealed. A vacuum is a very good thermal insulator, and there is little heat loss from the cooling solution.

After freezing the sample, the entire apparatus is evacuated by opening the stopcock. When the evacuation is complete, the stopcock is closed, and the Dewar flask is removed. The sample is allowed to thaw and then it is frozen again. This freeze–thaw–freeze cycle removes any air or gases that were trapped in the frozen sample. Next the stopcock is opened to evacuate the system again. When the second evacuation is complete, the stopcock is closed, and the Dewar flask is moved to the other arm to cool the empty container. As the sample warms, it will vaporize, travel to the other side, and be frozen or liquefied by the cooling solution. This transfer of the liquid from one arm to the other may take quite a while, but *no heating is required.*

The bulb-to-bulb distillation is most effective when liquid nitrogen is used as coolant and when the vacuum system can achieve a pressure of 10^{-3} mm Hg or lower. This requires a vacuum pump; an aspirator cannot be used.

Part D. Semimicroscale Distillation

16.9 VACUUM DISTILLATION—SEMIMICROSCALE METHODS

There are times when the amount of material to be distilled is too large to use a microscale apparatus with a Hickman head, yet too small to use traditional macroscale equipment. If you have a distillation head and a vacuum adapter in your microscale kit, an apparatus similar to that in Figure 16.11 may be used. (The distillation head and the vacuum adapter are optional microscale pieces, and many standard microscale kits do not include them. However, most manufacturers have these pieces in their catalogs, and some kits may include them.)

The semimicroscale apparatus shown in Figure 16.11 (without a condenser) works well if the liquid to be distilled has a relatively high boiling point. For lower boiling liquids, a condenser may be added yielding a standard distillation apparatus such as that shown in Figure 14.1, page 201. If foaming or bumping is a serious concern, a Claisen head may be added, yielding a standard vacuum distillation apparatus similar to that shown in Figure 16.1, but using microscale pieces.

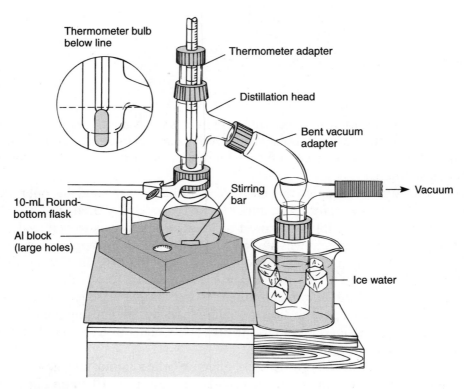

Figure 16.11 A semimicroscale vacuum distillation.

Part E. Vacuum Pump Methods

16.10 THE MECHANICAL VACUUM PUMP

The aspirator is not capable of yielding pressures below about 5 mm Hg. This is the vapor pressure of water of 0°C, and water freezes at this temperature. A more realistic value of pressure for an aspirator is about 20 mm Hg. When pressures below 20 mm Hg are required, a vacuum pump will have to be employed. Figure 16.12 illustrates a mechanical vacuum pump and its associated glassware. The vacuum pump operates on a principle similar to that of the aspirator, but the vacuum pump uses a high-boiling oil, rather than water, to remove air from the attached system. The oil used in a vacuum pump, a silicone oil or a high-molecular-weight hydrocarbon-based oil, has a very low vapor pressure, and very low system pressures can be achieved. A good vacuum pump, with new oil, can achieve pressures of 10^{-3} or 10^{-4} mm Hg. Instead of discarding the oil as it is used, it is recycled continuously through the system.

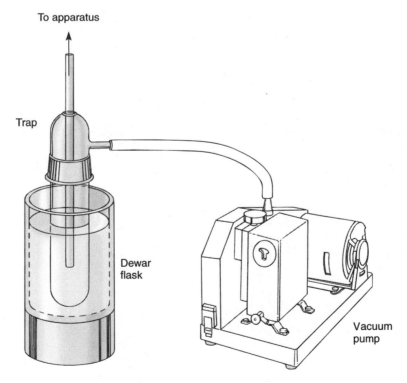

Figure 16.12 A vacuum pump and its trap.

A cooled trap is required when using a vacuum pump. This trap protects the oil in the pump from any vapors that may be present in the system. If vapors from organic solvents, or from the organic compounds being distilled, dissolve in the oil, the vapor pressure of the oil will increase, rendering it less effective. A special type of vacuum trap is illustrated in Figure 16.12. It is designed to fit into an insulated Dewar flask so that the coolant will last for a long period. At a minimum, this flask should be filled with ice water, but a dry ice–acetone mixture or liquid nitrogen is required to achieve lower temperatures and better protect the oil. Often two traps are used; the first trap contains ice water and the second trap dry ice–acetone or liquid nitrogen. The first trap liquefies low-boiling vapors that might freeze or solidify in the second trap and block it.

Part F. Manometers

16.11 THE CLOSED-END MANOMETER

The principal device used to measure pressures in a vacuum distillation is the **closed-end manometer.** Two basic types are shown in Figures 16.13 and 16.14. The manometer shown in Figure 16.13 is widely used because it is relatively easy to construct. It consists of a U-tube that is closed at one end and mounted on a wooden support. You can construct the manometer from 9-mm glass capillary tubing and fill it, as shown in Figure 16.15.

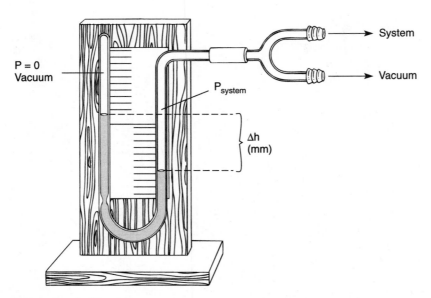

Figure 16.13 A simple U-tube manometer.

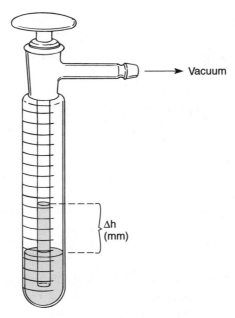

Figure 16.14 A commercial "stick" manometer.

CAUTION Mercury is a very toxic metal with cumulative effects. Because mercury has a high vapor pressure, it must not be spilled in the laboratory. You must not touch it with your skin. Seek immediate help from an instructor in case of a spill or if you break a manometer. Spills must be cleaned up immediately.

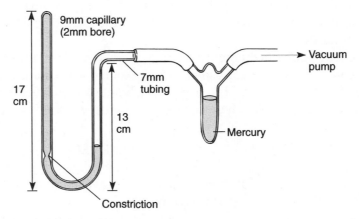

Figure 16.15 Filling a U-tube manometer.

A small filling device is connected to the U-tube with pressure tubing. The U-tube is evacuated with a good vacuum pump; then the mercury is introduced by tilting the mercury reservoir.

> **NOTE:** The entire filling operation should be conducted in a shallow pan in order to contain any spills that might occur.

Enough mercury should be added to form a column about 20 cm in total length. When the vacuum is interrupted by admitting air, the mercury is forced by atmospheric pressure to the end of the evacuated tube. The manometer is then ready for use. The constriction shown in Figure 16.15 helps to protect the manometer against breakage when the pressure is released. Be sure that the column of mercury is long enough to pass through this constriction.

When an aspirator or any other vacuum source is used, a manometer can be connected into the system. As the pressure is lowered, the mercury rises in the right tube and drops in the left tube until Δh corresponds to the approximate pressure of the system (see Fig. 16.13).

$$\Delta h = (P_{\text{system}} - P_{\text{reference arm}}) = (P_{\text{system}} - 10^{-3} \text{ mm Hg}) \approx P_{\text{system}}$$

A short piece of metric ruler or a piece of graph paper ruled in millimeter squares is mounted on the support board to allow Δh to be read. No addition or subtraction is necessary, because the reference pressure (created by the initial evacuation when filling) is approximately zero (10^{-3} mm Hg) when referred to readings in the 10- to 50-mm Hg range. To determine the pressure, count the number of millimeter squares beginning at the top of the mercury column on the left and continuing downward to the top of the mercury column on the right. This is the height difference Δh, and it gives the pressure in the system directly.

A commercial counterpart to the U-tube manometer is shown in Figure 16.14. With this manometer, the pressure is given by the difference in the mercury levels in the inner and outer tubes.

The manometers described here have a range of about 1–150 mm Hg in pressure. They are convenient to use when an aspirator is the source of vacuum. For high-vacuum systems (pressures below 1 mm Hg), a more elaborate manometer or an electronic measuring device must be used. These devices will not be discussed here.

16.12 CONNECTING AND USING A MANOMETER

The most common use of a closed-end manometer is to monitor pressure during a reduced-pressure distillation. The manometer is placed in a vacuum distillation system, as shown in Figure 16.16. Generally an aspirator is the source of vacuum. Both the manometer and the distillation apparatus should be protected by a trap from possible

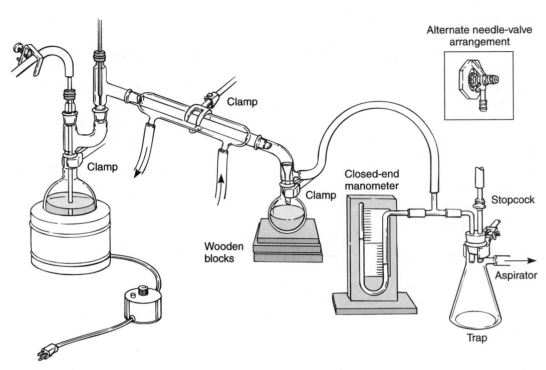

Figure 16.16 Connecting a manometer to the system. To construct a "bleed," the needle valve may replace the stopcock.

backups in the water line. Alternatives to the trap arrangements shown in Figure 16.16 appear in Figures 16.5 and 16.8. Notice in each case that the trap has a device (screw clamp or stopcock) for opening the system to the atmosphere. This is especially important in using a manometer, because you should always make pressure changes slowly. If this is not done, there is a danger of spraying mercury throughout the system, breaking the manometer, or spurting mercury into the room. In the closed-end manometer, if the system is opened suddenly, the mercury rushes to the closed end of the U-tube. The mercury rushes with such speed and force that the end will be broken out of the manometer. Air should be admitted *slowly* by opening the valve cautiously. In a similar fashion, the valve should be closed slowly when the vacuum is being started, or mercury may be forcefully drawn into the system through the open end of the manometer.

If the pressure in a reduced-pressure distillation is lower than desired, it is possible to adjust it by means of a **bleed valve.** The stopcock can serve this function in Figure 16.16 if it is opened only a small amount. In those systems with a screw clamp on the trap (Figs. 16.3 and 16.4), remove the screw clamp from the trap valve and attach the base of a Tirrill-style Bunsen burner. The needle valve in the base of the burner can be used to adjust precisely the amount of air that is admitted (bled) to the system and hence control the pressure.

PROBLEMS

1. Give some reasons that would lead you to purify a liquid by using vacuum distillation rather than by using simple distillation.
2. When using an aspirator as a source of vacuum in a vacuum distillation, do you turn off the aspirator before venting the system? Explain.
3. A compound was distilled at atmospheric pressure and had a boiling range of 310–325°C. What would be the approximate boiling range of this liquid if it was distilled under vacuum at 20 mm Hg?
4. Boiling stones generally do not work when performing a vacuum distillation. What substitutes may be used?
5. What is the purpose of the trap that is used during a vacuum distillation performed with an aspirator?

CHAPTER 17

Sublimation

In Chapter 13, the influence of temperature on the change in vapor pressure of a liquid was considered (see Fig. 13.1, page 190). It was shown that the vapor pressure of a liquid increases with temperature. Because the boiling point of a liquid occurs when its vapor pressure is equal to the applied pressure (normally atmospheric pressure), the vapor pressure of a liquid equals 760 mm Hg at its boiling point. The vapor pressure of a solid also varies with temperature. Because of this behavior, some solids can pass directly into the vapor phase without going through a liquid phase. This process is called **sublimation.** Because the vapor can be resolidified, the overall vaporization–solidification cycle can be used as a purification method. The purification can be successful only if the impurities have significantly lower vapor pressures than the material being sublimed.

Part A. Theory

17.1 VAPOR PRESSURE BEHAVIOR OF SOLIDS AND LIQUIDS

In Figure 17.1, vapor pressure curves for solid and liquid phases for two different substances are shown. Along lines *AB* and *DF*, the sublimation curves, the solid and vapor are at equilibrium. To the left of these lines, the solid phase exists, and to the right of these lines, the vapor phase is present. Along lines *BC* and *FG*, the liquid and vapor are at equilibrium. To the left of these lines, the liquid phase exists, and to the right, the vapor is present. The two substances vary greatly in their physical properties, as shown in Figure 17.1.

In the first case (Fig. 17.1A), the substance shows normal change-of-state behavior on being heated, going from solid to liquid to gas. The dashed line, which represents an atmospheric pressure of 760 mm Hg, is located *above* the melting point *B* in Figure 17.1A. Thus, the applied pressure (760 mm Hg) is *greater* than the vapor pressure of the solid–liquid phase at the melting point. Starting at *A*, as the temperature of the solid is raised, the vapor pressure increases along *AB* until the solid is observed to melt at *B*. At *B* the vapor pressures of *both* the solid and liquid are identical. As the temperature continues to rise, the vapor pressure will increase along *BC* until the

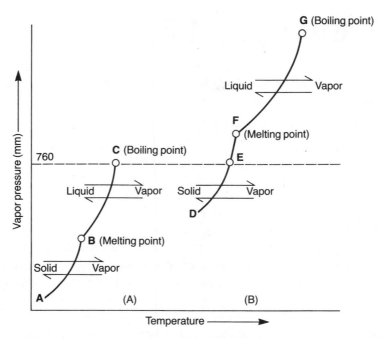

Figure 17.1 Vapor pressure curves for solids and liquids. **(A)** This substance shows normal solid-to-liquid-to-gas transitions at 760 mm Hg pressure. **(B)** This substance shows a solid-to-gas transition at 760 mm Hg pressure.

liquid is observed to boil at *C*. The description given is for the "normal" behavior expected for a solid substance. All three states (solid, liquid, and gas) are observed sequentially during the change in temperature.

In the second case (Fig. 17.1B), the substance develops enough vapor pressure to vaporize completely at a temperature below its melting point. The substance shows a solid-to-gas transition only. The dashed line is now located *below* the melting point *F* of this substance. Thus, the applied pressure (760 mm Hg) is *less* than the vapor pressure of the solid–liquid phase at the melting point. Starting at *D*, the vapor pressure of the solid rises as the temperature increases along line *DF*. However, the vapor pressure of the solid reaches atmospheric pressure (point *E*) *before* the melting point at *F* is attained. Therefore, sublimation occurs at *E*. No melting behavior will be observed at atmospheric pressure for this substance. For a melting point to be reached and the behavior along line *FG* to be observed, an applied pressure greater than the vapor pressure of the substance at point *F* would be required. This could be achieved by using a sealed pressure apparatus.

The sublimation behavior just described is relatively rare for substances at atmospheric pressure. Several compounds exhibiting this behavior—carbon dioxide, perfluorocyclohexane, and hexachloroethane—are listed in Table 17.1. Notice that these compounds have vapor pressures *above* 760 mm Hg at their melting points. In

Table 17.1 Vapor Pressures of Solids at Their Melting Points

Compound	Vapor Pressure of Solid at MP (mm Hg)	Melting Point (°C)
Carbon dioxide	3876 (5.1 atm)	−57
Perfluorocyclohexane	950	59
Hexachloroethane	780	186
Camphor	370	179
Iodine	90	114
Naphthalene	7	80
Benzoic acid	6	122
p-Nitrobenzaldehyde	0.009	106

other words, their vapor pressures reach 760 mm Hg below their melting points and they sublime rather than melt. Anyone trying to determine the melting point of hexachloroethane at atmospheric pressure will see vapor pouring from the end of the melting point tube! With a sealed capillary tube, the melting point of 186°C is observed.

17.2 SUBLIMATION BEHAVIOR OF SOLIDS

Sublimation is usually a property of relatively nonpolar substances that also have highly symmetrical structures. Symmetrical compounds have relatively high melting points and high vapor pressures. The ease with which a substance can escape from the solid state is determined by the strength of intermolecular forces. Symmetrical molecular structures have a relatively uniform distribution of electron density and a small dipole moment. A smaller dipole moment means a higher vapor pressure because of lower electrostatic attractive forces in the crystal.

Solids sublime if their vapor pressures are greater than atmospheric pressure at their melting points. Some compounds with the vapor pressures at their melting points are listed in Table 17.1. The first three entries in the table were discussed in Section 17.1. At atmospheric pressure they would sublime rather than melt, as shown in Figure 17.1B.

The next four entries in Table 17.1 (camphor, iodine, naphthalene, and benzoic acid) exhibit typical change-of-state behavior (solid, liquid, and gas) at atmospheric pressure, as shown in Figure 17.1A. These compounds sublime readily under reduced pressure, however. Vacuum sublimation is discussed in Section 17.3.

Compared with many other organic compounds, camphor, iodine, and naphthalene have relatively high vapor pressures at relatively low temperatures. For example, they have a vapor pressure of 1 mm Hg at 42, 39, and 53°C, respectively. Although this vapor pressure does not seem very large, it is high enough to lead, after a time, to **evaporation** of the solid from an open container. Mothballs (naphthalene and

1,4-dichlorobenzene) show this behavior. When iodine stands in a closed container over a period of time, you can observe movement of crystals from one part of the container to another.

Although chemists often refer to any solid–vapor transition as sublimation, the process described for camphor, iodine, and naphthalene is really an **evaporation** of a solid. Strictly speaking, a sublimation point is like a melting point or a boiling point. It is defined as the point at which the vapor pressure of the solid *equals* the applied pressure. Many liquids readily evaporate at temperatures far below their boiling points. It is, however, much less common for solids to evaporate. Solids that readily sublime (evaporate) must be stored in sealed containers. When the melting point of such a solid is being determined, some of the solid may sublime and collect toward the open end of the melting-point tube while the rest of the sample melts. To solve the sublimation problem, seal the capillary tube or rapidly determine the melting point. It is possible to use the sublimation behavior to purify a substance. For example, at atmospheric pressure, camphor can be readily sublimed, just below its melting point at 175°C. At 175°C the vapor pressure of camphor is 320 mm Hg. The vapor solidifies on a cool surface.

17.3 VACUUM SUBLIMATION

Many organic compounds sublime readily under reduced pressure. When the vapor pressure of the solid equals the applied pressure, sublimation occurs, and the behavior is identical to that shown in Figure 17.1B. The solid phase passes directly into the vapor phase. From the data given in Table 17.1, you should expect camphor, naphthalene, and benzoic acid to sublime at or below the respective applied pressures of 370, 7, and 6 mm Hg. In principle, you can sublime *p*-nitrobenzaldehyde (last entry in the table), but it would not be practical because of the low applied pressure required.

17.4 ADVANTAGES OF SUBLIMATION

One advantage of sublimation is that no solvent is used and therefore none needs to be removed later. Sublimation also removes occluded material, like molecules of solvent, from the sublimed substance. For instance, caffeine (sublimes at 178°C, melts at 236°C) absorbs water gradually from the atmosphere to form a hydrate. During sublimation, this water is lost, and anhydrous caffeine is obtained. If too much solvent is present in a sample to be sublimed, however, instead of becoming lost, it condenses on the cooled surface and thus interferes with the sublimation.

Sublimation is a faster method of purification than crystallization but not as selective. Similar vapor pressures are often a factor in dealing with solids that sublime; consequently, little separation can be achieved. For this reason, solids are far more often purified by crystallization. Sublimation is most effective in removing a volatile substance from a nonvolatile compound, particularly a salt or other inorganic material.

Sublimation is also effective in removing highly volatile bicyclic or other symmetrical molecules from less volatile reaction products. Examples of volatile bicyclic compounds are borneol, isoborneol, and camphor.

Borneol   Camphor   Isoborneol

Part B. Macroscale and Microscale Sublimation

17.5 SUBLIMATION—METHODS

Sublimation can be used to purify solids. The solid is warmed until its vapor pressure becomes high enough for it to vaporize and condense as a solid on a cooled surface placed closely above. Three types of apparatus are illustrated in Figure 17.2. Since all of the parts fit securely, they are all capable of holding a vacuum. Chemists usually perform vacuum sublimations because most solids undergo the solid-to-gas transition only at low pressures. Reduction of pressure also helps to prevent thermal decomposition of substances that would require high temperatures to sublime at ordinary pressures. One end of a piece of rubber pressure tubing is attached to the apparatus and the other end is attached to an aspirator or to the house vacuum system or to a vacuum pump.

A sublimation is probably best carried out using one of the pieces of microscale equipment shown in Figures 17.2A and B. It is recommended that the laboratory instructor make available either one type or the other to be used on a communal basis. Each apparatus shown employs a central tube (closed on one end) filled with ice-cold water that serves as a condensing surface. The tube is filled with ice chips and a minimum of water. If the cooling water becomes warm before the sublimation is completed, a Pasteur pipet can be used to remove the warm water. The tube is then refilled with more ice-cold water. Warm water is undesirable because the vapor will not condense efficiently to form a solid as readily on a warm surface as it would on a cold surface. A poor recovery of solid results.

The apparatus shown in Figure 17.2C can be constructed from a sidearm test tube, a neoprene adapter, and a piece of glass tubing sealed at one end. Alternatively, a 15×125-mm test tube may be used instead of the piece of glass tubing. The test tube is inserted into a No. 1 neoprene adapter using a little water as a lubricant. All pieces must fit securely to obtain a good vacuum and to avoid water being drawn into the sidearm test tube around the rubber adapter. To achieve an adequate seal, the sidearm test tube may need to be flared somewhat.

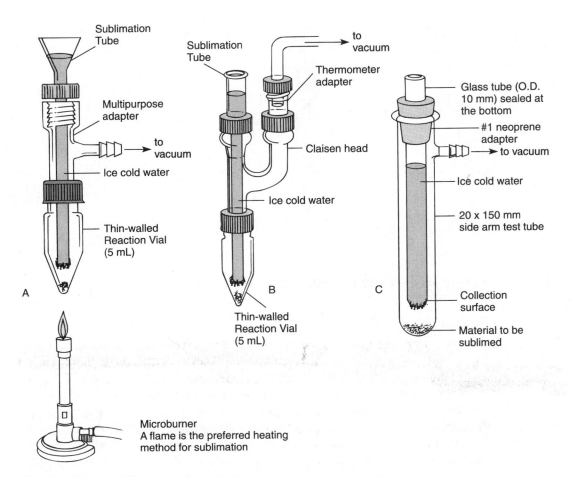

Figure 17.2 A sublimation apparatus.

A flame is the preferred heating device because the sublimation will occur more quickly than with other heating devices. The sublimation will be finished before the ice water warms significantly. The burner can be held by its cool base (not the hot barrel!) and moved up and down the sides of the outer tube to "chase" any solid that has formed on the sides toward the cold tube in the center. When using the apparatus shown in Figures 17.2A and B with a flame, you will need to use a thin-wall vial. Thicker glass can shatter when heated with a flame.

Remember that while performing a sublimation, it is important to keep the temperature below the melting point of the solid. After sublimation, the material that has collected on the cooled surface is recovered by removing the central tube (cold-finger) from the apparatus. Take care in removing this tube to avoid dislodging the crystals that have collected. The deposit of crystals is scraped from the inner tube with a spatula. If reduced pressure has been used, the pressure must be released carefully to keep a blast of air from dislodging the crystals.

17.6 SUBLIMATION—SPECIFIC DIRECTIONS

A. Microscale Apparatus

Assemble a sublimation apparatus as shown in Figure 17.2A.[1] Place your impure compound in a small Erlenmeyer flask. Add approximately 0.5 mL of methylene chloride[2] to the Erlenmeyer flask, swirl to dissolve the solid, and transfer the solution of your compound to a clean 5-mL thin-walled conical vial, using a clean and dry Pasteur pipet. Add a few more drops of methylene chloride to the flask in order to rinse the compound out completely. Transfer this liquid to the conical vial. Evaporate the methylene chloride from the conical vial by gentle heating in a warm water bath under a stream of dry air or nitrogen.

Insert the cold finger into the sublimation apparatus. If you are using the sublimator with the multipurpose adapter, adjust it so that the tip of the cold finger will be positioned about 1 cm above the bottom of the conical vial. Be sure that the inside of the assembled apparatus is clean and dry. If you are using an aspirator, install a trap between the aspirator and the sublimation apparatus. Turn on the vacuum and check to make sure that all joints in the apparatus are sealed tightly. Place *ice-cold* water in the inner tube of the apparatus. Heat the sample gently and carefully with a microburner to sublime your compound. Hold the burner in your hand (hold it at its base, *not* by the hot barrel), and apply the heat by moving the flame back and forth under the conical vial and up the sides. If the sample begins to melt, remove the flame for a few seconds before you resume heating. When sublimation is complete, discontinue heating. Remove the cold water and remaining ice from the inner tube and allow the apparatus to cool while continuing to apply the vacuum.

When the apparatus is at room temperature, slowly vent the vacuum and *carefully* remove the inner tube. If this operation is done carelessly, the sublimed crystals may be dislodged from the inner tube and fall back into the conical vial. Scrape the sublimed compound onto a tared piece of smooth paper, and determine the weight of your compound recovered.

B. Sidearm Test Tube Apparatus

Assemble a sublimation apparatus as shown in Figure 17.2C. Insert a 15 × 125-mm test tube into a No. 2 neoprene adapter, using a *little* water as a lubricant, until the tube is fully inserted. Place the crude compound into a 20 × 150-mm sidearm test tube. Next place the 15 × 120-mm test tube into the sidearm test tube, making sure they fit together tightly. Turn on the aspirator or house vacuum and make sure a good seal is obtained. At the point at which a good seal has been achieved, you should hear

---

[1]If you are using another type of sublimation apparatus, your instructor will provide you with specific instructions on how to assemble it correctly.

[2]If your compound does not dissolve freely in methylene chloride, use some other appropriate low-boiling solvent, such as ether, acetone, or pentane.

or observe a change in the water velocity in the aspirator. At this time, also make sure that the central tube is centered in the sidearm test tube; this will allow for optimal collection of the purified compound. Once the vacuum has been established, place small chips of ice in the test tube to fill it.[3] When a good vacuum seal has been obtained and ice has been added to the inner test tube, heat the sample gently and carefully with a microburner to sublime your compound. Hold the burner in your hand (hold it at the *base*, not by the hot barrel) and apply heat by moving the flame back and forth under the outer tube and up the sides. If the sample begins to melt, remove the flame for a few seconds before you resume heating. When sublimation is complete, remove the burner and allow the apparatus to cool. As the apparatus is cooling, and before you disconnect the vacuum, remove the water and ice from the inner tube using a Pasteur pipet.

When the apparatus has cooled and the water has been removed from the tube, you may disconnect the vacuum. The vacuum should be removed carefully to avoid dislodging the crystals from the inner tube by the sudden rush of air into the apparatus. *Carefully* remove the inner tube of the sublimation apparatus. If this operation is done carelessly, the sublimed crystals may be dislodged from the inner tube and fall back into the residue. Scrape the sublimed compound onto tared weighing paper, using a small spatula. Determine the weight of this purified compound.

PROBLEMS

1. Why is solid carbon dioxide called dry ice? How does it differ from solid water in behavior?
2. Under what conditions can you have *liquid* carbon dioxide?
3. A solid substance has a vapor pressure of 800 mm Hg at its melting point (80°C). Describe how the solid behaves as the temperature is raised from room temperature to 80°C, while the atmospheric pressure is held constant at 760 mm Hg.
4. A solid substance has a vapor pressure of 100 mm Hg at the melting point (100°C). Assuming an atmospheric pressure of 760 mm Hg, describe the behavior of this solid as the temperature is raised from room temperature to its melting point.
5. A substance has a vapor pressure of 50 mm Hg at the melting point (100°C). Describe how you would experimentally sublime this substance.

---

[3]It is very important that ice not be added to the inner test tube until the vacuum has been established. If the ice is added before the vacuum is turned on, condensation on the outer walls of the inner tube will contaminate the sublimed compound.

CHAPTER 18

Steam Distillation

The simple, fractional, and vacuum distillations described in Chapters 14, 15, and 16 are applicable to completely soluble (miscible) mixtures only. When liquids are *not* mutually soluble (immiscible), they can also be distilled, but with a somewhat different result. A mixture of immiscible liquids will boil at a lower temperature than the boiling points of any of the separate components as pure compounds. When steam is used to provide one of the immiscible phases, the process is called **steam distillation**. The advantage of this technique is that the desired material distills at a temperature below 100°C. Thus, if unstable or very high-boiling substances are to be removed from a mixture, decomposition is avoided. Because all gases mix, the two substances can mix in the vapor and codistill. Once the distillate is cooled, the desired component, which is not miscible, separates from the water. Steam distillation is used widely in isolating liquids from natural sources. It is also used in removing a reaction product from a tarry reaction mixture.

Part A. Theory

18.1 DIFFERENCES BETWEEN DISTILLATION OF MISCIBLE AND IMMISCIBLE MIXTURES

$$\text{Miscible Liquids} \qquad P_{total} = P_A^0 N_A + P_B^0 N_B \qquad [1]$$

Two liquids A and B that are mutually soluble (miscible), and that do not interact, form an ideal solution and follow Raoult's Law, as shown in Equation 1. Note that the vapor pressures of pure liquids P_A^0 and P_B^0 are not added directly to give the total pressure P_{total} but are reduced by the respective mole fractions N_A and N_B. The total pressure above a miscible or homogeneous solution will depend on P_A^0 and P_B^0 and also on N_A and N_B. Thus, the composition of the vapor will depend on *both* the vapor pressures and the mole fractions of each component.

$$\text{Immiscible Liquids} \qquad P_{total} = P_A^0 + P_B^0 \qquad [2]$$

In contrast, when two mutually insoluble (immiscible) liquids are "mixed" to give a heterogeneous mixture, each exerts its own vapor pressure, independently of the other, as shown in Equation 2. The mole fraction term does not appear in this equation, because the compounds are not miscible. You simply add the vapor pressures of the pure liquids P_A^0 and P_B^0 at a given temperature to obtain the total pressure above the mixture. When the total pressure equals 760 mm Hg, the mixture boils. The composition of the vapor from an immiscible mixture, in contrast to the miscible mixture, is determined only by the vapor pressures of the two substances codistilling. Equation 3 defines the composition of the vapor from an immiscible mixture. Calculations involving this equation are given in Section 18.2.

$$\frac{\text{Moles A}}{\text{Moles B}} = \frac{P_A^0}{P_B^0} \qquad [3]$$

A mixture of two immiscible liquids boils at a lower temperature than the boiling points of either component. The explanation for this behavior is similar to that given for minimum-boiling-point azeotropes (Chapter 15, Section 15.8). Immiscible liquids behave as they do because an extreme incompatibility between the two liquids leads to higher combined vapor pressure than Raoult's Law would predict. The higher combined vapor pressures cause a lower boiling point for the mixture than for either single component. Thus, you may think of steam distillation as a special type of azeotropic distillation in which the substance is completely insoluble in water.

The differences in behavior of miscible and immiscible liquids, where it is assumed that P_A^0 equals P_B^0, are shown in Figure 18.1. Note that with miscible liquids, the composition of the vapor depends on the relative amounts of A and B present (Fig. 18.1A). Thus, the composition of the vapor must change during a distillation. In contrast, the composition of the vapor with immiscible liquids is independent of the

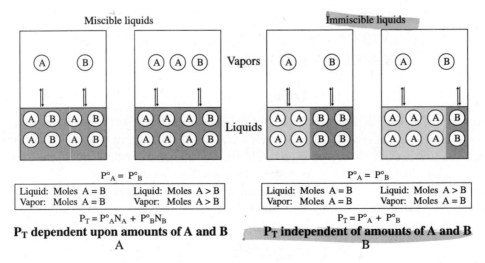

Figure 18.1 Total pressure behavior for miscible and immiscible liquids. **(A)** Ideal miscible liquids follow Raoult's Law: P_T depends on the mole fractions and vapor pressures of A and B. **(B)** Immiscible liquids do not follow Raoult's Law: P_T depends only on the vapor pressures of A and B.

amounts of A and B present (Fig. 18.1B). Hence, the vapor composition must remain *constant* during the distillation of such liquids, as predicted by Equation 3. Immiscible liquids act as if they were being distilled simultaneously from separate compartments, as shown in Figure 18.1B, even though in practice they are "mixed" during a steam distillation. Because all gases mix, they do give rise to a homogeneous vapor and codistill.

18.2 IMMISCIBLE MIXTURES: CALCULATIONS

The composition of the distillate is constant during a steam distillation, as is the boiling point of the mixture. The boiling points of steam-distilled mixtures will always be below the boiling point of water (bp 100°C) as well as the boiling point of any of the other substances distilled. Some representative boiling points and compositions of steam distillates are given in Table 18.1. Note that the higher the boiling point of a pure substance, the more closely the temperature of the steam distillate approaches, but does not exceed, 100°C. This is a reasonably low temperature, and it avoids the decomposition that might result at high temperatures with a simple distillation.

For immiscible liquids, the molar proportions of two components in a distillate equal the ratio of their vapor pressures in the boiling mixture, as given in Equation 3. When Equation 3 is rewritten for an immiscible mixture involving water, Equation 4 results. Equation 4 can be modified by substituting the relation moles = (weight/molecular weight) to give Equation 5.

$$\frac{\text{Moles substance}}{\text{Moles water}} = \frac{P^0_{\text{substance}}}{P^0_{\text{water}}} \qquad [4]$$

$$\frac{\text{Wt substance}}{\text{Wt water}} = \frac{(P^0_{\text{substance}})(\text{Molecular weight}_{\text{substance}})}{(P^0_{\text{water}})(\text{Molecular weight}_{\text{water}})} \qquad [5]$$

A sample calculation using this equation is given in Figure 18.2. Notice that the result of this calculation is very close to the experimental value given in Table 18.1.

Table 18.1 Boiling Points and Compositions of Steam Distillates

Mixture	Boiling Point of Pure Substance (°C)	Boiling Point of Mixture (°C)	Composition (% Water)
Benzene–water	80.1	69.4	8.9%
Toluene–water	110.6	85.0	20.2%
Hexane–water	69.0	61.6	5.6%
Heptane–water	98.4	79.2	12.9%
Octane–water	125.7	89.6	25.5%
Nonane–water	150.8	95.0	39.8%
1-Octanol–water	195.0	99.4	90.0%

Problem How many grams of water must be distilled to steam distill 1.55 g of 1-octanol from an aqueous solution? What will be the composition (wt%) of the distillate? The mixture distills at 99.4°C.

Answer The vapor pressure of water at 99.4°C must be obtained from the *CRC Handbook* (= 744 mm Hg).

(a) Obtain the partial pressure of 1-octanol.

$$P°_{1\text{-octanol}} = P_{total} - P°_{water}$$
$$P°_{1\text{-octanol}} = (760 - 744) = 16 \text{ mm Hg}$$

(b) Obtain the composition of the distillate.

$$\frac{\text{wt 1-octanol}}{\text{wt water}} = \frac{(16)(130)}{(744)(18)} = 0.155 \text{ g/g-water}$$

(c) Clearly, 10 g of water must be distilled.

$$(0.155 \text{ g/g-water})(10 \text{ g-water}) = 1.55 \text{ g 1-octanol}$$

(d) Calculate the **weight** percentages.

$$1\text{-octanol} = 1.55 \text{ g}/(10 \text{ g} + 1.55 \text{ g}) = 13.4\%$$
$$\text{water} = 10 \text{ g}/(10 \text{ g} + 1.55 \text{ g}) = 86.6\%$$

Figure 18.2 Sample calculations for a steam distillation.

Part B. Macroscale Distillation

18.3 STEAM DISTILLATION—MACROSCALE METHODS

Two methods for steam distillation are in general use in the laboratory: the **direct method** and the **live steam method.** In the first method, steam is generated *in situ* (in place) by heating a distillation flask containing the compound and water. In the second method, steam is generated outside and is passed into the distillation flask using an inlet tube.

A. Direct Method

A macroscale direct method steam distillation is illustrated in Figure 18.3. Although a heating mantle may be used, it is probably best to use a flame with this method, because a large volume of water must be heated rapidly. A boiling stone must be used to prevent bumping. The separatory funnel allows more water to be added during the course of the distillation.

Distillate is collected as long as it is either cloudy or milky white in appearance. Cloudiness indicates that an immiscible liquid is separating. When the distillate runs clear in the distillation, it is usually a sign that only water is distilling. However, there

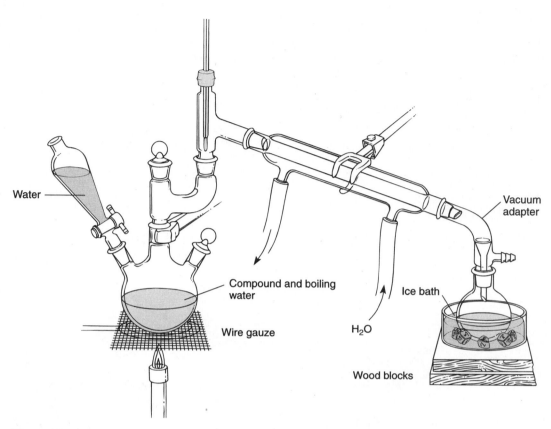

Labels: Water; Compound and boiling water; Wire gauze; H₂O; Ice bath; Wood blocks; Vacuum adapter

Figure 18.3 A macroscale direct steam distillation.

are some steam distillations where the distillate is never cloudy, even though material has codistilled. You must observe carefully, and be sure to collect enough distillate so that all of the organic material codistills.

B. Live Steam Method

A macroscale steam distillation using the live steam method is shown in Figure 18.4. If steam lines are available in the laboratory, they may be attached directly to the steam trap (purge them first to drain water). If steam lines are not available, an external steam generator (see inset) must be prepared. The external generator usually will require a flame to produce steam at a rate fast enough for the distillation. When the distillation is first started, the clamp at the bottom of the steam trap is left open. The steam lines will have a large quantity of condensed water in them until they are well heated. When the lines become hot and condensation of steam ceases, the clamp may be closed. Occasionally, the clamp will have to be reopened to remove condensate. In this method, the steam agitates the mixture as it enters the bottom of the flask, and a stirrer or boiling stone is not required.

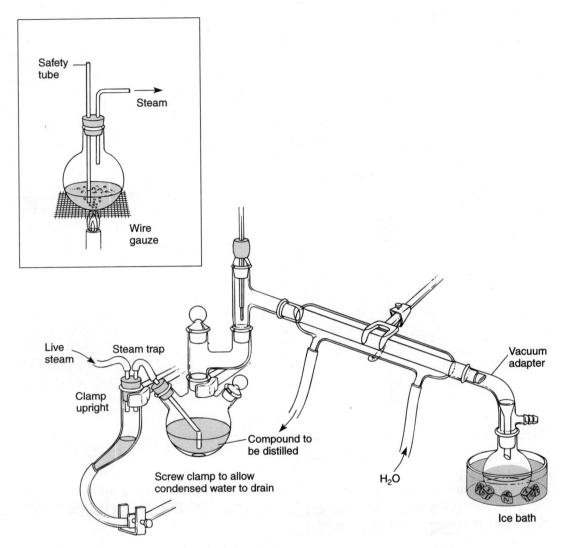

Figure 18.4 A macroscale steam distillation using live steam.

▰▰▰ **CAUTION** Hot steam can produce very severe burns.

Sometimes it is helpful to heat the three-necked distilling flask with a heating mantle (or flame) to prevent excessive condensation at that point. Steam must be admitted at a fast enough rate for you to see the distillate condensing as a milky white fluid in the condenser. The vapors that codistill will separate on cooling to give this cloudiness. When the condensate becomes clear, the distillation is near the end. The flow of water through the condenser should be faster than in other types of distillation to help cool the vapors. Make sure the vacuum adapter remains cool to the touch. An ice bath may be used to cool the receiving flask if desired. When the distillation is to be stopped, the screw clamp on the steam trap should be opened, and the steam inlet tube must be removed from the three-necked flask. If this is not done, liquid will back up into the tube and steam trap.

Part C. Microscale Distillation

18.4 STEAM DISTILLATION—MICROSCALE METHODS

The direct method of steam distillation is the only one suitable for microscale experiments. Steam is produced in the conical vial or distillation flask (*in situ*) by heating water to its boiling point in the presence of the compound to be distilled. This method works well for small amounts of materials. A microscale steam distillation apparatus is shown in Figure 18.5. Water and the compound to be distilled are placed in the flask and heated. A stirring bar or a boiling stone should be used to prevent bumping. The vapors of the water and the desired compound codistill when they are heated. They are condensed and collect in the Hickman head. When the Hickman head fills, the distillate is removed with a Pasteur pipet and placed in another vial for storage. For the typical microscale experiment, it will be necessary to fill the well and remove the distillate three or four times. All of these distillate fractions are placed in the same storage container. The efficiency in collecting the distillate can sometimes be improved if the inside walls of the Hickman head are rinsed several times into the well. A Pasteur pipet is used to perform the rinsing. Distillate is withdrawn from the well, and then it is used to wash the walls of the Hickman head all the way around the head. After the walls have been washed and when the well is full, the distillate can be withdrawn and transferred to the storage container. It may be necessary to add more water during the course of the distillation. More water is added (remove the condenser if used) through the center of the Hickman head by using a Pasteur pipet.

Part D. Semimicroscale Distillation

18.5 STEAM DISTILLATION—SEMIMICROSCALE METHODS

The apparatus shown in Figure 14.13, page 218, may also be used to perform a steam distillation at the microscale level or slightly above. This apparatus avoids the need to empty the collected distillate during the course of the distillation as is required when a Hickman head is used.

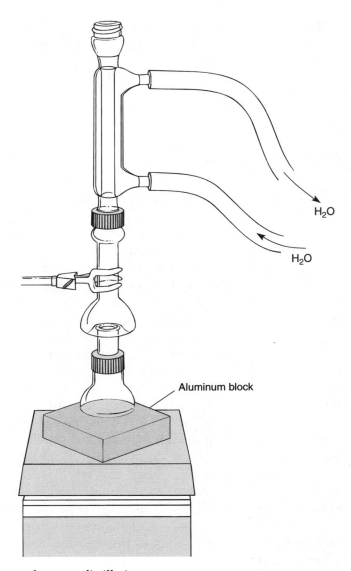

Figure 18.5 Microscale steam distillation.

PROBLEMS

1. Calculate the weight of benzene codistilled with each gram of water and the percentage composition of the vapor produced during a steam distillation. The boiling point of the mixture is 69.4°C. The vapor pressure of water at 69.4°C is 227.7 mm Hg. Compare the result with the data in Table 18.1.

2. Calculate the approximate boiling point of a mixture of bromobenzene and water at atmospheric pressure. A table of vapor pressure of water and bromobenzene at various temperatures is given.

Temperature (°C)	Vapor Pressures (mm Hg)	
	Water	*Bromobenzene*
93	588	110
94	611	114
95	634	118
96	657	122
97	682	127
98	707	131
99	733	136

3. Calculate the weight of nitrobenzene that codistills (bp 99°C of mixture) with each gram of water during a steam distillation. You may need the data given in Problem 2.

4. A mixture of *p*-nitrophenol and *o*-nitrophenol can be separated by steam distillation. The *o*-nitrophenol is steam volatile and the *para* isomer is not volatile. Explain. Base your answer on the ability of the isomers to form hydrogen bonds internally.

CHAPTER 19

Column Chromatography

The most modern and sophisticated methods of separating mixtures available to the organic chemist all involve **chromatography.** Chromatography is defined as the separation of a mixture of two or more different compounds or ions by distribution between two phases, one of which is stationary and the other of which is moving. Various types of chromatography are possible, depending on the nature of the two phases involved: **solid–liquid** (column, thin-layer, and paper), **liquid–liquid** (high-performance liquid), and **gas–liquid** (vapor-phase) chromatographic methods are common.

All chromatography works on much the same principle as solvent extraction (Chapter 12). Basically, the methods depend on the differential solubilities or adsorptivities of the substances to be separated relative to the two phases between which they are to be partitioned. In this chapter, column chromatography, a solid–liquid method, is considered. Thin-layer chromatography is examined in Chapter 20; high-performance liquid chromatography is discussed in Chapter 21; and gas chromatography, a gas–liquid method, is discussed in Chapter 22.

19.1 ADSORBENTS

Column chromatography is a technique based on both adsorptivity and solubility. It is a solid–liquid phase-partitioning technique. The solid may be almost any material that does not dissolve in the associated liquid phase; the solids used most commonly are silica gel $SiO_2 \cdot xH_2O$, also called silicic acid, and alumina $Al_2O_3 \cdot xH_2O$. These compounds are used in their powdered or finely ground forms (usually 200 to 400 mesh).[1]

Most alumina used for chromatography is prepared from the impure ore bauxite $Al_2O_3 \cdot xH_2O + Fe_2O_3$. The bauxite is dissolved in hot sodium hydroxide and filtered to remove the insoluble iron oxides; the alumina in the ore forms the soluble amphoteric

---

[1] The term "mesh" refers to the number of openings per linear inch found in a screen. A large number refers to a fine screen (finer wires more closely spaced). When particles are sieved through a series of these screens, they are classified by the smallest mesh screen that they will pass through. Mesh 5 would represent a coarse gravel, while mesh 800 would be a fine powder.

hydroxide $Al(OH)_4^-$. The hydroxide is precipitated by CO_2, which reduces the pH, as $Al(OH)_3$. When heated, the $Al(OH)_3$ loses water to form pure alumina Al_2O_3.

$$Bauxite\ (crude) \xrightarrow{\text{hot NaOH}} Al(OH)_4^-(aq) + Fe_2O_3\ (insoluble)$$

$$Al(OH)_4^-(aq) + CO_2 \longrightarrow Al(OH)_3 + HCO_3^-$$

$$2Al(OH)_3 \xrightarrow{\text{heat}} Al_2O_3\ (s) + 3H_2O$$

Alumina prepared in this way is called **basic alumina** because it still contains some hydroxides. Basic alumina cannot be used for chromatography of compounds that are base sensitive. Therefore, it is washed with acid to neutralize the base, giving **acid-washed alumina.** This material is unsatisfactory unless it has been washed with enough water to remove *all* the acid; on being so washed, it becomes the best chromatographic material, called **neutral alumina.** If a compound is acid sensitive, either basic or neutral alumina must be used. You should be careful to ascertain what type of alumina is being used for chromatography. Silica gel is not available in any form other than that suitable for chromatography.

19.2 INTERACTIONS

If powdered or finely ground alumina (or silica gel) is added to a solution containing an organic compound, some of the organic compound will **adsorb** onto or adhere to the fine particles of alumina. Many kinds of intermolecular forces cause organic molecules to bind to alumina. These forces vary in strength according to their type. Nonpolar compounds bind to the alumina using only van der Waals forces. These are weak forces, and nonpolar molecules do not bind strongly unless they have extremely high molecular weights. The most important interactions are those typical of polar organic compounds. Either these forces are of the dipole–dipole type or they involve some direct interaction (coordination, hydrogen bonding, or salt formation). These types of interactions are illustrated in Figure 19.1, which for convenience shows only a portion of the alumina structure. Similar interactions occur with silica gel. The strengths of such interactions vary in the following approximate order:

Salt formation > coordination > hydrogen bonding > dipole–dipole > van der Waals

Strength of interaction varies among compounds. For instance, a strongly basic amine would bind more strongly than a weakly basic one (by coordination). In fact, strong bases and strong acids often interact so strongly that they **dissolve** alumina to some extent. You can use the following rule of thumb:

NOTE: The more polar the functional group, the stronger the bond to alumina (or silica gel).

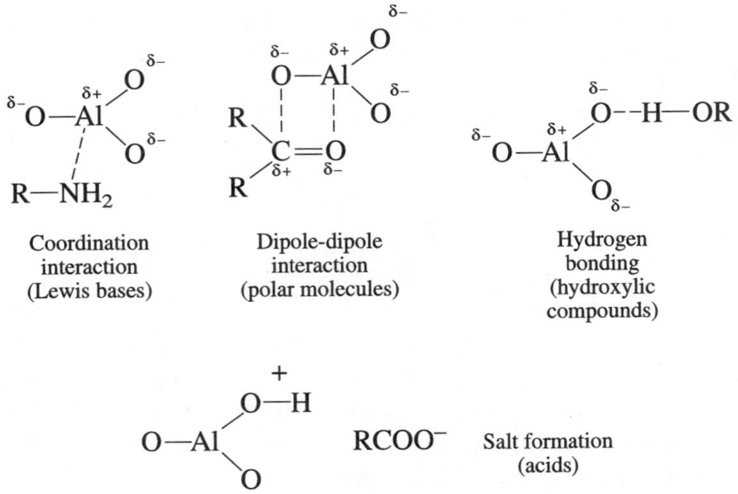

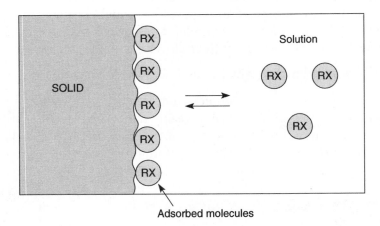

Figure 19.1 Possible interactions of organic compounds with alumina.

A similar rule holds for solubility. Polar solvents dissolve polar compounds more effectively than nonpolar solvents; nonpolar compounds are dissolved best by nonpolar solvents. Thus, the extent to which any given solvent can wash an adsorbed compound from alumina depends almost directly on the relative polarity of the solvent. For example, although a ketone adsorbed on alumina might not be removed by hexane, it might be removed completely by chloroform. For any adsorbed material, a kind of **distribution** equilibrium can be envisioned between the adsorbent material and the solvent. This is illustrated in Figure 19.2.

The distribution equilibrium is **dynamic,** with molecules constantly **adsorbing** from the solution and **desorbing** into it. The average number of molecules remaining adsorbed on the solid particles at equilibrium depends both on the particular molecule (RX) involved and the dissolving power of the solvent with which the adsorbent must compete.

Figure 19.2 Dynamic adsorption equilibrium.

19.3 PRINCIPLE OF COLUMN CHROMATOGRAPHIC SEPARATION

The dynamic equilibrium mentioned previously and the variations in the extent to which different compounds adsorb on alumina or silica gel underlie a versatile and ingenious method for **separating** mixtures of organic compounds. In this method, the mixture of compounds to be separated is introduced onto the top of a cylindrical glass column (Fig. 19.3) **packed** or filled with fine alumina particles (stationary solid phase). The adsorbent is continuously washed by a flow of solvent (moving phase) passing through the column.

Initially, the components of the mixture adsorb onto the alumina particles at the top of the column. The continuous flow of solvent through the column **elutes,** or washes, the solutes off the alumina and sweeps them down the column. The solutes (or materials to be separated) are called **eluates** or **elutants,** and the solvents are called **eluents.** As the solutes pass down the column to fresh alumina, new equilibria are established among the adsorbent, the solutes, and the solvent. The constant equilibration means that different compounds will move down at differing rates depending on their relative affinity for the adsorbent on one hand, and for the solvent on the other. Because the number of alumina particles is large, because they are closely packed, and because fresh solvent is being added continuously, the number of equilibrations between adsorbent and solvent that the solutes experience is enormous.

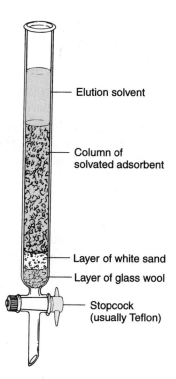

— Elution solvent

— Column of
solvated adsorbent

— Layer of white sand
— Layer of glass wool

— Stopcock
(usually Teflon)

Figure 19.3 A chromatographic column.

As the components of the mixture are separated, they begin to form moving bands (or zones), with each band containing a single component. If the column is long enough and the other parameters (column diameter, adsorbent, solvent, and flow rate) are correctly chosen, the bands separate from one another, leaving gaps of pure solvent in between. As each band (solvent and solute) passes out from the bottom of the column, it can be collected before the next band arrives. If the parameters mentioned are poorly chosen, the various bands either overlap or coincide, in which case either a poor separation or no separation is the result. A successful chromatographic separation is illustrated in Figure 19.4.

19.4 PARAMETERS AFFECTING SEPARATION

The versatility of column chromatography results from the many factors that can be adjusted. These include

1. Adsorbent chosen
2. Polarity of the solvents chosen
3. Size of the column (both length and diameter) relative to the amount of material to be chromatographed
4. Rate of elution (or flow)

By carefully choosing the conditions, almost any mixture can be separated. This technique has even been used to separate optical isomers. An optically active solid-phase adsorbent was used to separate the enantiomers.

Two fundamental choices for anyone attempting a chromatographic separation are the kind of adsorbent and the solvent system. In general, nonpolar compounds pass through the column faster than polar compounds, because they have a smaller affinity for the adsorbent. If the adsorbent chosen binds all the solute molecules (both polar and nonpolar) strongly, they will not move down the column. On the other hand, if too polar a solvent is chosen, all the solutes (polar and nonpolar) may simply be washed through the column, with no separation taking place. The adsorbent and the solvent should be chosen so that neither is favored excessively in the equilibrium competition for solute molecules.[2]

A. Adsorbents

In Table 19.1, various kinds of adsorbents (solid phases) used in column chromatography are listed. The choice of adsorbent often depends on the types of compounds to be separated. Cellulose, starch, and sugars are used for polyfunctional plant and animal materials (natural products) that are very sensitive to acid–base interactions.

---

[2]Often the chemist uses thin-layer chromatography (TLC), which is described in Chapter 20, to arrive at the best choices of solvents and adsorbents for the best separation. The TLC experimentation can be performed quickly and with extremely small amounts (microgram quantities) of the mixture to be separated. This saves significant time and materials. Chapter 20, Section 20.10, describes this use of TLC.

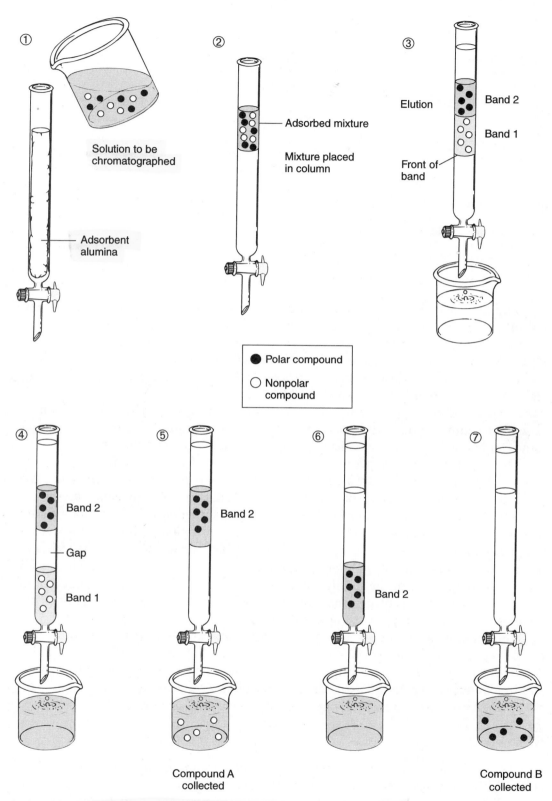

Figure 19.4 Sequences of steps in a chromatographic separation.

Table 19.1 Solid Adsorbents for Column Chromatography

Paper Cellulose Starch Sugars Magnesium silicate Calcium sulfate Silicic acid Florisil Magnesium oxide Aluminum oxide (alumina)[a] Activated charcoal (Norit)	Increasing strength of binding interactions toward polar compounds

[a]Basic, acid washed, and neutral.

Magnesium silicate is often used for separating acetylated sugars, steroids, and essential oils. Silica gel and Florisil are relatively mild toward most compounds and are widely used for a variety of functional groups—hydrocarbons, alcohols, ketones, esters, acids, azo compounds, and amines. Alumina is the most widely used adsorbent and is obtained in the three forms mentioned in Section 19.1: acidic, basic, and neutral. The pH of acidic or acid-washed alumina is approximately 4. This adsorbent is particularly useful for separating acidic materials such as carboxylic acids and amino acids. Basic alumina has a pH of 10 and is useful in separating amines. Neutral alumina can be used to separate a variety of nonacidic and nonbasic materials.

The approximate strength of the various adsorbents listed in Table 19.1 is also given. The order is only approximate and therefore it may vary. For instance, the strength, or separating abilities, of alumina and silica gel largely depends on the amount of water present. Water binds very tightly to either adsorbent, taking up sites on the particles that could otherwise be used for equilibration with solute molecules. If water is added to the adsorbent, it is said to have been **deactivated.** Anhydrous alumina or silica gel is said to be highly **activated.** High activity is usually avoided with these adsorbents. Use of the highly active forms of either alumina or silica gel, or of the acidic or basic forms of alumina, can often lead to molecular rearrangement or decomposition in certain types of solute compounds.

The chemist can select the degree of activity that is appropriate to carry out a particular separation. To accomplish this, highly activated alumina is mixed thoroughly with a precisely measured quantity of water. The water partially hydrates the alumina and thus reduces its activity. By carefully determining the amount of water required, the chemist can have available an entire spectrum of possible activities.

B. Solvents

In Table 19.2, some common chromatographic solvents are listed along with their relative ability to dissolve polar compounds. Sometimes a single solvent can be found that will separate all the components of a mixture. Sometimes a mixture of solvents

Table 19.2 Solvents (Eluents) for Chromatography

Petroleum ether	
Cyclohexane	
Carbon tetrachloride[a]	
Toluene	
Chloroform[a]	
Methylene chloride	Increasing polarity and
Diethyl ether	"solvent power" toward
Ethyl acetate	polar functional groups
Acetone	
Pyridine	
Ethanol	
Methanol	
Water	
Acetic acid	

[a]Suspected carcinogens.

can be found that will achieve separation. More often you must start elution with a nonpolar solvent to remove relatively nonpolar compounds from the column and then gradually increase the solvent polarity to force compounds of greater polarity to come down the column, or elute. The approximate order in which various classes of compounds elute by this procedure is given in Table 19.3. In general, nonpolar compounds travel through the column faster (elute first), and polar compounds travel more slowly (elute last). However, molecular weight is also a factor in determining the order of elution. A nonpolar compound of high molecular weight travels more slowly than a nonpolar compound of low molecular weight, and it may even be passed by some polar compounds.

Solvent polarity functions in two ways in column chromatography. First, a polar solvent will better dissolve a polar compound and move it down the column faster. Therefore, as already mentioned, the polarity of the solvent is usually increased during column chromatography to wash down compounds of increasing polarity. Second, as

Table 19.3 Elution Sequence for Compounds

Hydrocarbons	Fastest (will elute with nonpolar solvent)
Olefins	
Ethers	
Halocarbons	
Aromatics	
Ketones	Order of elution
Aldehydes	
Esters	
Alcohols	
Amines	
Acids, strong bases	Slowest (needs a polar solvent)

the polarity of the solvent increases, the solvent itself will displace adsorbed molecules from the alumina or silica and take their place on the column. Because of this second effect, a polar solvent will move **all types of compounds,** both polar and nonpolar, down the column at a faster rate than a nonpolar solvent.

When the polarity of the solvent has to be changed during a chromatographic separation, some precautions must be taken. Rapid changes from one solvent to another are to be avoided (especially when silica gel or alumina is involved). Usually, small percentages of a new solvent are mixed slowly into the one in use until the percentage reaches the desired level. If this is not done, the column packing often "cracks" as a result of the heat liberated when alumina or silica gel is mixed with a solvent. The solvent solvates the adsorbent, and the formation of a weak bond generates heat.

$$\text{Solvent} + \text{alumina} \rightarrow (\text{alumina} \cdot \text{solvent}) + \text{heat}$$

Often enough heat is generated locally to evaporate the solvent. The formation of vapor creates bubbles, which forces a separation of the column packing; this is called **cracking.** A cracked column does not produce a good separation, because it has discontinuities in the packing. The way in which a column is packed or filled is also very important in preventing cracking.

Certain solvents should be avoided with alumina or silica gel, especially with the acidic, basic, and highly active forms. For instance, with any of these adsorbents, acetone dimerizes via an aldol condensation to give diacetone alcohol. Mixtures of esters **transesterify** (exchange their alcoholic portions) when ethyl acetate or an alcohol is the eluent. Finally, the most active solvents (pyridine, methanol, water, and acetic acid) dissolve and elute some of the adsorbent itself. Generally, try to avoid solvents more polar than diethyl ether or methylene chloride in the eluent series (Table 19.2).

C. Column Size and Adsorbent Quantity

The column size and the amount of adsorbent must also be selected correctly to separate a given amount of sample well. As a rule of thumb, the amount of adsorbent should be 25 to 30 times, by weight, the amount of material to be separated by chromatography. Furthermore, the column should have a height-to-diameter ratio of about 8:1. Some typical relations of this sort are given in Table 19.4.

Table 19.4 Size of Column and Amount of Adsorbent for Typical Sample Sizes

Amount of Sample (g)	Amount of Adsorbent (g)	Column Diameter (mm)	Column Height (mm)
0.01	0.3	3.5	30
0.10	3.0	7.5	60
1.00	30.0	16.0	130
10.00	300.0	35.0	280

Note, as a caution, that the difficulty of the separation is also a factor in determining the size and length of the column to be used and the amount of adsorbent needed. Compounds that do not separate easily may require longer columns and more adsorbent than specified in Table 19.4. For easily separated compounds, a shorter column and less adsorbent may suffice.

D. Flow Rate

The at rate which solvent flows through the column is also significant in the effectiveness of a separation. In general, the time the mixture to be separated remains on the column is directly proportional to the extent of equilibration between stationary and moving phases. Thus, similar compounds eventually separate if they remain on the column long enough. The time a material remains on the column depends on the flow rate of the solvent. If the flow is too slow, however, the dissolved substances in the mixture may diffuse faster than the rate at which they move down the column. Then the bands grow wider and more diffuse, and the separation becomes poor.

19.5 PACKING THE COLUMN: TYPICAL PROBLEMS

The most critical operation in column chromatography is packing (filling) the column with adsorbent. The **column packing** must be evenly packed and free of irregularities, air bubbles, and gaps. As a compound travels down the column, it moves in an advancing zone, or **band.** It is important that the leading edge, or **front,** of this band be horizontal, or perpendicular to the long axis of the column. If two bands are close together and do not have horizontal band fronts, it is impossible to collect one band while completely excluding the other. The leading edge of the second band begins to elute before the first band has finished eluting. This condition can be seen in Figure 19.5. There are two main reasons for this problem. First, if the top surface edge of the adsorbent packing is not level, nonhorizontal bands result. Second, bands may be nonhorizontal if the column is not held in an exactly vertical position in both planes (front to back and side to side). When preparing a column, you must watch both these factors carefully.

Another phenomenon, called **streaming** or **channeling,** occurs when part of the band front advances ahead of the major part of the band. Channeling occurs if there are any cracks or irregularities in the adsorbent surface or any irregularities caused by air bubbles in the packing. A part of the advancing front moves ahead of the rest of the band by flowing through the channel. Two examples of channeling are shown in Figure 19.6.

The methods outlined in Sections 19.6, 19.7, and 19.8 are used to avoid problems resulting from uneven packing and column irregularities. These procedures should be followed carefully in preparing a chromatography column. Failure to pay close attention to the preparation of the column may affect the quality of the separation.

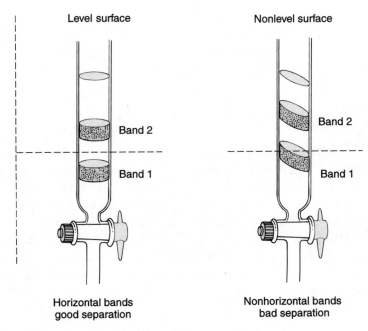

Figure 19.5 Comparison of horizontal and nonhorizontal band fronts.

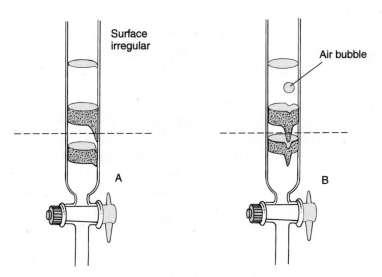

Figure 19.6 Channeling complications.

19.6 PACKING THE COLUMN: PREPARING THE SUPPORT BASE

Preparation of a column involves two distinct stages. In the first stage, a support base on which the packing will rest is prepared. This must be done so that the packing, a finely divided material, does not wash out of the bottom of the column. In the second stage, the column of adsorbent is deposited on top of the supporting base.

A. Macroscale Columns

For large-scale applications, a chromatography column is clamped upright (vertically). The column (Fig. 19.3) is a piece of cylindrical glass tubing with a stopcock attached at one end. The stopcock usually has a Teflon plug, because stopcock grease (used on glass plugs) dissolves in many of the organic solvents used as eluents. Stopcock grease in the eluent will contaminate the eluates.

Instead of a stopcock, a piece of flexible tubing may be attached to the bottom of the column, with a screw clamp used to stop or regulate the flow (Fig. 19.7). When a screw clamp is used, care must be taken that the tubing used is not dissolved by the solvents that will pass through the column during the experiment. Rubber, for instance, dissolves in chloroform, benzene, methylene chloride, toluene, or tetrahydrofuran (THF). Tygon tubing dissolves (actually, the plasticizer is removed) in many solvents, including benzene, methylene chloride, chloroform, ether, ethyl acetate, toluene, and THF. Polyethylene tubing is the best choice for use at the end of a column, because it is inert with most solvents.

Figure 19.7 Tubing with screw clamp to regulate solvent flow on a chromatography column.

Next the column is partially filled with a quantity of solvent, usually a nonpolar solvent such as hexane, and a support for the finely divided adsorbent is prepared in the following way. A loose plug of glass wool is tamped down into the bottom of the column with a long glass rod until all entrapped air is forced out as bubbles. Take care not to plug the column totally by tamping the glass wool too hard. A small layer of clean white sand is formed on top of the glass wool by pouring sand into the column. The column is tapped to level the surface of the sand. Any sand adhering to the side of the column is washed down with a small quantity of solvent. The sand forms a base that supports the column of adsorbent and prevents it from washing through the stopcock. The column is packed in one of two ways: by the slurry method or by the dry pack method.

B. Semimicroscale Columns

An alternative apparatus for macroscale column chromatography on a smaller scale is a commercial column, such as the one shown in Figure 19.8. This type of column is made of glass and has a solvent-resistant plastic stopcock at the

Figure 19.8 A commercial semimicroscale chromatography column. (The column shown is equipped with an optional solvent reservoir.)

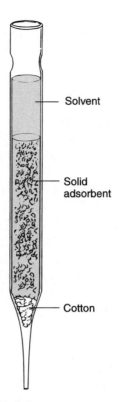

— Solvent

— Solid adsorbent

— Cotton

Figure 19.9 A microscale chromatography column.

bottom.[3] The stopcock assembly contains a filter disc to support the adsorbent column. An optional upper fitting, also made of solvent-resistant plastic, serves as a solvent reservoir. The column shown in Figure 19.8 is equipped with the solvent reservoir. This type of column is available in a variety of lengths, ranging from 100 to 300 mm. Because the column has a built-in filter disc, it is not necessary to prepare a support base before the adsorbent is added.

C. Microscale Columns

For microscale applications, a Pasteur pipet (5¾ inch) is used; it is clamped upright (vertically). To reduce the amount of solvent needed to fill the column, most of the tip of the pipet may be broken off. A small ball of cotton is placed in the pipet and tamped into position using a glass rod or a piece of wire. Take care not to plug the column totally by tamping the cotton too hard. The correct position of the cotton is shown in Figure 19.9. A microscale chromatography column is packed by one of the dry pack methods described in Sections 19.7 and 19.8.

---

[3]*Note to the Instructor*: With certain organic solvents, we have found that the "solvent-resistant" plastic stopcock may tend to dissolve! We recommend that instructors test their equipment with the solvent that they intend to use before the start of the laboratory class.

19.7 PACKING THE COLUMN: DEPOSITING THE ADSORBENT— DRY PACK METHODS

A. Dry Pack Method 1

Macroscale Columns.   In the first of the dry pack methods introduced here, the column is filled with solvent and allowed to drain *slowly*. The dry adsorbent is added, a little at a time, while the column is tapped gently with a pencil, finger, or glass rod.

A plug of cotton is placed at the base of the column, and an even layer of sand is formed on top (see page 289). The column is filled about half-full with solvent, and the solid adsorbent is added carefully from a beaker, while the solvent is allowed to flow slowly from the column. As the solid is added, the column is tapped as described for the slurry method (see page 302) to ensure that the column is packed evenly. When the column has the desired length, no more adsorbent is added. This method produces an evenly packed column. Solvent should be cycled through this column (for macroscale applications) several times before each use. The same portion of solvent that has drained from the column during the packing is used to cycle through the column.

Semimicroscale Columns.   The procedure to fill a commercial semimicroscale column is essentially the same as that used to fill a Pasteur pipet (see the following paragraph). The commercial column has the advantage that it is much easier to control the flow of solvent from the column during the filling process, because the stopcock can be adjusted appropriately. It is not necessary to use a cotton plug or to deposit a layer of sand before adding the adsorbent. The presence of the fritted disc at the base of the column prevents adsorbent from escaping from the column.

Microscale Columns.   To fill a microscale column, fill the Pasteur pipet (with the cotton plug, prepared as described in Section 19.6) about half-full with solvent. Using a microspatula, add the solid adsorbent slowly to the solvent in the column. As you add the solid, tap the column *gently* with a pencil, a finger, or a glass rod. The tapping promotes even settling and mixing and gives an evenly packed column free of air bubbles. As the adsorbent is added, solvent flows out of the Pasteur pipet. Because the adsorbent must not be allowed to dry during the packing process, you must use a means of controlling the solvent flow. If a piece of small-diameter plastic tubing is available, it can be fitted over the narrow tip of the Pasteur pipet. The flow rate can then be controlled using a screw clamp. A simple approach to controlling the flow rate is to use a finger over the top of the Pasteur pipet, much as you control the flow of liquid in a volumetric pipet. Continue adding the adsorbent slowly, with constant tapping, until the level of the adsorbent has reached the desired level. As you pack the column, be careful not to let the column run dry. The final column should appear as shown in Figure 19.9.

B. Dry Pack Method 2

Macroscale Columns. Macroscale columns can also be packed by a dry pack method that is commonly used in the packing of microscale columns (see Microscale Columns, below). In this method, the column is filled with dry adsorbent without any solvent. When the desired amount of adsorbent has been added, solvent is allowed to percolate through the column. The disadvantages described for the microscale method also apply to the macroscale method. This method is not recommended for use with silica gel or alumina, because the combination leads to uneven packing, air bubbles, and cracking, especially if a solvent that has a highly exothermic heat of solvation is used.

Semimicroscale Columns. The dry pack method 2 for semimicroscale columns is similar to that described for Pasteur pipets (see below), except that the plug of cotton is not required. The flow rate of solvent through the column can be controlled using the stopcock, which is part of the column assembly (see Fig. 19.7).

Microscale Columns. An alternative dry pack method for microscale columns is to fill the Pasteur pipet with *dry* adsorbent, without any solvent. Position a plug of cotton in the bottom of the Pasteur pipet. The desired amount of adsorbent is added slowly, and the pipet is tapped constantly, until the level of adsorbent has reached the desired height. Figure 19.9 can be used as a guide to judge the correct height of the column of adsorbent. When the column is packed, added solvent is allowed to percolate through the adsorbent until the entire column is moistened. The solvent is not added until just before the column is to be used.

This method is useful when the adsorbent is alumina, but it does not produce satisfactory results with silica gel. Even with alumina, poor separations can arise due to uneven packing, air bubbles, and cracking, especially if a solvent that has a highly exothermic heat of solvation is used.

19.8 PACKING THE COLUMN: DEPOSITING THE ADSORBENT— THE SLURRY METHOD

The slurry method is not recommended as a microscale method for use with Pasteur pipets. On a very small scale, it is too difficult to pack the column with the slurry without losing the solvent before the packing has been completed. Microscale columns should be packed by one of the dry pack methods, as described in Section 19.7.

In the slurry method, the adsorbent is packed into the column as a mixture of a solvent and an undissolved solid. The slurry is prepared in a separate container (Erlenmeyer flask) by adding the solid adsorbent, a little at a time, to a quantity of the solvent. This order of addition (adsorbent added to solvent) should be followed strictly, because the adsorbent solvates and liberates heat. If the solvent is added to the adsorbent, it may boil away almost as fast as it is added due to heat evolved. This will be especially true if ether or another low-boiling solvent is used. When this happens, the final mixture will be uneven and lumpy. Enough adsorbent is added to the solvent, and mixed by swirling the container, to form a thick, but flowing, slurry. The container should be swirled until the mixture is homogeneous and relatively free of entrapped air bubbles.

For a macroscale column, the procedure is as follows. When the slurry has been prepared, the column is filled about half-full with solvent, and the stopcock is opened to allow solvent to drain slowly into a large beaker. The slurry is mixed by swirling and is then poured in portions into the top of the draining column (a wide-necked funnel may be useful here). Be sure to swirl the slurry thoroughly before each addition to the column. The column is tapped constantly and *gently* on the side, during the pouring operation, with the fingers or with a pencil fitted with a rubber stopper. A short piece of large-diameter pressure tubing may also be used for tapping. The tapping promotes even settling and mixing and gives an evenly packed column free of air bubbles. Tapping is continued until all the material has settled, showing a well-defined level at the top of the column. Solvent from the collecting beaker may be readded to the slurry if it becomes too thick to be poured into the column at one time. In fact, the collected solvent should be cycled through the column several times to ensure that settling is complete and that the column is firmly packed. The downward flow of solvent tends to compact the adsorbent. Take care never to let the column "run dry" during packing. There should always be solvent on top of the absorbent column.

19.9 APPLYING THE SAMPLE TO THE COLUMN

The solvent (or solvent mixture) used to pack the column is normally the least polar elution solvent that can be used during chromatography. The compounds to be chromatographed are not highly soluble in the solvent. If they were, they would probably have a greater affinity for the solvent than for the adsorbent and would pass right through the column without equilibrating with the stationary phase.

The first elution solvent, however, is generally not a good solvent to use in preparing the sample to be placed on the column. Because the compounds are not highly soluble in nonpolar solvents, it takes a large amount of the initial solvent to dissolve the compounds, and it is difficult to get the mixture to form a narrow band on top of the column. A narrow band is ideal for an optimum separation of components. For the best separation, therefore, the compound is applied to the top of the column undiluted if it is a liquid, or in a *very small* amount of polar solvent if it is a solid. Water must not be used to dissolve the initial sample being chromatographed, because it reacts with the column packing.

In adding the sample to the column, use the following procedure. Lower the solvent level to the top of the adsorbent column by draining the solvent from the column. Add the sample (either a pure liquid or a solution) to form a small layer on top of the adsorbent. A Pasteur pipet is convenient for adding the sample to the column. Take care not to disturb the surface of the adsorbent. This is best accomplished by touching the pipet to the inside of the glass column and slowly draining it so as to allow the sample to spread into a thin film, which slowly descends to cover the entire adsorbent surface. Drain the pipet close to the surface of the adsorbent. When all the sample has been added, drain this small layer of liquid into the column until the top surface of the column *just begins* to dry. Then add a small layer of the chromatographic solvent carefully with a Pasteur pipet, again being careful not to disturb the surface. Drain this small layer of solvent into the column until the top surface of the column just dries. Add another small layer of fresh solvent, if necessary, and repeat the process until it is clear that the sample is strongly adsorbed on the top of the column. If the sample is colored and the fresh layer of solvent acquires some of this color, the sample has not been properly adsorbed. Once the sample has been properly applied, you can protect the level surface of the adsorbent by carefully filling the top of the column with solvent and sprinkling clean, white sand into the column so as to form a small protective layer on top of the adsorbent. For microscale applications, this layer of sand is not required.

Separations are often better if the sample is allowed to stand a short time on the column before elution. This allows a true equilibrium to be established. In columns that stand for too long, however, the adsorbent often compacts or even swells, and the flow can become annoyingly slow. Diffusion of the sample to widen the bands also becomes a problem if a column is allowed to stand over an extended period. For small-scale chromatography, using Pasteur pipets, there is no stopcock, and it is not possible to stop the flow. In this case, it is not necessary to allow the column to stand.

19.10 ELUTION TECHNIQUES

Solvents for analytical and preparative chromatography should be pure reagents. Commercial-grade solvents often contain small amounts of residue, which remain when the solvent is evaporated. For normal work, and for relatively easy separations that take only small amounts of solvent, the residue usually presents few problems. For large-scale work, commercial-grade solvents may have to be redistilled before use. This is especially true for hydrocarbon solvents, which tend to have more residue than other solvent types.

Elution of the products is usually begun with a nonpolar solvent, such as hexane or petroleum ether. The polarity of the elution solvent can be increased gradually by adding successively greater percentages of ether or toluene (for instance, 1, 2, 5, 10, 15, 25, 50, or 100%) or some other solvent of greater solvent power (polarity) than hexane. The transition from one solvent to another should not be too rapid in most solvent changes. If the two solvents to be changed differ greatly in their heats of solvation in binding to the adsorbent, enough heat can be generated to crack the column. Ether is especially troublesome in this respect, as it has both a low boiling point and a relatively high heat of solvation. Most organic compounds can be separated on silica gel or alumina using hexane–ether or hexane–toluene combinations for elution, and following these by pure methylene chloride. Solvents of greater polarity are usually avoided for the various reasons mentioned previously. In microscale work, the usual procedure is to use only one solvent for the chromatography.

The flow of solvent through the column should not be too rapid, or the solutes will not have time to equilibrate with the adsorbent as they pass down the column. If the rate of flow is too low, or stopped for a period, diffusion can become a problem—the solute band will diffuse, or spread out, in all directions. In either of these cases, separation will be poor. As a general rule (and only an approximate one), most macroscale columns are run with flow rates ranging from 5 to 50 drops of effluent per minute; a steady flow of solvent is usually avoided. Microscale columns made from Pasteur pipets do not have a means of controlling the solvent flow rate, but commercial microscale columns are equipped with stopcocks. The solvent flow rate in this type of column can be adjusted in a manner similar to that used with larger columns. To avoid diffusion of the bands, do not stop the column and do not set it aside overnight.

In some cases, the chromatography may proceed too slowly; the rate of solvent flow can be accelerated by attaching a rubber dropper bulb to the top of the Pasteur pipet column and squeezing *gently*. The additional air pressure forces the solvent through the column more rapidly. If this technique is used, however, care must be taken to remove the rubber bulb from the column before releasing it. Otherwise, air may be drawn up through the bottom of the column, destroying the column packing.

19.11 RESERVOIRS

When large quantities of solvent are used in a chromatographic separation, it is often convenient to use a solvent reservoir to forestall having to add small portions of fresh solvent continually. The simplest type of reservoir, a feature of many columns, is created by fusing the top of the column to a round-bottom flask (Fig. 19.10A). If the column has a standard-taper joint at its top, a reservoir can be created by joining a standard-taper separatory funnel to the column (Fig. 19.10B). In this arrangement, the stopcock is left open, and no stopper is placed in the top of the separatory funnel. A third common arrangement is shown in Figure 19.10C. A separatory funnel is filled with solvent; its stopper is wetted with solvent and put *firmly* in place. The funnel is inserted into the empty filling space at the top of the chromatographic column, and the stopcock is opened. Solvent flows out of the funnel, filling the space at the top of the column until the solvent level is well above the outlet of the separatory funnel. As solvent drains from the column, this arrangement automatically refills the space at the top of the column by allowing air to enter through the stem of the separatory funnel.

Some semimicroscale columns, such as that shown in Figure 19.8, are equipped with a solvent reservoir that fits onto the top of the column. It functions just like the reservoirs described in this section.

For a microscale chromatography, the portion of the Pasteur pipet above the adsorbent is used as a reservoir of solvent. Fresh solvent, as needed, is added by means of another Pasteur pipet. When it is necessary to change solvent, the new solvent is also added in this manner.

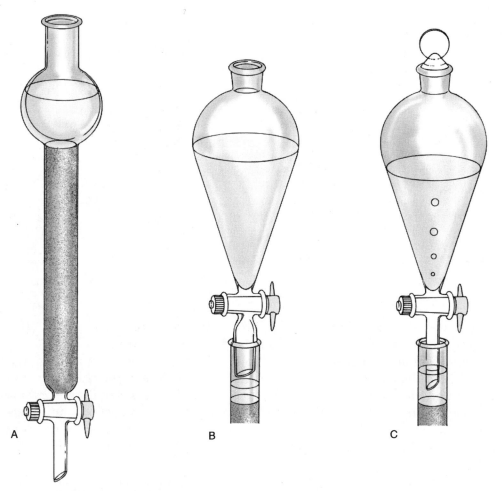

Figure 19.10 Various types of solvent-reservoir arrangements for chromatographic columns.

19.12 MONITORING THE COLUMN

It is a happy instance when the compounds to be separated are colored. The separation can then be followed visually and the various bands collected separately as they elute from the column. For the majority of organic compounds, however, this lucky circumstance does not exist, and other methods must be used to determine the positions of the bands. The most common method of following a separation of colorless compounds is to collect *fractions* of constant volume in preweighed flasks, or test tubes, to evaporate the solvent from each fraction, and to reweigh the container plus any residue. A plot of fraction number versus the weight of the residues after evaporation of solvent gives a plot similar to that in Figure 19.11. Clearly, fractions 2 through 7 (Peak 1) may be combined as a single compound, and so can fractions 8 through 11

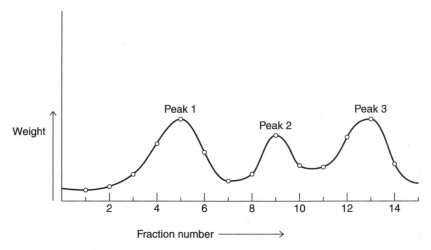

Figure 19.11 A typical elution graph.

(Peak 2) and 12 through 15 (Peak 3). The size of the fractions collected (1, 10, 100, or 500 mL) depends on the size of the column and the ease of separation.

Another common method of monitoring the column is to mix an inorganic phosphor into the adsorbent used to pack the column. When the column is illuminated with an ultraviolet light, the adsorbent treated in this way fluoresces. However, many solutes have the ability to **quench** the fluorescence of the indicator phosphor. In areas in which solutes are present, the adsorbent does not fluoresce, and a dark band is visible. In this type of column, the separation can also be followed visually.

Thin-layer chromatography is often used to monitor a column. This method is described in Chapter 20 (Section 20.10, page 323). Several sophisticated instrumental and spectroscopic methods, which we shall not detail, can also monitor a chromatographic separation.

19.13 TAILING

When a single solvent is used for elution, an elution curve (weight versus fraction) such as that shown as a solid line in Figure 19.12 is often observed. An ideal elution curve is shown by dashed lines. In the nonideal curve, the compound is said to be **tailing.** Tailing can interfere with the beginning of a curve or a peak of a second component and lead to a poor separation. One way to avoid this is to increase the polarity of the solvent constantly while eluting. In this way, at the tail of the peak, where the solvent polarity is increasing, the compound will move slightly faster than at the front and allow the tail to squeeze forward, forming a more nearly ideal band.

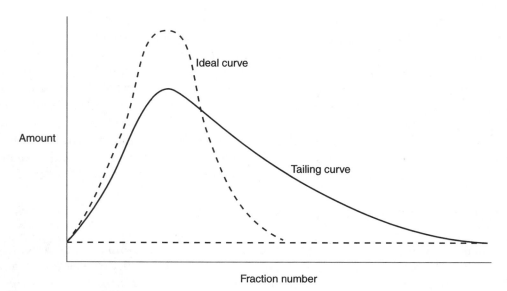

Figure 19.12 Elution curves: one ideal and one that "tails."

19.14 Recovering the Separated Compounds

In recovering each of the separated compounds of a chromatographic separation when they are solids, the various correct fractions are combined and evaporated. If the combined fractions contain sufficient material, they may be purified by recrystallization. If the compounds are liquids, the correct fractions are combined, and the solvent is evaporated. If sufficient material has been collected, liquid samples can be purified by distillation. The combination of chromatography–crystallization or chromatography–distillation usually yields very pure compounds. For microscale applications, the amount of sample collected is too small to allow a purification by crystallization or distillation. The samples that are obtained after the solvent has been evaporated are considered to be sufficiently pure, and no additional purification is attempted.

19.15 Decolorization by Column Chromatography

A common outcome of organic reactions is the formation of a product that is contaminated by highly colored impurities. Very often these impurities are highly polar, and they have a high molecular weight, as well as being colored. The purification of the desired product requires that these impurities be removed. Section 11.7 of Chapter 11 (pages 157–158) details methods of decolorizing an organic product. In most cases, these methods involve the use of a form of activated charcoal, or Norit.

An alternative, which is applied conveniently in microscale experiments, is to remove the colored impurity by column chromatography. Because of the polarity of the impurities, the colored components are strongly adsorbed on the stationary phase of the column, and the less polar desired product passes through the column and is collected.

Microscale decolorization of a solution on a chromatography column requires that a column be prepared in a Pasteur pipet, using either alumina or silica gel as the adsorbent (Sections 19.6 and 19.7). The sample to be decolorized is diluted to the point where crystallization within the column will not take place, and it is then passed through the column in the usual manner. The desired compound is collected as it exits the column, and the excess solvent is removed by evaporation (Chapter 7, Section 7.10, page 95).

19.16   GEL CHROMATOGRAPHY

The stationary phase in gel chromatography consists of a cross-linked polymeric material. Molecules are separated according to their *size* by their ability to penetrate a sievelike structure. Molecules permeate the porous stationary phase as they move down the column. Small molecules penetrate the porous structure more easily than large ones. Thus, the large molecules move through the column faster than the smaller ones and elute first. The separation of molecules by gel chromatography is depicted in Figure 19.13. With adsorption chromatography using materials such as alumina or silica, the

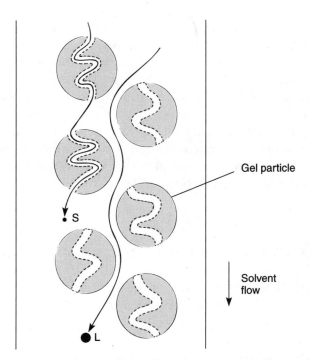

Figure 19.13 Gel chromatography. Comparison of the paths of large (L) and small (S) molecules through the column during the same interval of time.

order is usually the reverse. Small molecules (of low molecular weight) pass through the column *faster* than large molecules (of high molecular weight) because large molecules are more strongly attracted to the polar stationary phase.

Equivalent terms used by chemists for the gel chromatography technique are **gel-filtration chromatography** (biochemistry term), **gel-permeation chromatography** (polymer chemistry term), and **molecular-sieve chromatography. Size-exclusion chromatography** is a general term for the technique, and it is perhaps the most descriptive term for what occurs on a molecular level.

Sephadex is one of the most popular materials for gel chromatography. It is widely used by biochemists for separating proteins, nucleic acids, enzymes, and carbohydrates. Most often, water or aqueous solutions of buffers are used as the moving phase. Chemically, Sephadex is a polymeric carbohydrate that has been cross-linked. The degree of cross-linking determines the size of the "holes" in the polymer matrix. In addition, the hydroxyl groups on the polymer can adsorb water, which causes the material to swell. As it expands, "holes" are created in the matrix. Several different gels are available from manufacturers, each with its own set of characteristics. For example, a typical Sephadex gel, such as G-75, can separate molecules in the molecular weight (MW) range 3000 to 70,000. Assume a four-component mixture containing compounds with molecular weights of 10,000, 20,000, 50,000, and 100,000. The 100,000-MW compound would pass through the column first, because it cannot penetrate the polymer matrix. The 50,000-, 20,000-, and 10,000-MW compounds penetrate the matrix to varying degrees and would be separated. The molecules would elute in the order given (decreasing order of molecular weights). The gel separates on the basis of molecular size and configuration, rather than molecular weight.

Sephadex LH-20 has been developed for nonaqueous solvents. Some of the hydroxyl groups have been alkylated, and thus the material can swell under both aqueous and nonaqueous conditions (it now has "organic" character). This material can be used with several organic solvents, such as alcohol, acetone, methylene chloride, and aromatic hydrocarbons.

Another type of gel is based on a polyacrylamide structure (Bio-Gel P and Poly-Sep AA). A portion of a polyacrylamide chain is shown here:

$$-CH_2-CH-CH_2-CH-CH_2-CH-$$
$$C=O \qquad C=O \qquad C=O$$
$$NH_2 \qquad NH_2 \qquad NH_2$$

Gels of this type can also be used in water and some polar organic solvents. They tend to be more stable than Sephadex, especially under acidic conditions. Polyacrylamides can be used for many biochemical applications involving macromolecules. For separating synthetic polymers, cross-linked polystyrene beads (copolymer of styrene and divinylbenzene) find common application. Again, the beads are swollen before use. Common organic solvents can be used to elute the polymers. As with other gels, the higher-molecular-weight compounds elute before the lower-molecular-weight compounds.

19.17 FLASH CHROMATOGRAPHY

One of the drawbacks to column chromatography is that for large-scale preparative separations, the time required to complete a separation may be very long. Furthermore, the resolution that is possible for a particular experiment tends to deteriorate as the time for the experiment grows longer. This latter effect arises because the bands of compounds that move very slowly through a column tend to "tail."

A technique that can be useful in overcoming these problems has been developed. This technique, called **flash chromatography,** is actually a very simple modification of an ordinary column chromatography. In flash chromatography, the adsorbent is packed into a relatively short glass column, and air pressure is used to force the solvent through the adsorbent.

The apparatus used for flash chromatography is shown in Figure 19.14. The glass column is fitted with a Teflon stopcock at the bottom to control the flow rate of solvent. A plug of glass wool is placed in the bottom of the column to act as a support for the adsorbent. A layer of sand may also be added on top of the glass wool. The column is filled with adsorbent using the dry pack method. When the column has been filled, a fitting is attached to the top of the column, and the entire apparatus is connected to

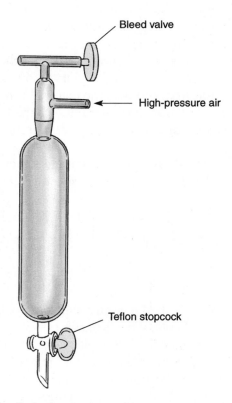

Figure 19.14 Apparatus for flash chromatography.

a source of high-pressure air or nitrogen. The fitting is designed so that the pressure applied to the top of the column can be adjusted precisely. The source of the high-pressure air is often a specially adapted air pump.

A typical column would use silica gel adsorbent (particle size = 40 to 63 μm) packed to a height of 5 inches in a glass column of 20-mm diameter. The pressure applied to the column would be adjusted to achieve a solvent flow rate such that the solvent level in the column would decrease by about 2 inches/minute. This system would be appropriate to separate the components of a 250-mg sample.

The high-pressure air forces the solvent through the column of adsorbent at a rate that is much greater than what would be achieved if the solvent flowed through the column under the force of gravity. Because the solvent is made to flow faster, the time required for substances to pass through the column is reduced. By itself, simply applying air pressure to the column might reduce the clarity of the separation, because the components of the mixture would not have time to establish themselves into distinctly separate bands. However, in flash chromatography, you can use a much finer adsorbent than would be used in ordinary chromatography. With a much smaller particle size for the adsorbent, the surface area is increased, and the resolution possible thereby improves.

A simple variation on this idea does not use air pressure. Instead, the lower end of the column is inserted into a stopper, which is fitted into the top of a suction flask. Vacuum is applied to the system, and the vacuum acts to draw the solvent through the adsorbent column. The overall effect of this variation is similar to that obtained when air pressure is applied to the top of the column.

REFERENCES

Deyl, Z., Macek, K., and Janák, J. *Liquid Column Chromatography*. Amsterdam: Elsevier, 1975.

Heftmann, E. *Chromatography*. 3rd ed. New York: Van Nostrand Reinhold, 1975.

Jacobson, B. M. "An Inexpensive Way to Do Flash Chromatography." *Journal of Chemical Education 65* (May 1988): 459.

Still, W. C., Kahn, M., and Mitra, A. "Rapid Chromatographic Technique for Preparative Separations with Moderate Resolution." *Journal of Organic Chemistry 43* (1978): 2923.

PROBLEMS

1. A sample was placed on a chromatography column. Methylene chloride was used as the eluting solvent. All of the components eluted off the column, but no separation was observed. What must have been happening during this experiment? How would you change the experiment to overcome this problem?

2. You are about to purify an impure sample of naphthalene by column chromatography. What solvent should you use to elute the sample?

3. Consider a sample that is a mixture composed of biphenyl, benzoic acid, and benzyl alcohol. Predict the order of elution of the components in this mixture. Assume that the chromatography uses a silica column, and the solvent system is based on cyclohexane, with an increasing proportion of methylene chloride added as a function of time.

4. An orange compound was added to the top of a chromatography column. Solvent was added immediately, and the entire volume of solvent in the solvent reservoir turned orange. No separation could be obtained from the chromatography experiment. What went wrong?

5. A yellow compound, dissolved in methylene chloride, is added to a chromatography column. The elution is begun using petroleum ether as the solvent. After 6 L of solvent had passed through the column, the yellow band still had not traveled down the column appreciably. What should be done to make this experiment work better?

6. You have 0.50 g of a mixture that you wish to purify by column chromatography. How much adsorbent should you use to pack the column? Estimate the appropriate column diameter and height.

7. In a particular sample, you wish to collect the component with the *highest* molecular weight as the *first* fraction. What chromatographic technique should you use?

8. A colored band shows an excessive amount of tailing as it passes through the column. What can you do to rectify this problem?

9. How would you monitor the progress of a column chromatography when the sample is colorless? Describe at least two methods.

CHAPTER 20

Thin-Layer Chromatography

Thin-layer chromatography (TLC) is a very important technique for the rapid separation and qualitative analysis of small amounts of material. It is ideally suited for the analysis of mixtures and reaction products in both macroscale and microscale experiments. The technique is closely related to column chromatography. In fact, TLC can be considered column chromatography *in reverse*, with the solvent ascending the adsorbent, rather than descending. Because of this close relationship to column chromatography, and because the principles governing the two techniques are similar, Chapter 19, on column chromatography, should be read first.

20.1 PRINCIPLES OF THIN-LAYER CHROMATOGRAPHY

Like column chromatography, TLC is a solid–liquid partitioning technique. However, the moving liquid phase is not allowed to percolate down the adsorbent; it is caused to *ascend* a thin layer of adsorbent coated onto a backing support. The most typical backing is a plastic material, but other materials are also used. A thin layer of the adsorbent is spread onto the plate and allowed to dry. A coated and dried plate is called a **thin-layer plate** or a **thin-layer slide.** (The reference to *slide* comes about because microscope slides were often used to prepare small thin-layer plates.) When a thin-layer plate is placed upright in a vessel that contains a shallow layer of solvent, the solvent ascends the layer of adsorbent on the plate by capillary action.

In TLC, the sample is applied to the plate before the solvent is allowed to ascend the adsorbent layer. The sample is usually applied as a small spot near the base of the plate; this technique is often referred to as **spotting.** The plate is spotted by repeated applications of a sample solution from a small capillary pipet. When the filled pipet touches the plate, capillary action delivers its contents to the plate, and a small spot is formed.

As the solvent ascends the plate, the sample is partitioned between the moving liquid phase and the stationary solid phase. During this process, you are **developing,** or **running,** the thin-layer plate. In development, the various components in the applied mixture are separated. The separation is based on the many equilibrations the solutes experience between the moving and the stationary phases. (The nature of these equilibrations was thoroughly discussed in Chapter 19, Sections 19.2 and 19.3, pages 287–290.) As in column chromatography, the least polar substances advance faster than the most polar substances. A separation results from the differences in the rates at which the individual components of the mixture advance upward on the plate. When many substances are present in a mixture, each has its own characteristic solubility and adsorptivity properties, depending on the functional groups in its structure. In general, the stationary phase is strongly polar and strongly binds polar substances. The moving liquid phase is usually less polar than the adsorbent and most easily dissolves substances that are less polar or even nonpolar. Thus, the most polar substances travel slowly upward, or not at all, and nonpolar substances travel more rapidly if the solvent is sufficiently nonpolar.

When the thin-layer plate has been developed, it is removed from the developing tank and allowed to dry until it is free of solvent. If the mixture that was originally spotted on the plate was separated, there will be a vertical series of spots on the plate. Each spot corresponds to a separate component or compound from the original mixture. If the components of the mixture are colored substances, the various spots will be clearly visible after development. More often, however, the "spots" will not be visible because they correspond to colorless substances. If spots are not apparent, they can be made visible only if a **visualization method** is used. Often, spots can be seen when the thin-layer plate is held under ultraviolet light; the ultraviolet lamp is a common visualization method. Also common is the use of iodine vapor. The plates are placed in a chamber containing iodine crystals and left to stand for a short time. The iodine reacts with the various compounds adsorbed on the plate to give colored complexes that are clearly visible. Because iodine often changes the compounds by reaction, the components of the mixture cannot be recovered from the plate when the iodine method is used. (Other methods of visualization are discussed in Section 20.7.)

20.2 COMMERCIALLY PREPARED TLC PLATES

The most convenient type of TLC plate is prepared commercially and sold in a ready-to-use form. Many manufacturers supply glass plates precoated with a durable layer of silica gel or alumina. More conveniently, plates are also available that have either a flexible plastic backing or an aluminum backing. The most common types of commercial TLC plates are composed of plastic sheets that are coated with silica gel and polyacrylic acid, which serves as a binder. A fluorescent indicator may be mixed with the silica gel. Due to the presence of compounds in the sample the indicator renders the spots visible under ultraviolet light (see Section 20.7). Although these plates are relatively expensive compared with plates prepared in the laboratory, they are far more convenient to use, and they provide more consistent results. The plates are manufactured quite uniformly. Because the plastic backing is flexible, an additional advantage is that the coating does not flake off the plates easily. The plastic sheets (usually 8 by 8 inch square) can also be cut with a pair of scissors or paper cutter to whatever size may be required.

If the package of commercially prepared TLC plates has been opened previously, or if the plates have not been purchased recently, they should be dried before use. Dry the plates by placing them in an oven at 100°C for 30 minutes, and store them in a desiccator until they are to be used.

20.3 PREPARATION OF THIN-LAYER SLIDES AND PLATES

Commercially prepared plates (Section 20.2) are the most convenient to use, and we recommend their use for most applications. If you must prepare your own slides or plates, this section provides directions for doing so. The two adsorbent materials used most often for TLC are alumina G (aluminum oxide) and silica gel G (silicic acid). The G designation stands for gypsum (calcium sulfate). Calcined gypsum $CaSO_4 \cdot \frac{1}{2}H_2O$ is better known as plaster of Paris. When exposed to water or moisture, gypsum sets in a rigid mass $CaSO_4 \cdot 2H_2O$, which binds the adsorbent together and to the glass plates used as a backing support. In the adsorbents used for TLC, about 10–13% by weight of gypsum is added as a binder. The adsorbent materials are otherwise similar to those used in column chromatography; the adsorbents used in column chromatography have a larger particle size, however. The material for thin-layer work is a fine powder. The small particle size, along with the added gypsum, makes it impossible to use silica gel G or alumina G for column work. In a column, these adsorbents generally set so rigidly that solvent virtually stops flowing through the column.

For separations involving large amounts of material, or for difficult separations, it may be necessary to use larger thin-layer plates. Under these circumstances, you may have to prepare your own plates. Plates with dimensions up to 200–250 cm^2 are common. With larger plates, it is desirable to have a somewhat durable coating, and a water slurry of the adsorbent should be used to prepare them. If silica gel is used, the slurry should be prepared in the ratio of about 1 g silica gel G to each 2 mL of water. The glass plate used for the thin-layer plate should be washed, dried, and placed on a sheet of newspaper. Place two strips of masking tape along two edges of the plate. Use more than one layer of masking tape if a thicker coating is desired on the plate. A slurry is prepared, shaken well, and poured along one of the untaped edges of the plate.

CAUTION Avoid breathing silica dust or methylene chloride, prepare and use the slurry in a hood, and avoid getting methylene chloride or the slurry mixture on your skin. Perform the coating operation under a hood.

A heavy piece of glass rod, long enough to span the taped edges, is used to level and spread the slurry over the plate. While the rod is resting on the tape, it is pushed along the plate from the end at which the slurry was poured toward the opposite end of the plate. This is illustrated in Figure 20.1. After the slurry is spread, the masking tape strips are removed, and the plates are dried in a 110°C oven for about 1 hour. Plates of 200–250 cm^2 are easily prepared by this method. Larger plates present more difficulties. Many laboratories have a commercially manufactured spreading machine that makes the entire operation simpler.

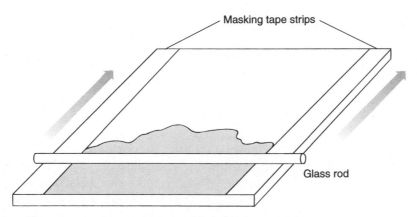

Figure 20.1 Preparing a large thin-layer chromatography plate.

20.4 SAMPLE APPLICATION: SPOTTING THE PLATES

A. Preparing a Micropipet

To apply the sample that is to be separated to the thin-layer plate, use a micropipet. A micropipet is easily made from a short length of thin-walled capillary tubing such as that used for melting-point determinations, but open at both ends. The capillary tubing is heated at its midpoint with a microburner and rotated until it is soft. When the tubing is soft, the heated portion of the tubing is drawn out until a constricted portion of tubing 4–5 cm long is formed. After cooling, the constricted portion of tubing is scored at its center with a file or scorer and broken. The two halves yield two capillary micropipets. Try to make a clean break without jagged or sharp edges. Figure 20.2 shows how to make such pipets.

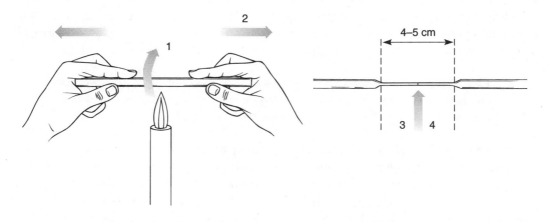

① Rotate in flame until soft.   ③ Score lightly in center of pulled section.
② Remove from flame and pull.   ④ Break in half to give two pipets.

Figure 20.2 The construction of two capillary micropipets.

B. Spotting the Plate

To apply a sample to the plate, begin by placing about 1 mg of a solid test substance, or one drop of a liquid test substance, in a small container such as a watch glass or a test tube. Dissolve the sample in a few drops of a volatile solvent. Acetone or methylene chloride is usually a suitable solvent. If a solution is to be tested, it can often be used directly (undiluted). The small capillary pipet, prepared as described, is filled by dipping the pulled end into the solution to be examined. Capillary action fills the pipet. Empty the pipet by touching it lightly to the thin-layer plate at a point about 1 cm from the bottom (Fig. 20.3). The spot must be high enough to not dissolve in the developing solvent. It is important to touch the plate very lightly and not to gouge a hole in the adsorbent. When the pipet touches the plate, the solution is transferred to the plate as a small spot. The pipet should be touched to the plate very briefly and then removed. If the pipet is held to the plate, its entire contents will be delivered to the plate. Only a small amount of material is needed. It is often helpful to blow gently on the plate as the sample is applied. This helps keep the spot small by evaporating the solvent before it can spread out on the plate. The smaller the spot formed, the better the separation obtainable. If needed, additional material can be applied to the plate by repeating the spotting procedure. You should repeat the procedure with several small amounts, rather than apply one large amount. The solvent should be allowed to evaporate between applications. If the spot is not small (about 2 mm in diameter), a new plate should be prepared. The capillary pipet may be used several times if it is rinsed between uses. It is repeatedly dipped into a small portion of solvent to rinse it and touched to a paper towel to empty it.

As many as three different spots may be applied to a one-inch-wide TLC plate. Each spot should be about 1 cm from the bottom of the plate, and all spots should be evenly spaced, with one spot in the center of the plate. Due to diffusion, spots often increase in diameter as the plate is developed. To keep spots containing different materials from merging, and to avoid confusing the samples, do not place more than three spots on a single plate. Larger plates can accommodate many more samples.

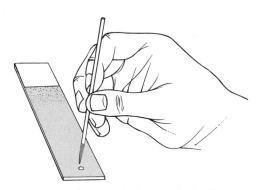

Figure 20.3 Spotting the thin-layer chromatography plate with a drawn capillary pipet.

20.5 DEVELOPING (RUNNING) TLC PLATES

A. Preparing a Development Chamber

A convenient development chamber for TLC plates can be made from a 4-oz wide-mouthed jar. An alternative development chamber can be constructed from a beaker, using aluminum foil to cover the opening. The inside of the jar or beaker should be lined with a piece of filter paper, cut so that it does not quite extend around the inside of the jar. A small vertical opening (2–3 cm) should be left in the filter paper to observe the development. Before development, the filter paper inside the jar or beaker should be thoroughly moistened with the development solvent. The solvent-saturated liner helps to keep the chamber saturated with solvent vapors, thereby speeding the development. Once the liner is saturated, the level of solvent in the bottom of the development chamber is adjusted to a depth of about 5 mm, and the chamber is capped (or covered with aluminum foil) and set aside until it is to be used. A correctly prepared development chamber (with TLC plate in place) is shown in Figure 20.4.

B. Developing the TLC Plate

Once the spot has been applied to the thin-layer plate and the solvent has been selected (see Section 20.6), the plate is placed in the chamber for development. The plate must be placed in the chamber carefully so that none of the coated portion touches the filter paper liner. In addition, the solvent level in the bottom of the chamber must not be above the spot that was applied to the plate, or the spotted material will dissolve in the pool of solvent instead of undergoing chromatography. Once the plate has been placed correctly, replace the cap on the developing chamber and wait

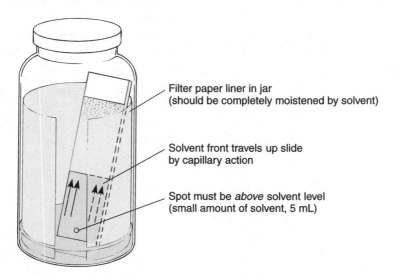

Filter paper liner in jar
(should be completely moistened by solvent)

Solvent front travels up slide
by capillary action

Spot must be *above* solvent level
(small amount of solvent, 5 mL)

Figure 20.4 A development chamber with a thin-layer chromatography plate undergoing development.

for the solvent to advance up the plate by capillary action. This generally occurs rapidly, and you should watch carefully. As the solvent rises, the plate becomes visibly moist. When the solvent has advanced to within 5 mm of the end of the coated surface, the plate should be removed, and the position of the solvent front should be marked immediately by scoring the plate along the solvent line with a pencil. The solvent front must not be allowed to travel beyond the end of the coated surface. The plate should be removed before this happens. The solvent will not actually advance beyond the end of the plate, but spots allowed to stand on a completely moistened plate on which the solvent is not in motion expand by diffusion. Once the plate has dried, any visible spots should be outlined on the plate with a pencil. If no spots are apparent, a visualization method (Section 20.7) may be needed.

20.6 CHOOSING A SOLVENT FOR DEVELOPMENT

The development solvent used depends on the materials to be separated. You may have to try several solvents before a satisfactory separation is achieved. Because small TLC plates can be prepared and developed rapidly, an empirical choice is usually not hard to make. A solvent that causes all the spotted material to move with the solvent front is too polar. One that does not cause any of the material in the spot to move is not polar enough. As a guide to the relative polarity of solvents, consult Table 19.2 in Chapter 19 (page 293).

Methylene chloride and toluene are solvents of intermediate polarity and good choices for a wide variety of functional groups to be separated. For hydrocarbon materials, good first choices are hexane, petroleum ether (ligroin), or toluene. Hexane or petroleum ether with varying proportions of toluene or ether gives solvent mixtures of moderate polarity that are useful for many common functional groups. Polar materials may require ethyl acetate, acetone, or methanol.

A rapid way to determine a good solvent is to apply several sample spots to a single plate. The spots should be placed a minimum of 1 cm apart. A capillary pipet is filled with a solvent and gently touched to one of the spots. The solvent expands outward in a circle. The solvent front should be marked with a pencil. A different solvent is applied to each spot. As the solvents expand outward, the spots expand as concentric rings. From the appearance of the rings, you can judge approximately the suitability of the solvent. Several types of behavior experienced with this method of testing are shown in Figure 20.5.

20.7 VISUALIZATION METHODS

It is fortunate when the compounds separated by TLC are colored because the separation can be followed visually. More often than not, however, the compounds are colorless. In that case, some reagent or some method must be used to make the separated materials visible. Reagents that give rise to colored spots are called **visualization reagents.** Methods of viewing that make the spots apparent are **visualization methods.**

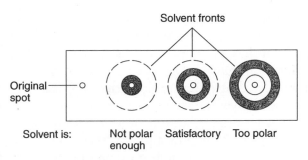

Figure 20.5 The concentric ring method of testing solvents.

The most common method of visualization is by an ultraviolet (UV) lamp. Under UV light, compounds often look like bright spots on the plate. This often suggests the structure of the compound. Certain types of compounds shine very brightly under UV light, because they fluoresce.

Plates can be purchased with a fluorescent indicator added to the adsorbent. A mixture of zinc and cadmium sulfides is often used. When treated in this way and held under UV light, the entire plate fluoresces. However, dark spots appear on the plate where the separated compounds are seen to quench this fluorescence.

Iodine is also used to visualize plates. Iodine reacts with many organic materials to form complexes that are either brown or yellow. In this visualization method, the developed and dried TLC plate is placed in a 4-oz wide-mouthed screw cap jar along with a few crystals of iodine. The jar is capped and gently warmed on a steam bath or a hot plate at low heat. The jar fills with iodine vapors, and the spots begin to appear. When the spots are sufficiently intense, the plate is removed from the jar and the spots are outlined with a pencil. The spots are not permanent. Their appearance results from the formation of complexes the iodine makes with the organic substances. As the iodine sublimes off the plate, the spots fade. Hence, they should be marked immediately. Nearly all compounds except saturated hydrocarbons and alkyl halides form complexes with iodine. The intensities of the spots do not accurately indicate the amount of material present, except in the crudest way.

In addition to the preceding methods, several chemical methods are available that either destroy or permanently alter the separated compounds through reaction. Many of these methods are specific for particular functional groups.

Alkyl halides can be visualized if a dilute solution of silver nitrate is sprayed on the plates. Silver halides are formed. These halides decompose if exposed to light, giving rise to dark spots (free silver) on the TLC plate.

Most organic functional groups can be made visible if they are charred with sulfuric acid. Concentrated sulfuric acid is sprayed on the plate, which is then heated in an oven at 110°C to complete the charring. Permanent spots are thus created.

Colored compounds can be prepared from colorless compounds by making derivatives before spotting them on the plate. An example of this is the preparation of 2,4-dinitrophenylhydrazones from aldehydes and ketones to produce yellow and orange compounds. You may also spray the 2,4-dinitrophenylhydrazine reagent on the plate after the ketones or aldehydes have separated. Red and yellow spots form where the compounds are located. Other examples of this method are the use of ferric chloride to visualize phenols and the use of bromocresol green to detect carboxylic acids. Chromium trioxide, potassium dichromate, and potassium permanganate can be used to visualize compounds that are easily oxidized. *p*-Dimethylaminobenzaldehyde easily detects amines. Ninhydrin reacts with amino acids to make them visible. Numerous other methods and reagents available from various supply outlets are specific for certain types of functional groups. These visualize only the class of compounds of interest.

20.8 PREPARATIVE PLATES

If you use large plates (Section 20.3), materials can be separated and the separated components can be recovered individually from the plates. Plates used in this way are called **preparative plates.** For preparative plates, a thick layer of adsorbent is generally used. Instead of being applied as a spot or a series of spots, the mixture to be separated is applied as a line of material about 1 cm from the bottom of the plate. As the plate is developed, the separated materials form bands. After development, you can observe the separated bands, usually by UV light, and outline the zones in pencil. If the method of visualization is destructive, most of the plate is covered with paper to protect it, and the reagent is applied only at the extreme edge of the plate.

Once the zones have been identified, the adsorbent in those bands is scraped from the plate and extracted with solvent to remove the adsorbed material. Filtration removes the adsorbent, and evaporation of the solvent gives the recovered component from the mixture.

20.9 THE R_f VALUE

Thin-layer chromatography conditions include

1. Solvent system
2. Adsorbent
3. Thickness of the adsorbent layer
4. Relative amount of material spotted

Under an established set of such conditions, a given compound always travels a fixed distance relative to the distance the solvent front travels. This ratio of the distance the compound travels to the distance the solvent travels is called the **R_f value.** The symbol **R_f** stands for "retardation factor," or "ratio to front," and it is expressed as a decimal fraction:

$$R_f = \frac{\text{distance traveled by substance}}{\text{distance traveled by solvent front}}$$

When the conditions of measurement are completely specified, the R_f value is constant for any given compound, and it corresponds to a physical property of that compound.

The R_f value can be used to identify an unknown compound, but like any other identification based on a single piece of data, the R_f value is best confirmed with some additional data. Many compounds can have the same R_f value, just as many compounds have the same melting point.

It is not always possible, in measuring an R_f value, to duplicate exactly the conditions of measurement another researcher has used. Therefore, R_f values tend to be of more use to a single researcher in one laboratory than they are to researchers in different laboratories. The only exception to this is when two researchers use TLC plates from the same source, as in commercial plates, or know the exact details of how the plates were prepared. Nevertheless, the R_f value can be a useful guide. If exact values cannot be relied on, the relative values can provide another researcher with useful information about what to expect. Anyone using published R_f values will find it a good idea to check them by comparing them with standard substances whose identity and R_f values are known.

To calculate the R_f value for a given compound, measure the distance that the compound has traveled from the point at which it was originally spotted. For spots that are not too large, measure to the center of the migrated spot. For large spots, the measurement should be repeated on a new plate, using less material. For spots that show tailing, the measurement is made to the "center of gravity" of the spot. This first distance measurement is then divided by the distance the solvent front has traveled from the same original spot. A sample calculation of the R_f values of two compounds is illustrated in Figure 20.6.

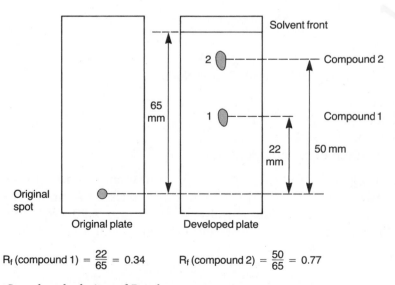

$$R_f \text{ (compound 1)} = \frac{22}{65} = 0.34 \qquad R_f \text{ (compound 2)} = \frac{50}{65} = 0.77$$

Figure 20.6 Sample calculation of R_f values.

20.10 THIN-LAYER CHROMATOGRAPHY APPLIED IN ORGANIC CHEMISTRY

Thin-layer chromatography has several important uses in organic chemistry. It can be used in the following applications:

1. To establish that two compounds are identical
2. To determine the number of components in a mixture
3. To determine the appropriate solvent for a column-chromatographic separation
4. To monitor a column-chromatographic separation
5. To check the effectiveness of a separation achieved on a column, by crystallization or by extraction
6. To monitor the progress of a reaction

In all these applications, TLC has the advantage that only small amounts of material are necessary. Material is not wasted. With many of the visualization methods, less than a tenth of a microgram (10^{-7} g) of material can be detected. On the other hand, samples as large as a milligram may be used. With preparative plates that are large (about 9 inches on a side) and have a relatively thick coating of adsorbent (>500 μm), it is often possible to separate from 0.2 to 0.5 g of material at one time. The main disadvantage of TLC is that volatile materials cannot be used, because they evaporate from the plates.

Thin-layer chromatography can establish that two compounds suspected to be identical are in fact identical. Simply spot both compounds side by side on a single plate, and develop the plate. If both compounds travel the same distance on the plate (have the same R_f value), they are probably identical. If the spot positions are not the same, the compounds are definitely not identical. It is important to spot compounds *on the same plate*. This is especially important with slides and plates that you prepare yourself. Because they vary widely from plate to plate, no two plates have exactly the same thickness of adsorbent. If you use commercial plates, this precaution is not necessary, although it is nevertheless strongly recommended.

Thin-layer chromatography can establish whether a sample is a single substance or a mixture. A single substance gives a single spot no matter what solvent is used to develop the plate. On the other hand, the number of components in a mixture can be established by trying various solvents on a mixture. A word of caution should be given. It may be difficult, in dealing with compounds of very similar properties, for example, isomers, to find a solvent that will separate the mixture. Inability to achieve a separation is not absolute proof that a sample is a single pure substance. Many compounds can be separated only by *multiple developments* of the TLC slide with a fairly nonpolar solvent. In this method, you remove the plate after the first development and allow it to dry. After being dried, it is placed in the chamber again and developed once more. This effectively doubles the length of the slide. At times, several developments may be necessary.

When a mixture is to be separated, you can use TLC to choose the best solvent to separate it if column chromatography is contemplated. You can try various solvents on a plate coated with the same adsorbent as will be used in the column. The solvent that resolves the components best will probably work well on the column. These small-scale experiments are quick, use very little material, and save time that would be wasted by attempting to separate the entire mixture on the column. Similarly, TLC plates can *monitor* a column. A hypothetical situation is shown in Figure 20.7. A solvent was found that would separate the mixture into four components (A–D). A column was run using this solvent, and 11 fractions of 15 mL each were collected. Thin-layer analysis of the various fractions showed that Fractions 1–3 contained Component A; Fractions 4–7, Component B; Fractions 8–9, Component C; and Fractions 10–11, Component D. A small amount of cross-contamination was observed in Fractions 3, 4, 7, and 9.

In another TLC example, a researcher found a product from a reaction to be a mixture. It gave two spots, A and B, on a TLC plate. After the product was crystallized, the crystals were found by TLC to be pure A, whereas the mother liquor was found to have a mixture of A and B. The crystallization was judged to have purified A satisfactorily.

Finally, it is often possible to monitor the progress of a reaction by TLC. At various points during a reaction, samples of the reaction mixture are taken and subjected to TLC analysis. An example is given in Figure 20.8. In this case, the desired reaction was the conversion of A to B. At the beginning of the reaction (0 hour), a TLC plate was prepared that was spotted with pure A, pure B, and the reaction mixture. Similar plates were prepared at 0.5, 1, 2, and 3 hours after the start of the reaction. The plates showed that the reaction was complete in 2 hours. When the reaction was run longer than 2 hours, a new compound, side product C, began to appear. Thus, the optimum reaction time was judged to be 2 hours.

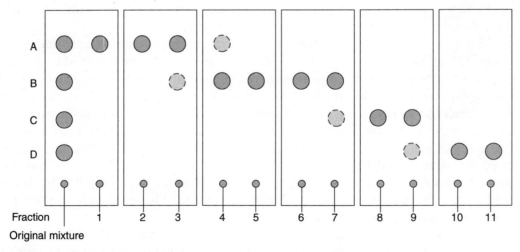

Figure 20.7 Monitoring column chromatography with TLC plates.

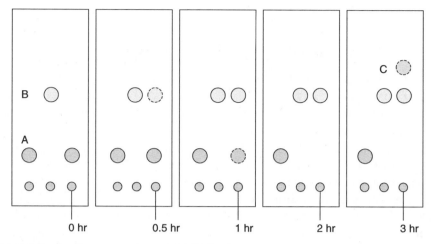

Figure 20.8 Monitoring a reaction with TLC plates.

20.11 PAPER CHROMATOGRAPHY

Paper chromatography is often considered to be related to thin-layer chromatography. The experimental techniques are somewhat like those of TLC, but the principles are more closely related to those of extraction. Paper chromatography is actually a liquid–liquid partitioning technique, rather than a solid–liquid technique. For paper chromatography, a spot is placed near the bottom of a piece of high-grade filter paper (Whatman No. 1 is often used). Then the paper is placed in a developing chamber. The development solvent ascends the paper by capillary action and moves the components of the spotted mixture upward at differing rates. Although paper consists mainly of pure cellulose, the cellulose itself does not function as the stationary phase. Rather, the cellulose absorbs water from the atmosphere, especially from an atmosphere saturated with water vapor. Cellulose can absorb up to about 22% of water. It is this water adsorbed on the cellulose that functions as the stationary phase. To ensure that the cellulose is kept saturated with water, many development solvents used in paper chromatography contain water as a component. As the solvent ascends the paper, the compounds are partitioned between the stationary water phase and the moving solvent. Because the water phase is stationary, the components in a mixture that are most highly water soluble, or those that have the greatest hydrogen-bonding capacity, are the ones that are held back and move most slowly. Paper chromatography applies mostly to highly polar compounds or to compounds that are polyfunctional. The most common use of paper chromatography is for sugars, amino acids, and natural pigments. Because filter paper is manufactured consistently, R_f values can often be relied on in paper chromatographic work. However, R_f values are customarily measured from the leading edge (top) of the spot—not from its center, as is customary in TLC.

PROBLEMS

1. A student spots an unknown sample on a TLC plate and develops it in dichloromethane solvent. Only one spot, for which the R_f value is 0.95, is observed. Does this indicate that the unknown material is a pure compound? Using thin-layer chromatography, what can be done to verify the purity of the sample?

2. You and another student were each given an unknown compound. Both samples contained colorless material. You each used the same brand of commercially prepared TLC plate and developed the plates using the same solvent. Each of you obtained a single spot of $R_f = 0.75$. Were the two samples necessarily the same substances? How could you prove unambiguously that they were identical using TLC?

3. Each of the solvents given should effectively separate one of the following mixtures by TLC. Match the appropriate solvent with the mixture that you would expect to separate well with that solvent. Select your solvent from the following: hexane, methylene chloride, or acetone. You may need to look up the structures of the solvents and compounds in a handbook.
   (a) 2-Phenylethanol and acetophenone
   (b) Bromobenzene and p-xylene
   (c) Benzoic acid, 2,4-dinitrobenzoic acid, and 2,4,6-trinitrobenzoic acid.

4. Consider a sample that is a mixture composed of biphenyl, benzoic acid, and benzyl alcohol. The sample is spotted on a TLC plate and developed in a dichloromethane–cyclohexane solvent mixture. Predict the relative R_f values for the three components in the sample. *Hint:* See Table 19.3.

5. Consider the following errors that could be made when running TLC. Indicate what should be done to correct the error.
   (a) A two-component mixture containing 1-octene and 1,4-dimethylbenzene gave only one spot with an R_f value of 0.95. The solvent used was acetone.
   (b) A two-component mixture containing a dicarboxylic acid and a tricarboxylic acid gave only one spot with an R_f value of 0.05. The solvent used was hexane.
   (c) When a TLC plate was developed, the solvent front ran off the top of the plate.

6. Calculate the R_f value of a spot that travels 5.7 cm, with a solvent front that travels 13 cm.

7. A student spots an unknown sample on a TLC plate and develops it in pentane solvent. Only one spot, for which the R_f value is 0.05, is observed. Is the unknown material a pure compound? Using thin-layer chromatography, what can be done to verify the purity of the sample?

8. A colorless unknown substance is spotted on a TLC plate and developed in the correct solvent. The spots do not appear when visualization with a UV lamp or iodine vapors is attempted. What could you do to visualize the spots if the compound is
   (a) An alkyl halide
   (b) A ketone
   (c) An amino acid
   (d) A sugar

CHAPTER 21

High-Performance Liquid Chromatography (HPLC)

The separation that can be achieved is greater if the column packing used in column chromatography is made more dense by using an adsorbent that has a smaller particle size. The solute molecules encounter a much larger surface area on which they can be adsorbed as they pass through the column packing. At the same time, the solvent spaces between the particles are reduced in size. As a result of this tight packing, equilibrium between the liquid and solid phases can be established very rapidly with a fairly short column, and the degree of separation is markedly improved. The disadvantage of making the column packing more dense is that the solvent flow rate becomes very slow or even stops. Gravity is not strong enough to pull the solvent through a tightly packed column.

A recently developed technique can be applied to obtain much better separations with tightly packed columns. A pump forces the solvent through the column packing. As a result, solvent flow rate is increased and the advantage of better separation is retained. This technique, called **high-performance liquid chromatography (HPLC)**, is becoming widely applied to problems in which separations by ordinary column chromatography are unsatisfactory. Because the pump often provides pressures in excess of 1000 pounds per square inch (psi), this method is also known as **high-pressure liquid chromatography**. High pressures are not required, however, and satisfactory separations can be achieved with pressures as low as 100 psi.

The basic design of an HPLC instrument is shown in Figure 21.1. The instrument contains the following essential components:

1. Solvent reservoir
2. Solvent filter and degasser
3. Pump
4. Pressure gauge
5. Sample injection system
6. Column
7. Detector
8. Amplifier and electronic controls
9. Chart recorder

There may be other variations on this simple design. Some instruments have heated ovens to maintain the column at a specified temperature, fraction collectors, and microprocessor-controlled data-handling systems. Additional filters for

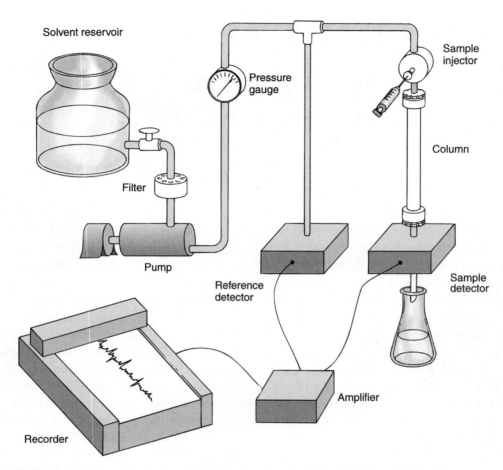

Figure 21.1 A schematic diagram of a high-performance liquid chromatograph.

the solvent and sample may also be included. You may find it interesting to compare this schematic diagram with Figure 22.2 in Chapter 22 (page 335) for a gas chromatography instrument. Many of the essential components are common to both types of instruments.

21.1 ADSORBENTS AND COLUMNS

The most important factor to consider when choosing a set of experimental conditions is the nature of the material packed into the column. You must also consider the size of the column that will be selected. The chromatography column is generally packed with silica or alumina adsorbents. Unlike column chromatography, however, the adsorbents used for HPLC have a much smaller particle size. Typically, particle size ranges from 5 to 20 μm in diameter for HPLC; it is on the order of 100 μm for column chromatography.

The adsorbent is packed into a column that can withstand the elevated pressures typical of this type of experiment. Generally, the column is constructed of stainless steel, although some columns that are constructed of a rigid polymeric material ("PEEK"—Poly Ether Ether Ketone) are available commercially. A strong column is required to withstand the high pressures that may be used. The columns are fitted with stainless steel connectors, which ensure a pressure-tight fit between the column and the tubing that connects the column to the other components of the instrument.

Columns that fulfill a large number of specialized purposes are available. In this chapter, we consider only the four most important types of columns:

1. Normal-phase chromatography
2. Reversed-phase chromatography
3. Ion-exchange chromatography
4. Size-exclusion chromatography

In most types of chromatography, the adsorbent is more polar than the mobile phase. For example, the solid packing material, which may be either silica or alumina, has a stronger affinity for polar molecules than does the solvent. As a result, the molecules in the sample adhere strongly to the solid phase, and their progress down the column is much slower than the rate at which solvent moves through the column. The time required for a substance to move through the column can be altered by changing the polarity of the solvent. In general, as the solvent becomes more polar, the faster substances move through the column. This type of behavior is known as **normal-phase chromatography.** In HPLC, you inject a sample onto a normal-phase column and elute it by varying the polarity of the solvent, much as you do with ordinary column chromatography. Disadvantages of normal-phase chromatography are that retention times tend to be long and bands have a tendency to "tail."

These disadvantages can be ameliorated by selecting a column in which the solid support is *less polar* than the moving solvent phase. This type of chromatography is known as **reversed-phase chromatography.** In this type of chromatography, the silica column packing is treated with alkylating agents. As a result, nonpolar alkyl groups are bonded to the silica surface, making the adsorbent nonpolar. The alkylating agents that are used most commonly can attach methyl ($—CH_3$), octyl ($—C_8H_{17}$), or octadecyl ($—C_{18}H_{37}$) groups to the silica surface. The latter variation, in which an 18-carbon chain is attached to the silica, is the most popular. This type of column is known as a **C_{18} column.** The bonded alkyl groups have an effect similar to what would be produced by an extremely thin organic solvent layer coating the surface of the silica particles. The interactions that take place between the substances dissolved in the solvent and the stationary phase thus become more like those observed in a liquid–liquid extraction. The solute particles distribute themselves between the two "solvents"—that is, between the moving solvent and the organic coating on the silica. The longer the chains of the alkyl groups that are bonded to the silica, the more effective the alkyl groups are as they interact with solute molecules.

Reversed-phase chromatography is widely used because the rate at which solute molecules exchange between moving phase and stationary phase is very rapid, which means that substances pass through the column relatively quickly. Furthermore, problems arising from the "tailing" of peaks are reduced. A disadvantage of this type of column, however, is that the chemically bonded solid phases tend to decompose. The organic groups are slowly hydrolyzed from the surface of the silica, which leaves a normal silica surface exposed. Thus, the chromatographic process that takes place on the column slowly shifts from a reversed-phase to a normal-phase separation mechanism.

Another type of solid support that is sometimes used in reversed-phase chromatography is organic polymer beads. These beads present a surface to the moving phase that is largely organic in nature.

For solutions of ions, select a column that is packed with an ion-exchange resin. This type of chromatography is known as **ion-exchange chromatography.** The ion-exchange resin that is chosen can be either an anion-exchange resin or a cation-exchange resin, depending upon the nature of the sample being examined.

A fourth type of column is known as a **size-exclusion column** or a **gel-filtration column.** The interaction that takes place on this type of column is similar to that described in Chapter 19, Section 19.16, page 308.

21.2 COLUMN DIMENSIONS

The dimensions of the column that you use depend upon the application. For analytical applications, a typical column is constructed of tubing that has an inside diameter of between 4 and 5 mm, although analytical columns with inside diameters of 1 or 2 mm are also available. A typical analytical column has a length of about 7.5 to 30 cm. This type of column is suitable for the separation of a 0.1- to 5-mg sample. With columns of smaller diameter, it is possible to perform an analysis with samples smaller than 1 *micro*gram.

High-performance liquid chromatography is an excellent analytical technique, but the separated compounds may also be isolated. The technique can be used for preparative experiments. Just as in column chromatography, the fractions can be collected into individual receiving containers as they pass through the column. The solvents can be evaporated from these fractions, allowing you to isolate separated components of the original mixture. Samples that range in size from 5 to 100 mg can be separated on a semipreparative, or **semiprep column.** The dimensions of a semiprep column are typically 8 mm inside diameter and 10 cm in length. A semiprep column is a practical choice when you wish to use the same column for both analytical and preparative separations. A semiprep column is small enough to provide reasonable sensitivity in analyses, but it is also capable of handling moderate-size samples when you need to isolate the components of a mixture. Even larger samples can be separated using a **preparative column.** This type of column is useful when you wish to collect the components of a mixture and then use the pure samples for additional study (e.g., for a subsequent chemical reaction or for spectroscopic analysis). A preparative column may be as large as 20 mm inside diameter and 30 cm in length. A preparative column can handle samples as large as 1 g per injection.

21.3 SOLVENTS

The choice of solvent used for an HPLC separation depends on the type of chromatographic process selected. For a normal-phase separation, the solvent is selected based on its polarity. The criteria described in Chapter 19, Section 19.4B, page 292, are used. A solvent of very low polarity might be pentane, petroleum ether, hexane, or carbon tetrachloride; a solvent of very high polarity might be water, acetic acid, methanol, or 1-propanol.

For a reversed-phase experiment, a less polar solvent causes solutes to migrate *faster*. For example, for a mixed methanol–water solvent, as the percentage of methanol in the solvent increases (solvent becomes less polar), the time required to elute the components of a mixture from a column decreases. The behavior of solvents as eluents in a reversed-phase chromatography would be the reverse of the order shown in Table 19.2 on page 293.

If a single solvent (or solvent mixture) is used for the entire separation, the chromatogram is said to be **isochratic.** Special electronic devices are available with HPLC instruments that allow you to program changes in the solvent composition from the beginning to the end of the chromatography. These are called **gradient elution systems.** With gradient elution, the time required for a separation may be shortened considerably.

The need for pure solvents is especially acute with HPLC. The narrow bore of the column and the very small particle size of the column packing require that solvents be particularly pure and free of insoluble residue. In most cases, the solvents must be filtered through ultrafine filters and **degassed** (have dissolved gases removed) before they can be used.

The solvent gradient is chosen so that the eluting power of the solvent increases over the duration of the experiment. The result is that components of the mixture that tend to move very slowly through the column are caused to move faster as the eluting power of the solvent gradually increases. The instrument can be programmed to change the composition of the solvent following a linear gradient or a nonlinear gradient, depending on the specific requirements of the separation.

21.4 DETECTORS

A flow-through **detector** must be provided to determine when a substance has passed through the column. In most applications, the detector detects either the change in index of refraction of the liquid as its composition changes or the presence of solute by its absorption of ultraviolet or visible light. The signal generated by the detector is amplified and treated electronically in a manner similar to that found in gas chromatography (Chapter 22, Section 22.6, page 339).

A detector that responds to changes in the index of refraction of the solution may be considered the most universal of the HPLC detectors. The refractive index of the liquid passing through the detector changes slightly, but significantly, as the liquid changes from pure solvent to a liquid where the solvent contains some type of organic solute. This change in refractive index can be detected and compared to the refractive index of pure solvent. The difference in index values is then recorded as a peak on a chart. A disadvantage of this type of detector is that it must respond to very small changes in refractive index. As a result, the detector tends to be unstable and difficult to balance.

When the components of the mixture have some type of absorption in the ultraviolet or visible regions of the spectrum, a detector that is adjusted to detect absorption at a particular wavelength of light can be used. This type of detector is much more stable, and the readings tend to be more reliable. Unfortunately, many organic compounds do not absorb ultraviolet light, and this type of detector cannot be used.

21.5 PRESENTATION OF DATA

The data produced by an HPLC instrument appear in the form of a chart, where detector response is the vertical axis and time is represented on the horizontal axis. These are recorded on a continuously moving strip of chart paper, although they may also be observed in graphic form on a computer display. In virtually all respects, the form of the data is identical to that produced by a gas chromatograph; in fact, in many cases, the data-handling system for the two types of instruments is essentially identical. To understand how to analyze the data from an HPLC instrument, read Sections 22.11 and 22.12 of Chapter 22.

REFERENCE

Rubinson, K. A. *Chemical Analysis.* Boston: Little, Brown and Co., 1987. Chapter 14, "Liquid Chromatography."

PROBLEMS

1. For a mixture of biphenyl, benzoic acid, and benzyl alcohol, predict the order of elution and describe any differences that you would expect for a normal-phase HPLC experiment (in hexane solvent) compared with a reversed-phase experiment (in tetrahydrofuran–water solvent).
2. How would the **gradient elution program** differ between normal-phase and reversed-phase chromatography?

CHAPTER 22

Gas Chromatography

Gas chromatography is one of the most useful instrumental tools for separating and analyzing organic compounds that can be vaporized without decomposition. Common uses include testing the purity of a substance and separating the components of a mixture. The relative amounts of the components in a mixture may also be determined. In some cases, gas chromatography can be used to identify a compound. In microscale work, it can also be used as a preparative method to isolate pure compounds from a small amount of a mixture.

Gas chromatography resembles column chromatography in principle, but it differs in three respects. First, the partitioning processes for the compounds to be separated are carried out between a **moving gas phase** and a **stationary liquid phase.** (Recall that in column chromatography the moving phase is a liquid and the stationary phase is a solid adsorbent.) Second, the temperature of the gas system can be controlled, because the column is contained in an insulated oven. And third, the concentration of any given compound in the gas phase is a function of its vapor pressure only. Because gas chromatography separates the components of a mixture primarily on the basis of their vapor pressures (or boiling points), this technique is also similar in principle to fractional distillation. In microscale work, it is sometimes used to separate and isolate compounds from a mixture; fractional distillation would normally be used with larger amounts of material.

Gas chromatography (GC) is also known as vapor-phase chromatography (VPC) and as gas–liquid partition chromatography (GLPC). All three names, as well as their indicated abbreviations, are often found in the literature of organic chemistry. In reference to the technique, the last term, GLPC, is the most strictly correct and is preferred by most authors.

22.1 THE GAS CHROMATOGRAPH

The apparatus used to carry out a gas–liquid chromatographic separation is generally called a **gas chromatograph.** A typical student-model gas chromatograph, the GOW-MAC model 69-350, is illustrated in Figure 22.1. A schematic block diagram of a basic gas chromatograph is shown in Figure 22.2. The basic elements of the apparatus are apparent. The sample is injected into the chromatograph, and it

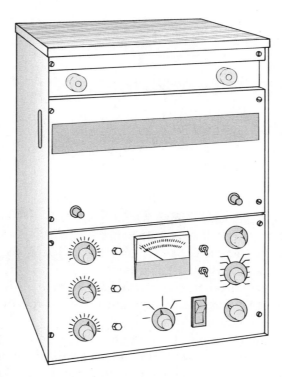

Figure 22.1 A gas chromatograph.

is immediately vaporized in a heated injection chamber and introduced into a moving stream of gas, called the **carrier gas.** The vaporized sample is then swept into a column filled with particles coated with a liquid adsorbent. The column is contained in a temperature-controlled oven. As the sample passes through the column, it is subjected to many gas–liquid partitioning processes, and the components are separated. As each component leaves the column, its presence is detected by an electrical detector that generates a signal that is recorded on a strip chart recorder.

Many modern instruments are also equipped with a microprocessor, which can be programmed to change parameters, such as the temperature of the oven, while a mixture is being separated on a column. With this capability, it is possible to optimize the separation of components and to complete a run in a relatively short time.

22.2 THE COLUMN

The heart of the gas chromatograph is the packed column. This column is usually made of copper or stainless steel tubing, but sometimes glass is used. The most common diameters of tubing are ⅛ inch (3 mm) and ¼ inch (6 mm). To construct a column, cut a piece of tubing to the desired length and attach the proper fittings on each of the two ends to connect to the apparatus. The most common length is 4–12 feet, but some columns may be up to 50 feet in length.

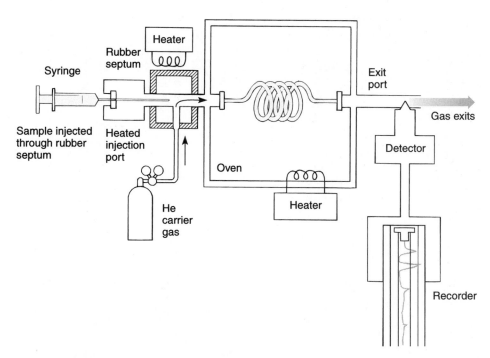

Figure 22.2 A schematic diagram of a gas chromatograph.

The tubing (column) is then packed with the **stationary phase.** The material chosen for the stationary phase is usually a liquid, a wax, or a low-melting solid. This material should be relatively nonvolatile; that is, it should have a low vapor pressure and a high boiling point. Liquids commonly used are high-boiling hydrocarbons, silicone oils, waxes, and polymeric esters, ethers, and amides. Some typical substances are listed in Table 22.1.

The liquid phase is usually coated onto a **support material.** A common support material is crushed firebrick. Many methods exist for coating the high-boiling liquid phase onto the support particles. The easiest is to dissolve the liquid (or low-melting wax or solid) in a volatile solvent like methylene chloride (bp 40°C). The firebrick (or other support) is added to this solution, which is then slowly evaporated (rotary evaporator) so as to leave each particle of support material evenly coated. Other support materials are listed in Table 22.2.

In the final step, the liquid-phase-coated support material is packed into the tubing as evenly as possible. The tubing is bent or coiled so that it fits into the oven of the gas chromatograph with its two ends connected to the gas entrance and exit ports.

Selection of a liquid phase usually revolves about two factors. First, most of them have an upper temperature limit above which they cannot be used. Above the specified limit of temperature, the liquid phase itself will begin to "bleed" off the column. Second, the materials to be separated must be considered. For polar samples, it is

Table 22.1 Typical Liquid Phases

	Type	Composition	Maximum Temperature (°C)	Typical Use
Apiezons (L, M, N, etc.)	Hydrocarbon greases (varying MW)	Hydrocarbon mixtures	250–300	Hydrocarbons
SE-30	Methyl silicone rubber	Like silicone oil, but cross-linked	350	General applications
DC-200	Silicone oil (R = CH$_3$)	$R_3Si-O-\left[\overset{R}{\underset{R}{Si}}-O\right]_n-SiR_3$	225	Aldehydes, ketones, halocarbons
DC-710	Silicone oil (R = CH$_3$) (R′ = C$_6$H$_5$)	$\left[\overset{R'}{\underset{R}{Si}}-O\right]_n$	300	General applications
Carbowaxes (400–20M)	Polyethylene glycols (varying chain lengths)	Polyether $HO-(CH_2CH_2-O)_m-CH_2CH_2OH$	Up to 250	Alcohols, ethers, halocarbons
DEGS	Diethylene glycol succinate	Polyester $\left[CH_2CH_2-O-\overset{O}{\underset{}{C}}-(CH_2)_2-\overset{O}{\underset{}{C}}-O\right]_n$	200	General applications

Increasing polarity →

Table 22.2 Typical Solid Supports

Crushed firebrick	Chromosorb T
Nylon beads	(Teflon beads)
Glass beads	Chromosorb P
Silica	(Pink diatomaceous earth,
Alumina	highly absorptive, pH 6–7)
Charcoal	Chromosorb W
Molecular sieves	(White diatomaceous earth,
	medium absorptivity, pH 8–10)
	Chromosorb G
	(like the above,
	low absorptivity, pH 8.5)

usually best to use a polar liquid phase; for nonpolar samples, a nonpolar liquid phase is indicated. The liquid phase performs best when the substances to be separated *dissolve* in it.

Most researchers today buy packed columns from commercial sources, rather than pack their own. A wide variety of types and lengths are available.

Alternatives to packed columns are Golay or glass capillary columns of diameters 0.1–0.2 mm. With these columns, no solid support is required, and the liquid is coated directly on the inner walls of the tubing. Liquid phases commonly used in glass capillary columns are similar in composition to those used in packed columns. They include DB-1 (similar to SE-30), DB-17 (similar to DC-710), and DB-WAX (similar to Carbowax 20M). The length of a capillary column is usually very long, typically 50–100 feet. Because of the length and small diameter, there is increased interaction between the sample and the stationary phase. Gas chromatographs equipped with these small-diameter columns are able to separate components more effectively than instruments using larger packed columns.

22.3 PRINCIPLES OF SEPARATION

After a column is selected, packed, and installed, the **carrier gas** (usually helium, argon, or nitrogen) is allowed to flow through the column supporting the liquid phase. The mixture of compounds to be separated is introduced into the carrier gas stream, where its components are equilibrated (or partitioned) between the moving gas phase and the stationary liquid phase (Figure 22.3). The latter is held stationary because it is adsorbed onto the surfaces of the support material.

The sample is introduced into the gas chromatograph by a microliter syringe. It is injected as a liquid or as a solution through a rubber septum into a heated chamber, called the **injection port,** where it is vaporized and mixed with the carrier gas. As this mixture reaches the column, which is heated in a controlled oven, it begins to equilibrate between the liquid and gas phases. The length of time required for a

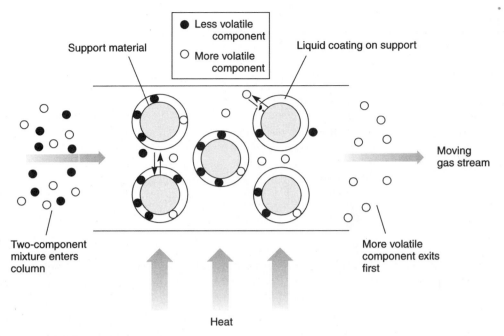

Figure 22.3 The separation process.

sample to move through the column is a function of how much time the sample spends in the vapor phase and how much time it spends in the liquid phase. The more time the sample spends in the vapor phase, the faster it gets to the end of the column. In most separations, the components of a sample have similar solubilities in the liquid phase. Therefore, the time the different compounds spend in the vapor phase is primarily a function of the vapor pressure of the compounds, and the more volatile component arrives at the end of the column first, as illustrated in Figure 22.3. By selecting the correct temperature of the oven and the correct liquid phase, the compounds in the injected mixture travel through the column at different rates and are separated.

22.4 FACTORS AFFECTING SEPARATION

Several factors determine the rate at which a given compound travels through a gas chromatograph. First, compounds with low boiling points will generally travel through the gas chromatograph faster than compounds with higher boiling points. This is because the column is heated, and low-boiling compounds always have higher vapor pressures than higher-boiling compounds. In general, therefore, for compounds with the same functional group, the higher the molecular weight the longer the retention time. For most molecules, the boiling point increases as the molecular weight increases. If the column is heated to a temperature that is too high, however, the entire mixture to be separated is flushed through the column at the same rate as the carrier gas, and no equilibration takes place with the liquid phase. On the other hand, at too low a temperature, the mixture dissolves in the liquid phase and never re-vaporizes. Thus, it is retained on the column.

The second factor is the rate of flow of the carrier gas. The carrier gas must not move so rapidly that molecules of the sample in the vapor phase cannot equilibrate with those dissolved in the liquid phase. This may result in poor separation between components in the injected mixture. If the rate of flow is too slow, however, the bands broaden significantly, leading to poor resolution (see Section 22.8).

The third factor is the choice of liquid phase used in the column. The molecular weights, functional groups, and polarities of the component molecules in the mixture to be separated must be considered when a liquid phase is being chosen. A different type of material is generally used for hydrocarbons, for instance, than for esters. The materials to be separated should dissolve in the liquid. The useful temperature limit of the liquid phase selected must also be considered.

The fourth factor is the length of the column. Compounds that resemble one another closely, in general, require longer columns than dissimilar compounds. Many kinds of isomeric mixtures fit into the "difficult" category. The components of isomeric mixtures are so much alike that they travel through the column at very similar rates. You need a longer column, therefore, to take advantage of any differences that may exist.

22.5 ADVANTAGES OF GAS CHROMATOGRAPHY

All factors that have been mentioned must be adjusted by the chemist for any mixture to be separated. Considerable preliminary investigation is often required before a mixture can be separated successfully into its components by gas chromatography. Nevertheless, the advantages of the technique are many.

First, many mixtures can be separated by this technique when no other method is adequate. Second, as little as 1–10 µL (1 µL = 10^{-6}L) of a mixture can be separated by this technique. This advantage is particularly important when working at the microscale level. Third, when gas chromatography is coupled with an electronic recording device (see following discussion), the amount of each component present in the separated mixture can be estimated quantitatively.

The range of compounds that can be separated by gas chromatography extends from gases, such as oxygen (bp –183°C) and nitrogen (bp –196°C), to organic compounds with boiling points over 400°C. The only requirement for the compounds to be separated is that they have an appreciable vapor pressure at a temperature at which they can be separated and that they be thermally stable at this temperature.

22.6 MONITORING THE COLUMN (THE DETECTOR)

To follow the separation of the mixture injected into the gas chromatograph, it is necessary to use an electrical device called a **detector**. Two types of detectors in common use are the **thermal conductivity detector (TCD)** and the **flame-ionization detector (FID)**.

The thermal conductivity detector is simply a hot wire placed in the gas stream at the column exit. The wire is heated by constant electrical voltage. When a steady stream of carrier gas passes over this wire, the rate at which it loses heat and its electrical conductance have constant values. When the composition of the vapor stream changes, the rate of heat flow from the wire, and hence its resistance, changes. Helium, which has a thermal conductivity higher than most organic substances, is a common carrier gas. Thus, when a substance elutes in the vapor stream, the thermal conductivity of the moving gases will be lower than with helium alone. The wire then heats up, and its resistance decreases.

A typical TCD operates by difference. Two detectors are used: one exposed to the actual effluent gas and the other exposed to a reference flow of carrier gas only. To achieve this situation, a portion of the carrier gas stream is diverted before it enters the injection port. The diverted gas is routed through a reference column into which no sample has been admitted. The detectors mounted in the sample and reference columns are arranged so as to form the arms of a Wheatstone bridge circuit, as shown in Figure 22.4. As long as the carrier gas alone flows over both detectors, the circuit is in balance. However, when a sample elutes from the sample column, the bridge circuit becomes unbalanced, creating an electrical signal. This signal can be amplified and used to activate a strip chart recorder. The recorder is an instrument that plots, by means of a moving pen, the unbalanced bridge current versus time on a continuously moving roll of chart paper. This record of detector response (current) versus time is called a **chromatogram.** A typical gas chromatogram is illustrated in Figure 22.5. Deflections of the pen are called **peaks.**

When a sample is injected, some air (CO_2, H_2O, N_2, and O_2) is introduced along with the sample. The air travels through the column almost as rapidly as the carrier

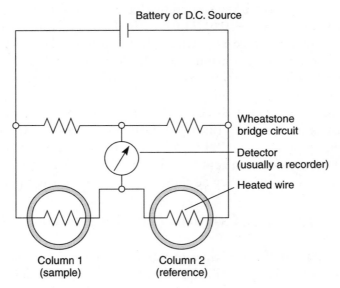

Figure 22.4 A typical thermal conductivity detector.

gas; as it passes the detector, it causes a small pen response, thereby giving a peak, called the **air peak.** At later times (t_1, t_2, t_3), the components also give rise to peaks on the chromatogram as they pass out of the column and past the detector.

In a flame-ionization detector, the effluent from the column is directed into a flame produced by the combustion of hydrogen, as illustrated in Figure 22.6. As organic compounds burn in the flame, ion fragments are produced that collect on the ring above the flame. The resulting electrical signal is amplified and sent to a recorder in a manner similar to a TCD, except that an FID does not produce an air peak. The main advantage of the FID is that it is more sensitive and can be used to analyze smaller quantities of sample. Also, because an FID does not respond to water, a gas chromatograph with this detector can be used to analyze aqueous solutions. Two disadvantages are that it is more difficult to operate and the detection process destroys the sample. Therefore, an FID gas chromatograph cannot be used to do preparative work.

22.7 RETENTION TIME

The period following injection that is required for a compound to pass through the column is called the **retention time** of that compound. For a given set of constant conditions (flow rate of carrier gas, column temperature, column length, liquid phase, injection port temperature, carrier), the retention time of any compound is always

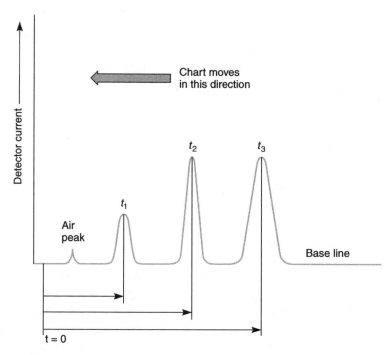

Figure 22.5 A typical gas chromatograph.

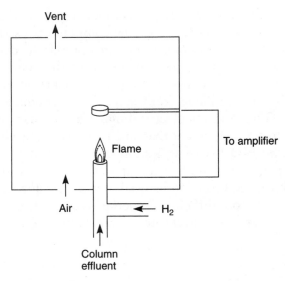

Figure 22.6 A flame-ionization detector.

constant (much like the R_f value in thin-layer chromatography, as described in Chapter 20, Section 20.9, page 321). The retention time is measured from the time of injection to the time of maximum pen deflection (detector current) for the component being observed. This value, when obtained under controlled conditions, can identify a compound by a direct comparison of it with values for known compounds determined under the same conditions. For easier measurement of retention times, most strip chart recorders are adjusted to move the paper at a rate that corresponds to time divisions calibrated on the chart paper. The retention times (t_1, t_2, t_3) are indicated in Figure 22.5 for the three peaks illustrated.

Most modern gas chromatographs are attached to a "data station," which uses a computer or a microprocessor to process the data. With these instruments, the chart often does not have divisions. Instead, the computer prints the retention time, usually to the nearest 0.01 minute, above each peak. A more complete discussion of the results obtained from a modern data station and how these data are treated may be found in Section 22.12.

22.8 POOR RESOLUTION AND TAILING

The peaks in Figure 22.5 are well **resolved.** That is, the peaks are separated from one another, and between each pair of adjacent peaks the tracing returns to the baseline. In Figure 22.7, the peaks overlap and the resolution is not good. Poor resolution is often caused by using too much sample, by a column that is too short, has too high a temperature, or has too large a diameter, by a liquid phase that does not discriminate well between the two components, or, in short, by almost any wrongly adjusted

Figure 22.7 Poor resolution or peaks overlap.

parameter. When peaks are poorly resolved, it is more difficult to determine the relative amount of each component. Methods for determining the relative percentages of each component are given in Section 22.11.

Another desirable feature illustrated by the chromatogram in Figure 22.5 is that each peak is symmetrical. A common example of an unsymmetrical peak is one in which **tailing** has occurred, as shown in Figure 22.8. Tailing usually results from injecting too much sample into the gas chromatograph. Another cause of tailing occurs with polar compounds, such as alcohols and aldehydes. These compounds may be temporarily adsorbed on column walls or areas of the support material that are not adequately coated by the liquid phase. Therefore they do not leave in a band, and tailing results.

22.9 QUALITATIVE ANALYSIS

A disadvantage of the gas chromatograph is that it gives no information about the identities of the substances it has separated. The little information it does provide is given by the retention time. It is hard to reproduce this quantity from day to day, however, and exact duplications of separations performed last month may be difficult to make this month. It is usually necessary to **calibrate** the column each time it is

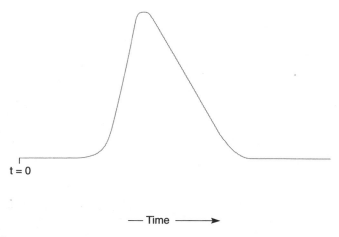

t = 0

— Time —→

Figure 22.8 Tailing.

used. That is, you must run pure samples of all known and suspected components of a mixture individually, just before chromatographing the mixture, to obtain the retention time of each known compound. As an alternative, each suspected component can be added, one by one, to the unknown mixture while the operator looks to see which peak has its intensity increased relative to the unmodified mixture. Another solution is to collect the components individually as they emerge from the gas chromatograph. Each component can then be identified by other means, such as by infrared or nuclear magnetic resonance spectroscopy or by mass spectrometry.

22.10 COLLECTING THE SAMPLE

For gas chromatographs with a thermal conductivity detector, it is possible to collect samples that have passed through the column. One method uses a gas-collection tube (see Figure 22.9), which is included in most microscale glassware kits. A collection tube is joined to the exit port of the column by inserting the ℱ 5/5 inner joint into a metal adapter, which is connected to the exit port. When a sample is eluted from the column in the vapor state, it is cooled by the connecting adapter and the gas-collection tube and condenses in the collection tube. The gas-collection tube is removed from the adapter when the recorder indicates that the desired sample has completely passed through the column. After the first sample has been collected, the process can be repeated with another gas-collection tube.

To isolate the liquid, the tapered joint of the collection tube is inserted into a 0.1-mL conical vial, which has a ℱ 5/5 outer joint. The assembly is placed into a test tube, as illustrated in Figure 22.10. During centrifugation, the sample is forced into the bottom of the conical vial. After disassembling the apparatus, the liquid can be re-

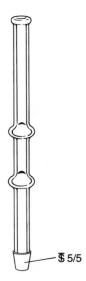

Figure 22.9 A gas chromatography collection tube.

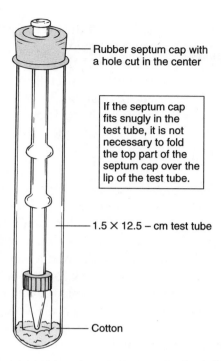

Rubber septum cap with
a hole cut in the center

If the septum cap
fits snugly in the
test tube, it is not
necessary to fold
the top part of the
septum cap over the
lip of the test tube.

1.5 X 12.5 – cm test tube

Cotton

Figure 22.10 A gas chromatography collection tube and a 0.1-mL conical vial.

moved from the vial with a syringe for a boiling-point determination or analysis by infrared spectroscopy. If a determination of the sample weight is desired, the empty
conical vial and cap should be tared and reweighed after the liquid has been collected.
It is advisable to dry the gas-collection tube and the conical vial in an oven before use,
to prevent contamination by water or other solvents used in cleaning this glassware.

Another method for collecting samples is to connect a cooled trap to the exit port of
the column. A simple trap, suitable for microscale work, is illustrated in Figure 22.11.
Suitable coolants include ice water, liquid nitrogen, or dry ice–acetone. For instance, if

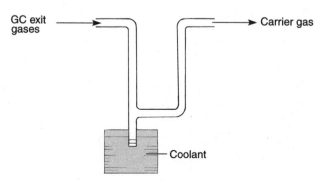

GC exit
gases

Carrier gas

Coolant

Figure 22.11 A collection trap.

the coolant is liquid nitrogen (bp –196°C) and the carrier gas is helium (bp –269°C), compounds boiling above the temperature of liquid nitrogen generally are condensed or trapped in the small tube at the bottom of the U-shaped tube. The small tube is scored with a file just below the point at which it is connected to the larger tube, the tube is broken off, and the sample is removed for analysis. To collect each component of the mixture, you must change the trap after each sample is collected.

22.11 QUANTITATIVE ANALYSIS

The area under a gas chromatograph peak is proportional to the amount (moles) of compound eluted. Hence, the molar percentage composition of a mixture can be approximated by comparing relative peak areas. This method of analysis assumes that the detector is equally sensitive to all compounds eluted and that it gives a linear response with respect to amount. Nevertheless, it gives reasonably accurate results.

The simplest method of measuring the area of a peak is by geometric approximation, or triangulation. In this method, you multiply the height h of the peak above the baseline of the chromatogram by the width of the peak at half of its height $w_{1/2}$. This is illustrated in Figure 22.12. The baseline is approximated by drawing a line between the two sidearms of the peak. This method works well only if the peak is symmetrical. If the peak has tailed or is unsymmetrical, it is best to cut out the peaks with scissors and weigh the pieces of paper on an **analytical balance.** Because the weight per area of a piece of good chart paper is reasonably constant from place to place, the ratio of the areas is the same as the ratio of the weights. To obtain a percentage composition for the mixture, first add all the peak areas (weights). Then, to calculate the percentage of any component in the mixture, divide its individual area by the total area and multiply the result by 100. A sample calculation is illustrated in Figure 22.13. If peaks overlap (see Fig. 22.7), either the gas chromatographic conditions must be readjusted to achieve better resolution of the peaks or the peak shape must be estimated.

There are various instrumental means, which are built into recorders, of detecting the amounts of each sample automatically. One method uses a separate pen that produces a trace that integrates the area under each peak. Another method employs an electronic device that automatically prints out the area under each peak and the percentage composition of the sample.

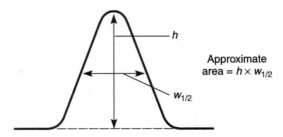

Approximate area = $h \times w_{1/2}$

Figure 22.12 Triangulation of a peak.

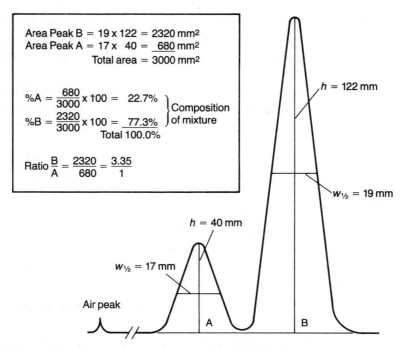

$$\text{Area Peak B} = 19 \times 122 = 2320 \text{ mm}^2$$
$$\text{Area Peak A} = 17 \times 40 = \underline{680} \text{ mm}^2$$
$$\text{Total area} = 3000 \text{ mm}^2$$

$$\%A = \frac{680}{3000} \times 100 = 22.7\%$$
$$\%B = \frac{2320}{3000} \times 100 = \underline{77.3\%}$$
$$\text{Total } 100.0\%$$

$$\left.\begin{array}{c} \\ \\ \end{array}\right\} \text{Composition of mixture}$$

$$\text{Ratio} \frac{B}{A} = \frac{2320}{680} = \frac{3.35}{1}$$

$h = 122$ mm

$w_{1/2} = 19$ mm

$h = 40$ mm

$w_{1/2} = 17$ mm

Air peak

A   B

Figure 22.13 Sample percentage composition calculation.

Most modern data stations (see Section 22.12) label the top of each peak with its retention time in minutes. When the trace is completed, the computer prints a table of all the peaks with their retention times, areas, and the percentage of the total area (sum of all the peaks) that each peak represents. Some caution should be used with these results because the computer often does not include smaller peaks, and occasionally does not resolve narrow peaks that are so close together that they overlap. If the trace has several peaks and you would like the ratio of only two of them, you will have to determine their percentages yourself using only their two areas or instruct the instrument to integrate only these two peaks.

For many applications, one assumes that the detector is equally sensitive to all compounds eluted. Compounds with different functional groups or with widely varying molecular weights, however, produce different responses with both TCD and FID gas chromatographs. With a TCD, the responses are different because not all compounds have the same thermal conductivity. Different compounds analyzed with an FID gas chromatograph also give different responses because the detector response varies with the type of ions produced. For both types of detectors, it is possible to calculate a **response factor** for each compound in a mixture. Response factors are usually determined by making up an equimolar mixture of two compounds, one of which is considered to be the reference. The mixture is separated on a gas chromatograph, and the relative percentages are calculated using one of the methods described previously. From these percentages you can determine a response factor for the compound being compared to the reference. If you do this for all the components in a mixture, you can then use these correction factors to make more accurate calculations of the relative percentages for the compounds in the mixture.

To illustrate how response factors are determined, consider the following example. An equimolar mixture of benzene, hexane, and ethyl acetate is prepared and analyzed using a flame-ionization gas chromatograph. The peak areas obtained are

Hexane	831158
Ethyl acetate	1449695
Benzene	966463

In most cases, benzene is taken as the standard, and its response factor is defined to be equal to 1.00. Calculation of the response factors for the other components of the test mixture proceeds as follows:

Hexane	$831158/966463 = 0.86$
Ethyl acetate	$1449695/966463 = 1.50$
Benzene	$966463/966463 = 1.00$ (by definition)

Notice that the response factors calculated in this example are molar response factors. It is necessary to correct these values by the relative molecular weights of each substance to obtain weight response factors.

When you use a flame-ionization gas chromatograph for quantitative analysis, it is first necessary to determine the response factors for each component of the mixture being analyzed, as just shown. For a quantitative analysis, it is likely that you will have to convert molar response factors into weight response factors. Next, the chromatography experiment using the unknown samples is performed. The observed peak areas for each component are corrected using the response factors in order to arrive at the correct weight percentage of each component in the sample. The application of response factors to correct the original results of a quantitative analysis will be illustrated in the following section.

22.12 TREATMENT OF DATA: CHROMATOGRAMS PRODUCED BY MODERN DATA STATIONS

A. Gas Chromatograms and Data Tables

Most modern gas chromatography instruments are equipped with computer-based data stations. Interfacing the instrument with a computer allows the operator to display and manipulate the results in whatever manner might be desired. The operator thus can view the output in a convenient form. Not only can the computer display the actual gas chromatogram, but it can also display the integration results. It can even display the result of two experiments simultaneously, making a comparison of parallel experiments convenient.

Figure 22.14 shows a gas chromatogram of a mixture of hexane, ethyl acetate, and benzene. The peaks corresponding to each peak can be seen; the peaks are labeled with their respective retention times:

	Retention Time (minutes)
Hexane	2.959
Ethyl acetate	3.160
Benzene	3.960

We can also see that there is a very small amount of an unspecified impurity, with a retention time of about 3.4 minutes.

Figure 22.15 shows part of the printed output that accompanies the gas chromatogram. It is this information that is used in the quantitative analysis of the mixture. According to the printout, the first peak has a retention time of 2.954 minutes (the difference between the retention times that appear as labels on the graph and those that appear in the data table are not significant). The computer has also determined the area under this peak (422373 counts). Finally, the computer has calculated the percentage of the first substance (hexane) by determining the total area of all the peaks in the chromatogram (1227054 counts) and dividing that into the area for the hexane peak. The result is displayed as 34.4217%. In a similar manner, the data table shows the retention times and peak areas for the other two peaks in the sample, along with a determination of the percentage of each substance in the mixture.

B. Application of Response Factors

If the detector responded with equal sensitivity to each of the components of the mixture, the data table shown in Figure 22.15 would contain the complete quantitative analysis of the sample. Unfortunately, as we have seen (Section 22.11), gas chromatography detectors respond more sensitively to some substances than they do to others. To correct for this discrepancy, it is necessary to apply corrections that are based on the **response factors** for each component of the mixture.

The method for determining the response factors was introduced in Section 22.11. In this section we will see how this information is applied in order to obtain a correct analysis. This example should serve to demonstrate the procedure for correcting raw gas chromatography results when response factors are known. According to the data table, the reported peak area for the first (hexane) peak is 422373 counts. The response factor for hexane was previously determined to be 0.86. The area of the hexane peak is thus corrected as follows:

$$422373/0.86 = 491000$$

Notice that the calculated result has been adjusted to reflect a reasonable number of significant figures.

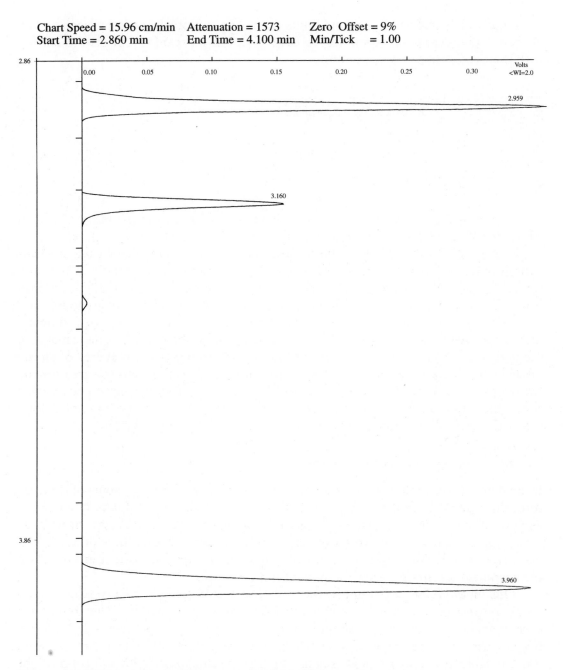

Figure 22.14 A sample gas chromatogram obtained from a data station.

```
Run Mode          : Analysis
Peak Measurement: Peak Area
Calculation Type: Percent

                           Ret.    Time                      Width
  Peak    Peak    Result   Time    Offset     Area    Sep.   1/2    Status
  No.     Name      ()     (min)   (min)    (counts)  Code  (sec)   Codes
  ----  ---------- -------  ------  ------  --------- ----- -----   ------
    1             34.4217  2.954   0.000    422373    BB    1.0
    2             16.6599  3.155   0.000    204426    BB    1.2
    3             48.9184  3.954   0.000    600255    BB    1.6
  ----  ---------- =======  ------  ======  ========= ----- -----   ------
        Totals:  100.0000                   0.000    1227054

Total Unidentified Counts :    1227054 counts

Detected Peaks: 8          Rejected Peaks: 5       Identified Peaks: 0

Multiplier: 1          Divisor: 1          Unidentified Peak Factor: 0

Baseline Offset: 1 microVolts

Noise (used): 28 microVolts - monitored before this run

Manual injection
```

****

Figure 22.15 A data table to accompany the gas chromatogram shown in Figure 22.14.

The areas for the other peaks in the gas chromatogram are corrected in a similar manner:

Hexane	422373/0.86 =	491000
Ethyl acetate	204426/1.50 =	136000
Benzene	600255/1.00 =	600000
Total peak area		1227000

Using these corrected areas, the true percentages of each component can be easily determined:

		Composition
Hexane	491000/1227000	40.0%
Ethyl acetate	136000/1227000	11.1%
Benzene	600000/1227000	48.9%
Total		100.0%

C. Determination of Relative Percentages of Components in a Complex Mixture

In some circumstances, one may wish to determine the relative percentages of two components when the mixture being analyzed may be more complex and may contain more than two components. Examples of this situation might include the analysis of a reaction product where the laboratory worker might be interested in the relative percentages of two isomeric products when the sample might also contain peaks arising from the solvent, unreacted starting material, or some other product or impurity.

The example provided in Figures 22.14 and 22.15 can be used to illustrate the method of determining the relative percentages of some, but not all, of the components in the sample. Assume we are interested in the relative percentages of hexane and ethyl acetate in the sample, but not in the percentage of benzene, which may be a solvent or an impurity. We know from the previous discussion that the *corrected* relative areas of the two peaks of interest are as follows:

	Relative Area
Hexane	491000
Ethyl acetate	136000
Total	627000

We can determine the relative percentages of the two components simply by dividing the area of each peak by the total area of the two peaks:

		Percentage
Hexane	491000/627000	78.3%
Ethyl acetate	136000/627000	21.7%
Total		100.0%

22.13 GAS CHROMATOGRAPHY–MASS SPECTROMETRY (GC–MS)

A recently developed variation on gas chromatography is **gas chromatography–mass spectrometry,** also known as **GC–MS.** In this technique a gas chromatograph is coupled to a mass spectrometer (see Chapter 28). In effect, the mass spectrometer acts as a detector. The gas stream emerging from the gas chromatograph is admitted through a valve into a tube, where it passes over the sample inlet system of the mass spectrometer. Some of the gas stream is thus admitted into the ionization chamber of the mass spectrometer.

The molecules in the gas stream are converted into ions in the ionization chamber, and thus the gas chromatogram is actually a plot of time versus **ion current,** a measure of the number of ions produced. At the same time that the molecules are converted into ions, they are also accelerated and passed through the **mass analyzer** of the instrument. The instrument, therefore, determines the mass spectrum of each fraction eluting from the gas chromatography column.

A drawback of this method involves the need for rapid scanning by the mass spectrometer. The instrument must determine the mass spectrum of each component in the mixture before the next component exits from the column, so that the spectrum of one substance is not contaminated by the spectrum of the next fraction.

Because high-efficiency capillary columns are used in the gas chromatograph, in most cases compounds are completely separated before the gas stream is analyzed. The typical GC–MS instrument has the capability of obtaining at least one scan per second in the range of 10 to 300 amu. Even more scans are possible if a narrow range of masses is analyzed. Using capillary columns, however, requires the user to take particular care to ensure that the sample does not contain any particles that might obstruct the flow of gases through the column. For this reason, the sample is carefully filtered through a very fine filter before the sample is injected into the chromatograph.

With a GC–MS system, a mixture can be analyzed and results obtained that resemble very closely those shown in Figures 22.14 and 22.15. A library search on each component of the mixture can also be conducted. The data stations of most instruments contain a library of standard mass spectra in their computer memory. If the components are known compounds, they can be identified tentatively by a comparison of their mass spectrum with the spectra of compounds found in the computer library. In this way, a "hit list" can be generated that reports on the probability that the compound in the library matches the known substance. A typical printout from a GC–MS instrument will list probable compounds that fit the mass spectrum of the component, the names of the compounds, their CAS Registry Numbers (see Chapter 30, Section 30.11, page 516), and a "quality" or "confidence" number. This last number provides an estimate of how closely the mass spectrum of the component matches the mass spectrum of the substance in the computer library.

A variation on the GC–MS technique includes coupling an infrared spectrometer (FT-IR) to a gas chromatograph. The substances that elute from the gas chromatograph are detected by determining their infrared spectra, rather than their mass spectra. A new technique that also resembles GC–MS is **high-performance liquid chromatography–mass spectrometry (HPLC–MS).** An HPLC instrument is coupled through a special interface to a mass spectrometer. The substances that elute from the HPLC column are detected by the mass spectrometer, and their mass spectra can be displayed, analyzed, and compared with standard spectra found in the computer library built into the instrument.

PROBLEMS

1. **(a)** A sample consisting of 1-bromopropane and 1-chloropropane is injected into a gas chromatograph equipped with a nonpolar column. Which compound has the shorter retention time? Explain your answer.

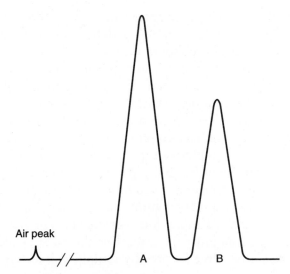

Figure 22.16 A chromatogram for Problem 2.

    (b) If the same sample were run several days later with the conditions as nearly the same as possible, would you expect the retention times to be identical to those obtained the first time? Explain.

2. Using triangulation, calculate the percentage of each component in a mixture composed of two substances, A and B. The chromatogram is shown in Figure 22.16.

3. Make a photocopy of the chromatogram in Figure 22.16. Cut out the peaks and weigh them on an analytical balance. Use the weights to calculate the percentage of each component in the mixture. Compare your answer to what you calculated in Problem 2.

4. What would happen to the retention time of a compound if the following changes were made?

    (a) Decrease the flow rate of the carrier gas

    (b) Increase the temperature of the column

    (c) Increase the length of the column

CHAPTER 23

Polarimetry

23.1 NATURE OF POLARIZED LIGHT

Light has a dual nature because it shows properties of both waves and particles. The wave nature of light can be demonstrated by two experiments: polarization and interference. Of the two, polarization is the more interesting to organic chemists, because polarization experiments can be used to learn something about the structure of an unknown molecule.

Ordinary white light consists of wave motion in which the waves have a variety of wavelengths and vibrate in all possible planes perpendicular to the direction of propagation. Light can be made to be **monochromatic** (of one wavelength or color) by using filters or special light sources. Frequently, a sodium lamp (sodium D line = 5893 Å) is used. Although the light from this lamp consists of waves of only one wavelength, the individual light waves still vibrate in all possible planes perpendicular to the beam. If we imagine that the beam of light is aimed directly at the viewer, ordinary light can be represented by showing the edges of the planes oriented randomly around the path of the beam, as in the left part of Figure 23.1.

A Nicol prism, which consists of a specially prepared crystal of Iceland spar (or calcite), has the property of serving as a screen that can restrict the passage of light waves. Waves that are vibrating in one plane are transmitted; those in all other planes are rejected (either refracted in another direction or absorbed). The light that passes through the prism is called **plane-polarized light,** and it consists of waves that vibrate in only one plane. A beam of plane-polarized light aimed directly at the viewer can be represented by showing the edges of the plane oriented in one particular direction, as in the right portion of Figure 23.1.

Figure 23.1 Ordinary versus plane-polarized light.

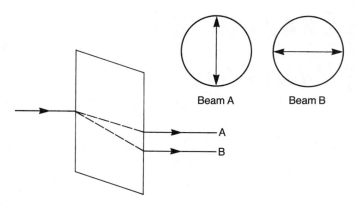

Figure 23.2 Double refraction.

Iceland spar has the property of **double refraction;** that is, it can split, or doubly refract, an entering beam of ordinary light into two separate emerging beams of light. Each of the two emerging beams (labeled A and B in Fig. 23.2) has only a single plane of vibration, and the plane of vibration in Beam A is perpendicular to the plane of Beam B. In other words, the crystal has separated the incident beam of ordinary light into two beams of plane-polarized light, with the plane of polarization of Beam A perpendicular to the plane of Beam B.

A single beam of plane-polarized light can be generated by taking advantage of the double-refracting property of Iceland spar. A Nicol prism, invented by the Scottish physicist William Nicol, consists of two crystals of Iceland spar cut to specified

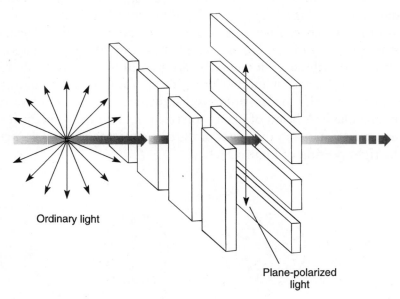

Figure 23.3 The picket-fence analogy.

angles and cemented by Canada balsam. This prism transmits one of the two beams of plane-polarized light while reflecting the other at a sharp angle so that it does not interfere with the transmitted beam. Plane-polarized light can also be generated by a Polaroid filter, a device invented by E. H. Land, an American physicist. Polaroid filters consist of certain types of crystals, embedded in transparent plastic and capable of producing plane-polarized light.

After passing through one Nicol prism, plane-polarized light can pass through a second Nicol prism, but only if the second prism has its axis oriented so that it is *parallel* to the incident light's plane of polarization. Plane-polarized light is *absorbed* by a Nicol prism that is oriented so that its axis is *perpendicular* to the incident light's plane of polarization. These situations can be illustrated by the picket-fence analogy, as shown in Figure 23.3. Plane-polarized light can pass through a fence whose slats are oriented in the proper direction but is blocked out by a fence whose slats are oriented perpendicularly.

An **optically active substance** interacts with polarized light to rotate the plane of polarization through some angle α. Figure 23.4 illustrates this phenomenon.

23.2 THE POLARIMETER

An instrument called a **polarimeter** is used to measure the extent to which a substance interacts with polarized light. A schematic diagram of a polarimeter is shown in Figure 23.5. The light from the source lamp is polarized by being passed through a fixed Nicol prism, called a **polarizer.** This light passes through the sample, with which it may or may not interact to have its plane of polarization rotated in one direction or the other. A second, rotatable Nicol prism, called the **analyzer,** is adjusted to allow the maximum amount of light to pass through. The number of degrees and the direction of rotation required for this adjustment are measured to give the **observed rotation** α.

So that data determined by several persons under different conditions can be compared, a standardized means of presenting optical rotation data is necessary. The most common way of presenting such data is by recording the **specific rotation** $[\alpha]_\lambda^t$,

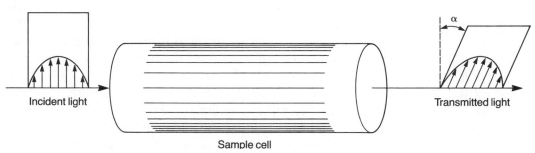

Figure 23.4 Optical activity.

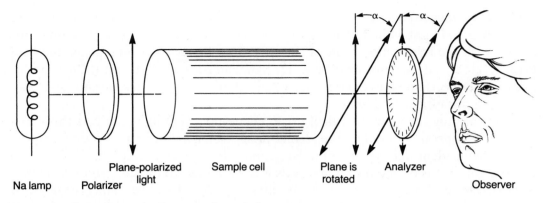

Na lamp   Polarizer   Plane-polarized light   Sample cell   Plane is rotated   Analyzer   Observer

Figure 23.5 A schematic diagram of a polarimeter.

which has been corrected for differences in concentration, cell path length, temperature, solvent, and wavelength of the light source. The equation defining the specific rotation of a compound in solution is

$$[\alpha]_\lambda^t = \frac{\alpha}{cl}$$

where α = observed rotation in degrees, c = concentration in grams per milliliter of solution, l = length of sample tube in decimeters, λ = wavelength of light (usually indicated as "D," for the sodium D line), and t = temperature in degrees Celsius. For pure liquids, the density d of the liquid in grams per milliliter replaces c in the preceding formula. You may occasionally want to compare compounds of different molecular weights, so a **molecular rotation,** based on moles instead of grams, is more convenient than a specific rotation. The molecular rotation M_λ^t is derived from the specific rotation $[\alpha]_\lambda^t$ by

$$M_\lambda^t = \frac{[\alpha]_\lambda^t \times \text{Molecular weight}}{100}$$

Usually, measurements are made at 25°C with the sodium D line as a light source; consequently, specific rotations are reported as $[\alpha]_D^{25}$.

Polarimeters that are now available incorporate electronics to determine the angle of rotation of chiral molecules. These instruments are essentially automatic. The only real difference between an automatic polarimeter and a manual one is that a light detector replaces the eye. No visual observation of any kind is made with an automatic instrument. A microprocessor adjusts the analyzer until the light reaching the detector is at a minimum. The angle of rotation is displayed digitally in an LCD window, including the sign of rotation. The simplest instrument is equipped with a sodium lamp that gives rotations based on the D line of sodium (589 nm). More expensive instruments use a tungsten lamp and filters so that wavelengths can be varied over a range of values. Using the latter instrument, a chemist can observe rotations at different wavelengths.

23.3 THE SAMPLE CELLS

It is important for the solution whose optical rotation is to be determined to contain no suspended particles of dust or dirt that might disperse the incident polarized light. Therefore, you must clean the sample cell carefully and make certain that there are no air bubbles trapped in the path of the light. The sample cells contain an enlarged ring near one end, in which the air bubbles may be trapped. Sample cells are available in various lengths, with 0.5 and 1.0 dm being the most common.

The sample is generally prepared by dissolving 0.1–0.5 g of the substance to be studied in 25 mL of solvent, usually water, ethanol, or methylene chloride (chloroform was used in the past). If the specific rotation of the substance is very high or very low, you may need to make the concentration of the solution respectively lower or higher, but usually this is determined after first trying a concentration range such as that suggested previously. The sample cell shown in Figure 23.6 is filled with solution. It is then tilted upward and tapped until the air bubbles move into the enlarged ring. It is important not to get fingerprints on the glass end plate in reassembling the cell.

23.4 OPERATION OF THE POLARIMETER

The procedures given here for preparing the cells and for operating the instrument are appropriate for the Zeiss polarimeter with the circular scale; other models of polarimeters are operated similarly. It is necessary before beginning the experiments to turn the power switch to the ON position and wait 5–10 minutes until the sodium lamp is properly warmed.

The instrument should be checked initially by making a zero reading with a sample cell filled only with solvent. If the zero reading does not correspond with the zero-degree calibration mark, then the difference in readings must be used to

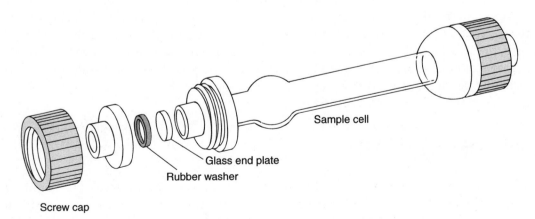

Figure 23.6 Polarimeter cell assembly.

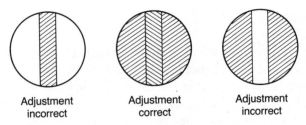

Adjustment incorrect   Adjustment correct   Adjustment incorrect

Figure 23.7 Split-field image in the polarimeter.

correct all subsequent readings. The reading is determined by laying the sample tube in the cradle, enlarged end up (making sure that there are no air bubbles in the light path), closing the cover, and turning the knob until the proper angle of the analyzer is reached. Most instruments, including the Zeiss polarimeter, are of the double-field type, in which the eye sees a split field whose sections must be matched in light intensity. The value of the angle through which the plane of polarized light has been rotated (if any) is read directly from the scale that can be seen through the eyepiece directly below the split-field image. Figure 23.7 shows how this split field might appear.

The cell containing the solution of the sample is then placed in the polarimeter, and the observed angle of rotation is measured in the same way. Be sure to record not only the numerical value of the angle of rotation in degrees but also the direction of rotation. Rotations clockwise are due to **dextrorotatory** substances and are indicated by the sign "+." Rotations counterclockwise are due to **levorotatory** substances and are indicated by the sign "−." It is best, in making a determination, to take several readings, including readings for which the actual value was approached from both sides. In other words, where the actual reading might be +75°, first approach this reading upward from a reading near zero; on the next measurement approach this reading downward from an angle greater than +75°. Duplicating readings and approaching the observed value from both sides reduce errors. The readings are then averaged to get the observed rotation α. This rotation is then corrected by the appropriate factors, according to the formulas in Section 23.2, to provide the specific rotation. The specific rotation is always reported as a function of temperature, indicating the wavelength by "D" if a sodium lamp is used and reporting the concentration and solvent used. For example $[\alpha]_D^{20} = +43.8°$ ($c = 7.5$ g/100 mL in absolute ethanol).

23.5 OPTICAL PURITY

When you prepare a sample of an enantiomer by a resolution method, the sample is not always 100% of a single enantiomer. It frequently is contaminated by residual amounts of the opposite stereoisomer. If you know the amount of each enantiomer in a mixture, you can calculate the **optical purity.** Some chemists prefer to use the term

enantiomeric excess (ee) rather than optical purity. The two terms can be used interchangeably. The percentage enantiomeric excess or optical purity is calculated as follows:

$$\% \text{ Optical purity} = \frac{\text{moles of one enantiomer} - \text{moles of other enantiomer}}{\text{total moles of both enantiomers}} \times 100$$

$$\% \text{ Optical purity} = \% \text{ Enantiomeric excess (ee)}$$

Often, it is difficult to apply the equation shown above because you do not know the exact amount of each enantiomer present in a mixture. It is far easier to calculate the optical purity (enantiomeric excess) by using the observed specific rotation of the mixture and dividing it by the specific rotation of the pure enantiomer. Values for the pure enantiomers can sometimes be found in literature sources.

$$\% \text{ Optical purity} = \% \text{ Enantiomeric excess} = \frac{\text{Observed specific rotation}}{\text{Specific rotation of pure enantiomer}} \times 100$$

This latter equation holds true only for mixtures of two chiral molecules that are mirror images of each other (enantiomers). If some other chiral substance is present in the mixture as an impurity, then the actual optical purity will deviate from the value calculated.

In a racemic (±) mixture, there is no excess enantiomer and the optical purity (enantiomeric excess) is zero; in a completely resolved material, the optical purity (enantiomeric excess) is 100%. A compound that is x% optically pure contains x% of one enantiomer and $(100 - x)$% of a racemic mixture.

Once the optical purity (enantiomeric excess) is known, the relative percentages of each of the enantiomers can be calculated easily. If the predominant form in the impure, optically active mixture is assumed to be the (+) enantiomer, the percentage of the (+) enantiomer is

$$\left[x + \left(\frac{100 - x}{2}\right)\right]\%$$

and the percentage of the (−) enantiomer is $[(100 - x)/2]\%$. The relative percentages of (+) and (−) forms in a partially resolved mixture of enantiomers can be calculated as shown next. Consider a partially resolved mixture of camphor enantiomers. The specific rotation for pure (+)-camphor is +43.8° in absolute ethanol, but the mixture shows a specific rotation of +26.3°.

$$\text{Optical purity} = \frac{+26.3°}{+43.8°} \times 100 = 60\% \text{ optically pure}$$

$$\% \text{ (+) enantiomer} = 60 + \left(\frac{100 - 60}{2}\right) = 80\%$$

$$\% \text{ (−) enantiomer} = \left(\frac{100 - 60}{2}\right) = 20\%$$

Notice that the difference between these two calculated values equals the optical purity or enantiomeric excess (ee).

PROBLEMS

1. Calculate the specific rotation of a substance that is dissolved in a solvent (0.4 g/mL) and that has an observed rotation of −10° as determined with a 0.5-dm cell.

2. Calculate the observed rotation for a solution of a substance (2.0 g/mL) that is 80% optically pure. A 2-dm cell is used. The specific rotation for the optically pure substance is +20°.

3. What is the optical purity of a partially racemized product if the calculated specific rotation is −8° and the pure enantiomer has a specific rotation of −10°? Calculate the percentage of each of the enantiomers in the partially racemized product.

CHAPTER 24
Refractometry

The **refractive index** is a useful physical property of liquids. A liquid can often be identified from a measurement of its refractive index. The refractive index can also provide a measure of the purity of the sample being examined. This is accomplished by comparing the experimentally measured refractive index with the value reported in the literature for an ultrapure sample of the compound. The closer the measured sample's value to the literature value, the purer the sample.

24.1 THE REFRACTIVE INDEX

The refractive index has as its basis the fact that light travels at a different velocity in condensed phases (liquids, solids) than in air. The refractive index n is defined as the ratio of the velocity of light in air to the velocity of light in the medium being measured:

$$n = \frac{V_{air}}{V_{liquid}} = \frac{\sin \theta}{\sin \phi}$$

It is not difficult to measure the ratio of the velocities experimentally. It corresponds to ($\sin \theta / \sin \phi$), where θ is the angle of incidence for a beam of light striking the surface of the medium and ϕ is the angle of refraction of the beam of light *within* the medium. This is illustrated in Figure 24.1.

The refractive index for a given medium depends on two variable factors. First, it is *temperature* dependent. The density of the medium changes with temperature; hence, the speed of light in the medium also changes. Second, the refractive index is *wavelength* dependent. Beams of light with different wavelengths are refracted to different extents in the same medium and give different refractive indices for that medium. It is usual to report refractive indices measured at 20°C, with a sodium discharge lamp as the source of illumination. The sodium lamp gives off yellow light of 589-nm wavelength, the so-called sodium D line. Under these conditions, the refractive index is reported in the following form:

$$n_D^{20} = 1.4892$$

The superscript indicates the temperature, and the subscript indicates that the sodium D line was used for the measurement. If another wavelength is used for the determination, the D is replaced by the appropriate value, usually in nanometers (1 nm = 10^{-9} m).

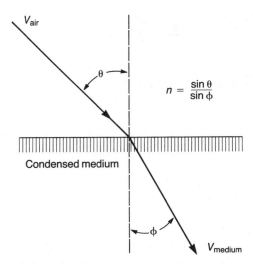

Figure 24.1 The refractive index.

Notice that the hypothetical value reported above has four decimal places. It is easy to determine the refractive index to within several parts in 10,000. Therefore, n_D is a very accurate physical constant for a given substance and can be used for identification. However, it is sensitive to even small amounts of impurity in the substance measured. Unless the substance is purified *extensively*, you will not usually be able to reproduce the last two decimal places given in a handbook or other literature source. Typical organic liquids have refractive index values between 1.3400 and 1.5600.

24.2 THE ABBÉ REFRACTOMETER

The instrument used to measure the refractive index is called a **refractometer.** Although many styles of refractometer are available, by far the most common instrument is the Abbé refractometer. This style of refractometer has the following advantages:

1. White light may be used for illumination; the instrument is compensated, however, so that the index of refraction obtained is actually that for the sodium D line.
2. The prisms can be temperature controlled.
3. Only a small sample is required (a few drops of liquid using the standard method, or about 5 µL using a modified technique).

A common type of Abbé refractometer is shown in Figure 24.2.

The optical arrangement of the refractometer is very complex; a simplified diagram of the internal workings is given in Figure 24.3. The letters *A, B, C,* and *D* label corresponding parts in both Figures 24.2 and 24.3. A complete description of refractometer optics is too difficult to attempt here, but Figure 24.3 gives a simplified diagram of the essential operating principles.

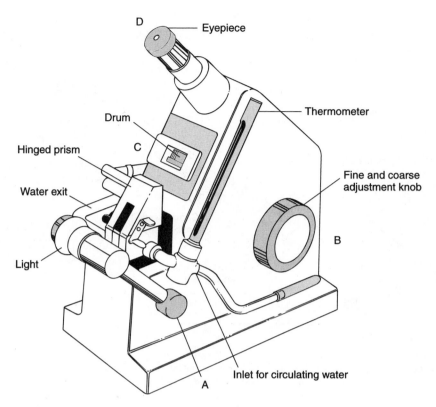

Figure 24.2 An Abbé refractometer (Bausch and Lomb Abbé 3L).

Using the standard method, the sample to be measured is introduced between the two prisms. If it is a free-flowing liquid, it may be introduced into a channel along the side of the prisms, injected from a Pasteur pipet. If it is a viscous sample, the prisms must be opened (they are hinged) by lifting the upper one; a few drops of liquid are applied to the lower prism with a Pasteur pipet or a wooden applicator. If a Pasteur pipet is used, take care not to touch the prisms, because they become scratched easily. When the prisms are closed, the liquid should spread evenly to make a thin film. With highly volatile samples, the remaining operations must be performed rapidly. Even when the prisms are closed, evaporation of volatile liquids can readily occur.

Next, you turn on the light and look into the eyepiece *D*. The hinged lamp is adjusted to give the maximum illumination to the visible field in the eyepiece. The light rotates at pivot *A*.

Rotate the coarse and fine adjustment knobs at *B* until the dividing line between the light and dark halves of the visual field coincide with the center of the crosshairs (Fig. 24.4). If the crosshairs are not in sharp focus, adjust the eyepiece to focus them. If the horizontal line dividing the light and dark areas appears as a colored

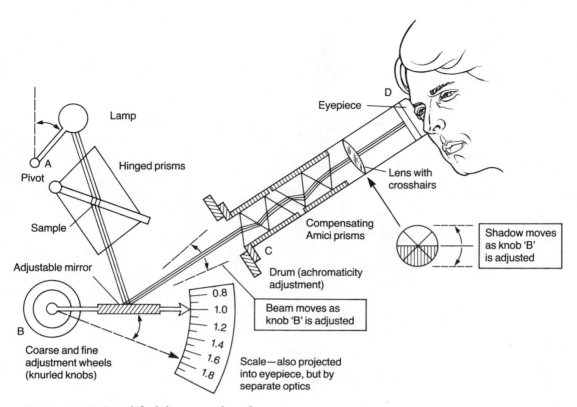

Figure 24.3 Simplified diagram of a refractometer.

band, as in Figure 24.5, the refractometer shows **chromatic aberration** (color dispersion). This can be adjusted with the knob labeled C (drum). This knurled knob rotates a series of prisms, called Amici prisms, that color compensate the refractometer and cancel out dispersion. Adjust the knob to give a sharp, uncolored division between the light and dark segments. When you have adjusted everything correctly (as in Fig. 24.4B), read the refractive index. In the instrument described here, press a small button on the left side of the housing to make the scale visible in the eyepiece. In other refractometers, the scale is visible at all times, frequently through a separate eyepiece.

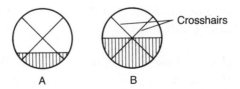

Figure 24.4 (A) Refractometer incorrectly adjusted. **(B)** Correct adjustment.

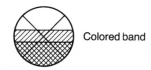

Colored band

Figure 24.5 A refractometer showing chromatic aberration (color dispersion). The dispersion is incorrectly adjusted.

Occasionally, the refractometer will be so far out of adjustment that it may be difficult to measure the refractive index of an unknown sample. When this happens, it is wise to place a pure sample of known refractive index in the instrument, set the scale to the correct value of refractive index, and adjust the controls for the sharpest line possible. Once this is done, it is easier to measure an unknown sample. It is especially helpful to perform this procedure prior to measuring the refractive index of a highly volatile sample.

> **NOTE:** There are many styles of refractometer, but most have adjustments similar to those described here.

In the procedure described above, several drops of liquid are required to obtain the refractive index. In some experiments, you may not have enough sample to use this standard method. It is possible to modify the procedure so that a reasonably accurate refractive index can be obtained on about 5 μL of liquid. Instead of placing the sample directly onto the prism, the sample is applied to a small piece of lens paper. The lens paper can be conveniently cut with a hand-held paper punch,[1] and the paper disc (0.6 cm diameter) is placed in the center of the bottom prism of the refractometer. To avoid scratching the prism, a forceps or tweezers with plastic tips should be used to handle the disc. About 5 μL of liquid is carefully placed on the lens paper using a microliter syringe. After closing the prisms, the refractometer is adjusted as described above and the refractive index is read. With this method, the horizontal line dividing the light and dark areas may not be as sharp as it is in the absence of the lens paper. It may also be impossible to eliminate color dispersion completely. Nonetheless, the refractive index values determined by this method are usually within 10 parts in 10,000 of the values determined by the standard procedure.

---

[1]To cut the lens paper more easily, place several sheets between two pieces of heavier paper, such as that used for file folders.

24.3 CLEANING THE REFRACTOMETER

In using the refractometer, always remember that if the prisms are scratched, the instrument will be ruined.

NOTE: Do not touch the prisms with any hard object.

This admonition includes Pasteur pipets and glass rods.

When measurements are completed, the prisms should be cleaned with ethanol or petroleum ether. *Soft* tissues are moistened with the solvent, and the prisms are wiped *gently*. When the solvent has evaporated from the prism surfaces, the prisms should be locked together. The refractometer should be left with the prisms closed to avoid collection of dust in the space between them. The instrument should also be turned off when it is no longer in use.

24.4 TEMPERATURE CORRECTIONS

Most refractometers are designed so that circulating water at a constant temperature can maintain the prisms at 20°C. If this temperature-control system is not used, or if the water is not at 20°C, a temperature correction must be made. Although the magnitude of the temperature correction may vary from one class of compound to another, a value of 0.00045 per degree Celsius is a useful approximation for most substances. The index of refraction of a substance *decreases* with *increasing* temperature. Therefore, add the correction to the observed n_D value for temperatures higher than 20°C and subtract it for temperatures lower than 20°C. For example, the reported n_D value for nitrobenzene is 1.5529. One would observe a value at 25°C of 1.5506. The temperature correction would be made as follows:

$$n_D^{20} = 1.5506 + 5(0.00045) = 1.5529$$

PROBLEMS

1. A solution consisting of isobutyl bromide and isobutyl chloride is found to have a refractive index of 1.3931 at 20°C. The refractive indices at 20°C of isobutyl bromide and isobutyl chloride are 1.4368 and 1.3785, respectively. Determine the molar composition (in percent) of the mixture by assuming a linear relation between the refractive index and the molar composition of the mixture.
2. The refractive index of a compound at 16°C is found to be 1.3982. Correct this refractive index to 20°C.

CHAPTER 25

Infrared Spectroscopy

Almost any compound having covalent bonds, whether organic or inorganic, will be found to absorb frequencies of electromagnetic radiation in the infrared region of the spectrum. The infrared region of the electromagnetic spectrum lies at wavelengths longer than those associated with visible light, which includes wavelengths from approximately 400 to 800 nm (1 nm = 10^{-9} m), but at wavelengths shorter than those associated with radio waves, which have wavelengths longer than 1 cm. For chemical purposes, we are interested in the *vibrational* portion of the infrared region. This portion includes radiations with wavelengths (λ) between 2.5 and 15 μm (1 μm = 10^{-6} m). The relation of the infrared region to other regions included in the electromagnetic spectrum is illustrated in Figure 25.1.

As with other types of energy absorption, molecules are excited to a higher energy state when they absorb infrared radiation. The absorption of the infrared radiation is, like other absorption processes, a quantized process. Only selected frequencies (energies) of infrared radiation are absorbed by a molecule. The absorption of infrared radiation corresponds to energy changes on the order of 8–40 kJ/mole (2–10 kcal/mole). Radiation in this energy range corresponds to the range encompassing the stretching and bending vibrational frequencies of the bonds in most covalent molecules. In the absorption process, those frequencies of infrared

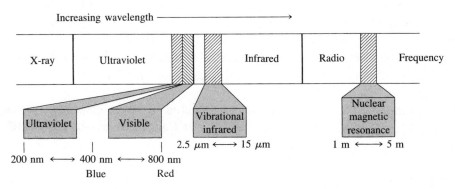

Figure 25.1 A portion of the electromagnetic spectrum, showing the relation of vibrational infrared radiation to other types of radiation.

radiation that match the natural vibrational frequencies of the molecule in question are absorbed, and the energy absorbed increases the *amplitude* of the vibrational motions of the bonds in the molecule.

Most chemists refer to the radiation in the vibrational infrared region of the electromagnetic spectrum by units called **wavenumbers** ($\overline{v}$). Wavenumbers are expressed in reciprocal centimeters (cm^{-1}) and are easily computed by taking the reciprocal of the wavelength (λ) expressed in centimeters. This unit has the advantage, for those performing calculations, of being directly proportional to energy. Thus, the vibrational infrared region of the spectrum extends from about 4000 to 650 cm^{-1} (or wavenumbers).

Wavelengths (μm) and wavenumbers (cm^{-1}) can be interconverted by the following relationships:

$$cm^{-1} = \frac{1}{(\mu m)} \times 10,000$$

$$\mu m = \frac{1}{(cm)^{-1}} \times 10,000$$

Part A. Sample Preparation and Recording the Spectrum

25.1 INTRODUCTION

To determine the infrared spectrum of the compound, one must place the compound in a sample holder or cell. In infrared spectroscopy this immediately poses a problem. Glass, quartz, and plastics absorb strongly throughout the infrared region of the spectrum (any compound with covalent bonds usually absorbs) and cannot be used to construct sample cells. Ionic substances must be used in cell construction. Metal halides (sodium chloride, potassium bromide, silver chloride) are commonly used for this purpose.

Sodium Chloride Cells.   Single crystals of sodium chloride are cut and polished to give plates that are transparent throughout the infrared region. These plates are then used to fabricate cells that can be used to hold *liquid* samples. Because sodium chloride is water soluble, samples must be *dry* before a spectrum can be obtained. In general, sodium chloride plates are preferred for most applications involving liquid samples. Potassium bromide plates may also be used in place of sodium chloride.

Silver Chloride Cells.   Cells may be constructed of silver chloride. These plates may be used for *liquid* samples that contain small amounts of water, because silver chloride is water insoluble. However, because water absorbs in the infrared region, as much water as possible should be removed, even when using silver chloride. Silver chloride plates must be stored in the dark. They darken when exposed to light, and they cannot be used with compounds that have an amino functional group. Amines react with silver chloride.

Solid Samples.   The easiest way to hold a *solid* sample in place is to dissolve the sample in a volatile organic solvent, place several drops of this solution on a salt plate, and allow the solvent to evaporate. This dry film method can be used only with modern FT-IR spectrometers. The other methods described here can be used with both FT-IR and dispersion spectrometers. A solid sample can also be held in place by making a potassium bromide pellet that contains a small amount of dispersed compound. A solid sample may also be suspended in mineral oil, which absorbs only in specific regions of the infrared spectrum. Another method is to dissolve the solid compound in an appropriate solvent and place the solution between two sodium chloride or silver chloride plates.

25.2 LIQUID SAMPLES—NaCl PLATES

The simplest method of preparing the sample, if it is a liquid, is to place a thin layer of the liquid between two sodium chloride plates that have been ground flat and polished. This is the method of choice when you need to determine the infrared spectrum of a pure liquid. A spectrum determined by this method is referred to as a **neat** spectrum. No solvent is used. The polished plates are expensive because they are cut from a large, single crystal of sodium chloride. Salt plates break easily, and they are water soluble.

Preparing the Sample.   Obtain two sodium chloride plates and a holder from the desiccator where they are stored. Moisture from fingers will mar and occlude the polished surfaces. Samples that contain water will destroy the plates.

> **NOTE:** The plates should be touched only on their edges. Be certain to use a sample that is dry or free from water.

Add one or two drops[1] of the liquid to the surface of one plate, then place the second plate on top. The pressure of this second plate causes the liquid to spread out and form a thin, capillary film between the two plates. As shown in Figure 25.2, set the plates between the bolts in a holder and place the metal ring carefully on the salt plates. Use the hex nuts to hold the salt plates in place.

> **NOTE:** Do not overtighten the nuts or the salt plates will cleave or split.

---

[1]Use a Pasteur pipet or a short length of microcapillary tubing. If you use the microcapillary tubing, it can be filled by touching it into the liquid sample. When you touch it (lightly) to the salt plate it will empty. Be careful not to scratch the plate.

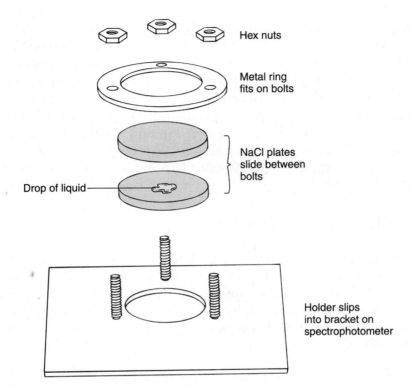

Figure 25.2 Salt plates and holder.

Tighten the nuts firmly, but do not use any force to turn them. Spin them with the fingers until they stop; then turn them just another fraction of a full turn, and they will be tight enough. If the nuts have been tightened carefully, you should observe a *transparent film of sample* (a uniform wetting of the surface). If a thin film has not been obtained, either loosen one or more of the hex nuts and adjust them so that a uniform film is obtained or add more sample.

The thickness of the film obtained between the two plates is a function of two factors: (1) the amount of liquid placed on the first plate (one drop, two drops, etc.) and (2) the pressure used to hold the plates together. If more than one or two drops of liquid has been used, it will probably be too much, and the resulting spectrum will show strong absorptions that are off the scale of the chart paper. Only enough liquid to wet both surfaces is needed.

If the sample has a very low viscosity, the capillary film may be too thin to produce a good spectrum. Another problem you may find is that the liquid is so volatile that the sample evaporates before the spectrum can be determined. In these cases, you may need to use the silver chloride plates discussed in Section 25.3, or a solution cell described in Section 25.6. Often you can obtain a reasonable spectrum by assembling the cell quickly and running the spectrum before the sample runs out of the salt plates or evaporates.

Determining the Infrared Spectrum.   Slide the holder into the slot in the sample beam of the spectrophotometer. Determine the spectrum according to the instructions provided by your instructor. In some cases, your instructor may ask you to calibrate your spectrum. If this is the case, refer to Section 25.8.

Cleaning and Storing the Salt Plates.   Once the spectrum has been determined, demount the holder and rinse the salt plates with methylene chloride (or *dry* acetone). (Keep the plates away from water!) Use a soft tissue, moistened with the solvent, to wipe the plates. If some of your compound remains on the plates, you may observe a shiny surface. Continue to clean the plates with solvent until no more compound remains on the surfaces of the plates.

 CAUTION   Avoid direct contact with methylene chloride. Return the salt plates and holder to the desiccator for storage.

25.3 LIQUID SAMPLES—AgCl PLATES

The minicell[2] shown in Figure 25.3 may also be used with liquids. The cell assembly consists of a two-piece threaded body, an O-ring, and two silver chloride plates. The plates are flat on one side, and there is a circular depression (0.025 mm or 0.10 mm deep) on the other side of the plate. An advantage of using silver chloride plates is that they may be used with wet samples or solutions. A disadvantage is that silver chloride darkens when exposed to light for extended periods. Silver chloride plates also scratch more easily than salt plates and react with amines.

Preparing the Sample.   Silver chloride plates should be handled in the same way as salt plates. Unfortunately, they are smaller and thinner (about like a contact lens) than salt plates, and care must be taken not to lose them! Remove them from the light-tight container with care. It is difficult to tell which side of the plate has the slight circular depression. Your instructor may have etched a letter on each plate to indicate which side is the flat one. To determine the infrared spectrum of a pure liquid (neat spectrum), select the flat side of each silver chloride plate. Insert the O-ring into the cell body as shown in Figure 25.3, place the plate into the cell body with the flat surface up, and add one drop or less of liquid to the plate.

NOTE: Do not use amines with AgCl plates.

---

[2]The Wilks Mini-Cell liquid sample holder is available from the Foxboro Company, 151 Woodward Avenue, South Norwalk, CT 06856. We recommend the AgCl cell windows with 0.10-mm depression, rather than the 0.025-mm depression.

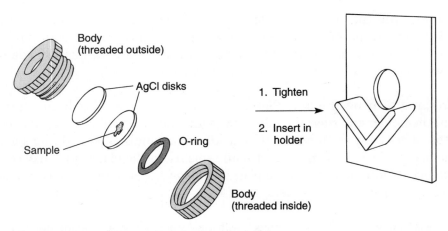

Figure 25.3 AgCl mini liquid cell and V-mount holder.

Place the second plate on top of the first with the flat side down. The orientation of the silver chloride plates is shown in Figure 25.4A. This arrangement is used to obtain a capillary film of your sample. Screw the top of the minicell into the body of the cell so that the silver chloride plates are held firmly together. A tight seal forms because AgCl deforms under pressure.

Other combinations may be used with these plates. For example, you may vary the sample path length by using the orientations shown in Figure 25.4B and C. If you add your sample and the 0.10-mm depression of one plate and cover it with the flat side of the other one, you obtain a path length of 0.10 mm (Fig. 25.4B). This arrangement is useful for analyzing volatile or low-viscosity liquids. Placement of the two plates with their depressions toward each other gives a path length of 0.20 mm (Fig. 25.4C). This orientation may be used for a solution of a solid (or liquid) in carbon tetrachloride (Section 25.6B).

Determining the Spectrum. Slide the V-mount holder shown in Figure 25.3 into the slot on the infrared spectophotometer. Set the cell assembly in the V-mount holder, and determine the infrared spectrum of the liquid.

Cleaning and Storing the AgCl Plates. Once the spectrum has been determined, the cell assembly holder should be demounted and the AgCl plates rinsed with methylene chloride or acetone. Do not use tissue to wipe the plates, as they scratch easily. AgCl plates are light sensitive. Store the plates in a light-tight container.

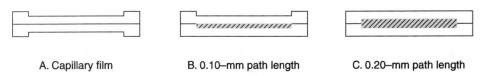

A. Capillary film   B. 0.10–mm path length   C. 0.20–mm path length

Figure 25.4 Path length variations for AgCl plates.

25.4 SOLID SAMPLES—DRY FILM

A simple method for determining the infrared spectrum of a solid sample is the **dry film** method. This method is easier than the other methods described here, it does not require any specialized equipment, and the spectra are excellent.[3] The disadvantage is that the dry film method can be used only with modern FT-IR spectrometers.

To use this method, place about 5 mg of your solid sample in a small, clean test tube. Add about five drops of methylene chloride (or diethyl ether or pentane), and stir the mixture to dissolve the solid. Using a Pasteur pipet (not a capillary tube), place several drops of the solution on the face of a salt plate. Allow the solvent to evaporate; a uniform deposit of your product will remain as a dry film coating the salt plate. Mount the salt plate on a V-shaped holder in the infrared beam. Note that only one salt plate is used; the second salt plate is not used to cover the first. Once the salt plate is positioned properly, you may determine the spectrum in the normal manner. With this method, it is *very important* that you clean your material off the salt plate. When you are finished, use methylene chloride or dry acetone to clean the salt plate.

25.5 SOLID SAMPLES—KBr PELLETS AND NUJOL MULLS

The methods described in this section can be used with both FT-IR and dispersion spectrometers.

A. KBr Pellets

One method of preparing a solid sample is to make a **potassium bromide (KBr) pellet.** When KBr is placed under pressure, it melts, flows, and seals the sample into a solid solution, or matrix. Because potassium bromide does not absorb in the infrared spectrum, a spectrum can be obtained on a sample without interference.

Preparing the Sample. Remove the agate mortar and pestle from the desiccator for use in preparing the sample. (Take care of them, they are expensive.) Grind 1 mg (0.001 g) of the solid sample for 1 minute in the agate mortar. At this point, the particle size will become so small that the surface of the solid appears shiny. Add 80 mg (0.080 g) of *powdered* potassium bromide, and grind the mixture for about 30 seconds with the pestle. Scrape the mixture into the middle with a spatula, and grind the mixture again for about 15 seconds. This grinding operation helps to mix the sample thoroughly with the KBr. You should work as rapidly as possible, because KBr absorbs water. The sample and KBr must be finely ground or the mixture will scatter the infrared radiation excessively. Using your spatula, heap the mixture in the center of the mortar. Return the bottle of potassium bromide to the desiccator where it is stored when it is not in use.

---

[3]P.L. Feist, *Journal of Chemical Education*, 78 (2001): 351.

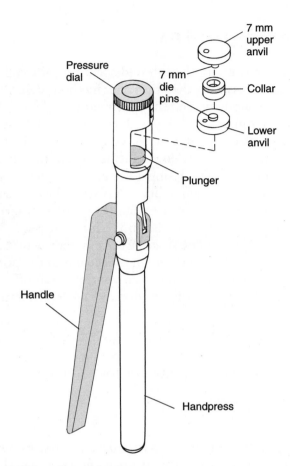

Figure 25.5 Making a KBr pellet with a hand press.

The sample and potassium bromide should be weighed on an analytical balance the first few times that a pellet is prepared. After some experience, you can estimate these quantities quite accurately by eye.

Making a Pellet Using a KBr Hand Press.   Two methods are commonly used to prepare KBr pellets. The first method uses the hand press apparatus shown in Figure 25.5.[4] Remove the die set from the storage container. Take extreme care to avoid scratching the polished surfaces of the die set. Place the anvil with the shorter die pin (lower anvil in Fig. 25.5) on a bench. Slip the collar over the pin. Remove about one-fourth of your KBr mixture with a spatula and transfer it into the collar. The powder may not cover the head of the pin completely but do not be

---

[4] KBr Quick Press unit is available from Wilmad Glass Company, Inc., Route 40 and Oak Road, Buena, NJ 08310.

concerned about this. Place the anvil with the longer die pin into the collar so that the die pin comes into contact with the sample. Never press the die set unless it contains a sample.

Lift the die set carefully by holding onto the lower anvil so that the collar stays in place. If you are careless with this operation, the collar may move enough to allow the powder to escape. Open the handle of the hand press slightly, tilt the press back a bit, and insert the die set into the press. Make sure that the die set is seated against the side wall of the chamber. Close the handle. It is imperative that the die set be seated against the side wall of the chamber so that the die is centered in the chamber. Pressing the die in an off-centered position can bend the anvil pins.

With the handle in the closed position, rotate the pressure dial so that the upper ram of the hand press just touches the upper anvil of the die assembly. Tilt the unit back so that the die set does not fall out of the hand press. Open the handle and rotate the pressure dial clockwise about one-half turn. Slowly compress the KBr mixture by closing the handle. The pressure should be no greater than that exerted by a very firm handshake. Do not apply excessive pressure or the dies may be damaged. If in doubt, rotate the pressure dial counterclockwise to lower the pressure. If the handle closes too easily, open the handle, rotate the pressure dial clockwise, and compress the sample again. Compress the sample for about 60 seconds.

After this time, tilt the unit back so that the die set does not fall out of the hand press. Open the handle, and carefully remove the die set from the unit. Turn the pressure dial counterclockwise about one full turn. Pull the die set apart and inspect the KBr pellet. Ideally, the pellet should appear clear like a piece of glass, but usually it will be translucent or somewhat opaque. There may be some cracks or holes in the pellet. The pellet will produce a good spectrum, even with imperfections, as long as light can travel through the pellet.

Making a Pellet with a KBr Minipress. The second method of preparing a pellet uses the minipress apparatus shown in Figure 25.6. Obtain a ground KBr mixture as described in "Preparing the Sample," and transfer a portion of the finely ground powder (usually not more than half) into a die that compresses it into a translucent pellet.

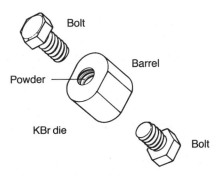

Figure 25.6 Making a KBr pellet with a minipress.

As shown in Figure 25.6, the die consists of two stainless steel bolts and a threaded barrel. The bolts have their ends ground flat. To use this die, screw one of the bolts into the barrel, but not all the way; leave one or two turns. Carefully add the powder with a spatula into the open end of the partly assembled die and tap it lightly on the bench top to give an even layer on the face of the bolt. While keeping the barrel upright, carefully screw the second bolt into the barrel until it is finger tight. Insert the head of the bottom bolt into the hexagonal hole in a plate bolted to the bench top. This plate keeps the head of one bolt from turning. The top bolt is tightened with a torque wrench to compress the KBr mixture. Continue to turn the torque wrench until you hear a loud click (the ratchet mechanism makes softer clicks) or until you reach the appropriate torque value (20 ft-lb). If you tighten the bolt beyond this point, you may twist the head off one of the bolts. Leave the die under pressure for about 60 seconds; then reverse the ratchet on the torque wrench or pull the torque wrench in the opposite direction to open the assembly. When the two bolts are loose, hold the barrel horizontally and carefully remove the two bolts. You should observe a clear or translucent KBr pellet in the center of the barrel. Even if the pellet is not totally transparent, you should be able to obtain a satisfactory spectrum as long as light passes through the pellet.

Determining the Infrared Spectrum. To obtain the spectrum, slide the holder appropriate for the type of die that you are using into the slot on the infrared spectrophotometer. Set the die containing the pellet in the holder so that the sample is centered in the optical path. Obtain the infrared spectrum. If you are using a double-beam instrument, you may be able to compensate (at least partially) for a marginal pellet by placing a wire screen or attenuator in the reference beam, thereby balancing the lowered transmittance of the pellet. An FT-IR instrument will automatically deal with the low intensity if you select the "autoscale" option.

Problems with an Unsatisfactory Pellet. If the pellet is unsatisfactory (too cloudy to pass light), one of several things may have been wrong:

1. The KBr mixture may not have been ground finely enough, and the particle size may be too big. The large particle size creates too much light scattering.
2. The sample may not be dry.
3. Too much sample may have been used for the amount of KBr taken.
4. The pellet may be too thick; that is, too much of the powdered mixture was put into the die.
5. The KBr may have been "wet" or have acquired moisture from the air while the mixture was being ground in the mortar.
6. The sample may have a low melting point. Low-melting solids not only are difficult to dry but also melt under pressure. You may need to dissolve the compound in a solvent and run the spectrum in solution (Section 25.6).

Cleaning and Storing the Equipment. After you have determined the spectrum, punch the pellet out of the die with a wooden applicator stick (a spatula should not be used as it may scratch the dies). Remember that the polished faces of the die set must not be scratched or they become useless. Pull a piece of Kimwipe through the die unit to remove all the sample. Also wipe any surfaces with a Kimwipe. *Do not wash the dies with water.* Check with your instructor to see if there are additional instructions for cleaning the die set. Return the dies to the storage container. Wash the mortar and pestle with water, dry them carefully with paper towels, and return them to the desiccator. Return the KBr powder to its desiccator.

B. Nujol Mulls

If an adequate KBr pellet cannot be obtained, or if the solid is insoluble in a suitable solvent, the spectrum of a solid may be determined as a **Nujol mull.** In this method, finely grind about 5 mg of the solid sample in an agate mortar with a pestle. Then add one or two drops of Nujol mineral oil (white) and grind the mixture to a very fine dispersion. The solid is not dissolved in the Nujol; it is actually a suspension. This mull is then placed between two salt plates using a rubber policeman. Mount the salt plates in the holder in the same way as for liquid samples (Section 25.2).

Nujol is a mixture of high-molecular-weight hydrocarbons. Hence it has absorptions in the C−H stretch and CH_2 and CH_3 bending regions of the spectrum (Fig. 25.7). Clearly, if Nujol is used, no information can be obtained in these portions of the spectrum. In interpreting the spectrum, you must ignore these Nujol peaks. It is important to label the spectrum immediately after it was determined, noting that it was determined as a Nujol mull. Otherwise you might forget that the C−H peaks belong to Nujol and not to the dispersed solid.

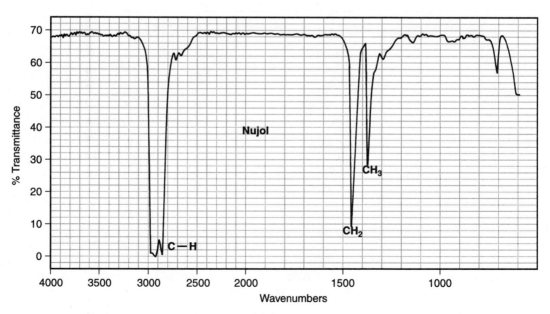

Figure 25.7 Infrared spectrum of Nujol (mineral oil).

25.6   SOLID SAMPLES—SOLUTION SPECTRA

A. Method A—Solution between Salt (NaCl) Plates

For substances that are soluble in carbon tetrachloride, a quick and easy method for determining the spectra of solids is available. Dissolve as much solid as possible in 0.1 mL of carbon tetrachloride. Place one or two drops of the solution between sodium chloride plates in precisely the same manner as used for pure liquids (Section 25.2). The spectrum is determined as described for pure liquids using salt plates (Section 25.2). You should work as quickly as possible. If there is a delay, the solvent will evaporate from between the plates before the spectrum is recorded. Because the spectrum contains the absorptions of the solute superimposed on the absorptions of carbon tetrachloride, it is important to remember that any absorption that appears near 800 cm^{-1} may be due to the stretching of the C—Cl bond of the solvent. Information contained to the right of about 900 cm^{-1} is not usable in this method. There are no other interfering bands for this solvent (see Fig. 25.8), and any other absorptions can be attributed to your sample. Chloroform solutions should not be studied by this method, because the solvent has too many interfering absorptions (see Fig 25.7).

▰▰▰ **CAUTION**   Carbon tetrachloride is a hazardous solvent. Work under the hood!

Carbon tetrachloride, besides being toxic, is suspected of being a carcinogen. In spite of the health problems associated with its use, there is no suitable alternative

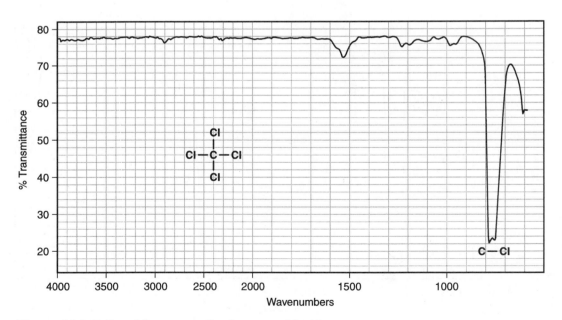

Figure 25.8 Infrared spectrum of carbon tetrachloride.

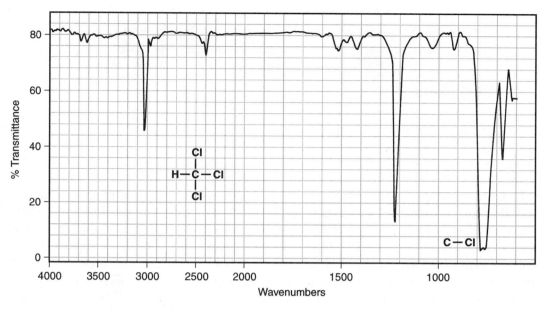

Figure 25.9 Infrared spectrum of chloroform.

solvent for infrared spectroscopy. Other solvents have too many interfering infrared absorption bands. Handle carbon tetrachloride very carefully to minimize the adverse health effects. The spectroscopic-grade carbon tetrachloride should be stored in a glass stoppered bottle in a hood. A Pasteur pipet should be attached to the bottle, possibly by storing it in a test tube taped to the side of the bottle. All sample preparation should be conducted in a hood. Rubber or plastic gloves should be worn. The cells should also be cleaned in the hood. All carbon tetrachloride used in preparing samples should be disposed of in an appropriately marked waste container.

B. Method B—AgCl Minicell

The AgCl minicell described in Section 25.3 may be used to determine the infrared spectrum of a solid dissolved in carbon tetrachloride. Prepare a 5–10% solution (5–10 mg in 0.1 mL) in carbon tetrachloride. If it is not possible to prepare a solution of this concentration because of low solubility, dissolve as much solid as possible in the solvent. Following the instructions given in Section 25.3, position the AgCl plates as shown in Figure 25.4C to obtain the maximum possible path length of 0.20 mm. When the cell is tightened firmly, the cell will not leak.

As indicated in Method A, the spectrum will contain the absorptions of the dissolved solid superimposed on the absorptions of carbon tetrachloride. A strong absorption appears near 800 cm^{-1} for C—Cl stretch in the solvent. No useful information may be obtained for the sample to the right of about 900 cm^{-1}, but other bands that appear in the spectrum will belong to your sample. Read the safety material provided in Method A. Carbon tetrachloride is toxic, and it should be used under a hood.

> **NOTE:** Care should be taken in cleaning the AgCl plates. Because AgCl plates scratch easily, they should not be wiped with tissue. Rinse them with methylene chloride, and keep them in a dark place. Amines will destroy the plates.

C. Method C—Solution Cells (NaCl)

The spectra of solids may also be determined in a type of permanent sample cell called a **solution cell**. (The infrared spectra of liquids may also be determined in this cell.) The solution cell, shown in Figure 25.10, is made from two salt plates, mounted with a Teflon spacer between them to control the thickness of the sample. The top sodium chloride plate has two holes drilled in it so that the sample can be introduced into the cavity between the two plates. These holes are extended through the face plate by two tubular extensions designed to hold Teflon plugs, which seal the internal chamber and prevent evaporation. The tubular extensions are tapered so that a syringe body (Luer lock without a needle) will fit snugly into them from the outside. The cells are thus filled from a syringe; usually, they are held upright and filled from the bottom entrance port.

These cells are very expensive, and you should try either Method A or B before using solution cells. If you do need them, obtain your instructor's permission and receive instruction before using the cells. The cells are purchased in matched pairs, with identical path lengths. Dissolve a solid in a suitable solvent, usually carbon tetrachloride, and add the solution to one of the cells (**sample cell**) as described in the previous paragraph. The pure solvent, identical to that used to dissolve the solid, is placed in the other cell (**reference cell**). The spectrum of the solvent is subtracted from the spectrum of the solution (not always completely), and a spectrum of the solute is thus provided. For the solvent compensation to be as exact as possible and to avoid contamination of the reference cell, it is essential that one cell be used as a reference and that the other cell be used as a sample cell without ever being interchanged. After the spectrum is determined, it is important to clean the cells by flushing them with clean solvent. They should be dried by passing dry air through the cell.

Solvents most often used in determining infrared spectra are carbon tetrachloride (Fig 25.8), chloroform (Fig 25.9), and carbon disulfide (Fig. 25.11). A 5–10% solution of solid in one of these solvents usually gives a good spectrum. Carbon tetrachloride and chloroform are suspected carcinogens; however, because there are no suitable alternative solvents, these compounds must be used in infrared spectroscopy. The procedure outlined on pages 380–381 for carbon tetrachloride should be followed. This procedure serves equally well for chloroform.

> **NOTE:** Before you use the solution cells, you must obtain the instructor's permission and instruction on how to fill and clean the cells.

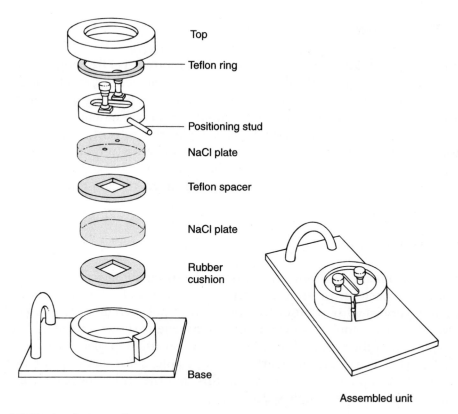

Figure 25.10 A solution cell.

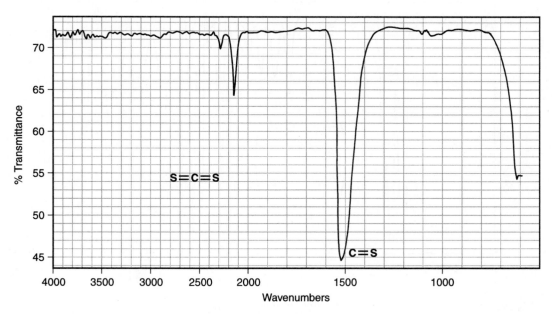

Figure 25.11 Infrared spectrum of carbon disulfide.

25.7 RECORDING THE SPECTRUM

The instructor will describe how to operate the infrared spectrophotometer, because the controls vary considerably, depending on the manufacturer, model of the instrument, and type. For example, some instruments involve pushing only a few buttons, whereas others use a more complicated computer interface system.

In all cases, it is important that the sample, the solvent, the type of cell or method used, and any other pertinent information be written on the spectrum immediately after the determination. This information may be important, and it is easily forgotten if not recorded. You may also need to calibrate the instrument (Section 25.8).

25.8 CALIBRATION

For some instruments, the frequency scale of the spectrum must be calibrated so that you know the position of each absorption peak precisely. You can recalibrate by recording a very small portion of the spectrum of polystyrene over the spectrum of your sample. The complete spectrum of polystyrene is shown in Figure 25.12. The most important of these peaks is at 1603 cm^{-1}; other useful peaks are at 2850 and 906 cm^{-1}. After you record the spectrum of your sample, substitute a thin film of polystyrene for the sample cell and record the **tips** (not the entire spectrum) of the most important peaks over the sample spectrum.

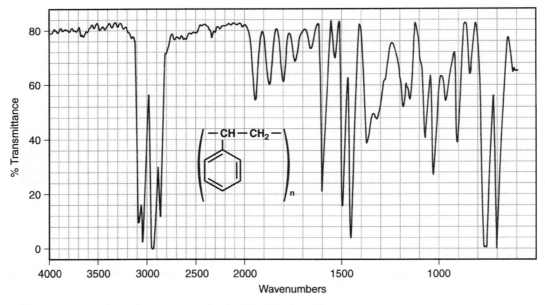

Figure 25.12 Infrared spectrum of polystyrene (thin film).

It is always a good idea to calibrate a spectrum when the instrument uses chart paper with a preprinted scale. It is difficult to align the paper properly so that the scale matches the absorption lines precisely. You often need to know the precise values for certain functional groups (for example, the carbonyl group). Calibration is essential in these cases.

With computer-interfaced instruments, the instrument does not need to be calibrated. With this type of instrument, the spectrum and scale are printed on blank paper at the same time. The instrument has an internal calibration that ensures that the positions of the absorptions are known precisely and that they are placed at the proper positions on the scale. With this type of instrument, it is often possible to print a list of the locations of the major peaks as well as to obtain the complete spectrum of your compound.

Part B.   Infrared Spectroscopy

25.9 Uses of the Infrared Spectrum

Because every type of bond has a different natural frequency of vibration, and because the same type of bond in two different compounds is in a slightly different environment, no two molecules of different structure have exactly the same infrared absorption pattern, or **infrared spectrum.** Although some of the frequencies absorbed in the two cases might be the same, in no case of two different molecules will their infrared spectra (the patterns of absorption) be identical. Thus, the infrared spectrum can be used to identify molecules much as a fingerprint can be used to identify people. Comparing the infrared spectra of two substances thought to be identical will establish whether or not they are in fact identical. If the infrared spectra of two substances coincide peak for peak (absorption for absorption), in most cases, the substances are identical.

A second and more important use of the infrared spectrum is that it gives structural information about a molecule. The absorptions of each type of bond (N—H, C—H, O—H, C—X, C=O, C—O, C—C, C=C, C≡C, C≡N, and so on) are regularly found only in certain small portions of the vibrational infrared region. A small range of absorption can be defined for each type of bond. Outside this range, absorptions will normally be due to some other type of bond. Thus, for instance, any absorption in the range 3000 ± 150 cm^{-1} will almost always be due to the presence of a CH bond in the molecule; an absorption in the range 1700 ± 100 cm^{-1} will normally be due to the presence of a C=O bond (carbonyl group) in the molecule. The same type of range applies to each type of bond. The way these are spread out over the vibrational infrared is illustrated schematically in Figure 25.13. It is a good idea to remember this general scheme for future convenience.

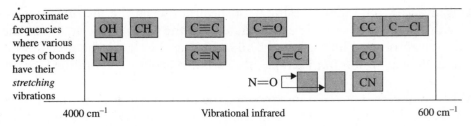

Figure 25.13 Approximate regions in which various common types of bonds absorb. (Bending, twisting, and other types of bond vibration have been omitted for clarity.)

25.10 MODES OF VIBRATION

The simplest types, or **modes,** of vibrational motion in a molecule that are **infrared active,** that is, give rise to absorptions, are the stretching and bending modes.

C—H   Stretching   Bending

Other, more complex types of stretching and bending are also active, however. To introduce several words of terminology, the normal modes of vibration for a methylene group are shown below.

In any group of three or more atoms—at least two of which are identical—there are *two* modes of stretching or bending: the symmetric mode and asymmetric

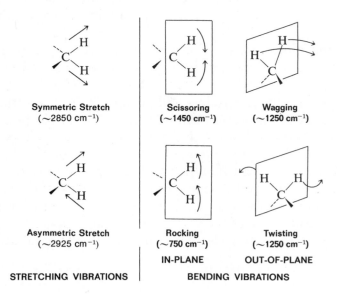

mode. Examples of such groupings are $-CH_3$, $-CH_2-$, $-NO_2$, $-NH_2$, and anhydrides $(CO)_2O$. For the anhydride, owing to asymmetric and symmetric modes of stretch, this functional group gives *two* absorptions in the C=O region. A similar phenomenon is seen for amino groups, where primary amines usually have *two* absorptions in the NH stretch region, whereas secondary amines R_2NH have only one absorption peak. Amides show similar bands. There are two strong N=O stretch peaks for a nitro group, which are caused by asymmetric and symmetric stretching modes.

25.11 WHAT TO LOOK FOR IN EXAMINING INFRARED SPECTRA

The instrument that determines the absorption spectrum for a compound is called an **infrared spectrophotometer.** The spectrophotometer determines the relative strengths and positions of all the absorptions in the infrared region and plots this information on a piece of paper. This plot of absorption intensity versus wavenumber or wavelength is referred to as the **infrared spectrum** of the compound. A typical infrared spectrum, that of methyl isopropyl ketone, is shown in Figure 25.14.

The strong absorption in the middle of the spectrum corresponds to C=O, the carbonyl group. Note that the C=O peak is quite intense. In addition to the characteristic position of absorption, the **shape** and **intensity** of this peak are also unique to the C=O bond. This is true for almost every type of absorption peak; both shape and

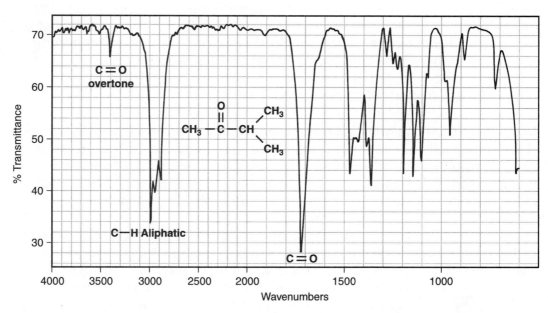

Figure 25.14 Infrared spectrum of methyl isopropyl ketone (neat liquid, salt plates).

intensity characteristics can be described, and these characteristics often make it possible to distinguish the peak in a confusing situation. For instance, to some extent both C=O and C=C bonds absorb the same region of the infrared spectrum:

$$C=O \quad 1850\text{--}1630 \text{ cm}^{-1}$$
$$C=C \quad 1680\text{--}1620 \text{ cm}^{-1}$$

However, the C=O bond is a strong absorber, whereas the C=C bond generally absorbs only weakly. Hence, a trained observer would not normally interpret a strong peak at 1670 cm^{-1} to be a carbon–carbon double bond nor a weak absorption at this frequency to be due to a carbonyl group.

The shape of a peak often gives a clue to its identity as well. Thus, although the NH and OH regions of the infrared overlap,

$$OH \quad 3650\text{--}3200 \text{ cm}^{-1}$$
$$NH \quad 3500\text{--}3300 \text{ cm}^{-1}$$

NH usually gives a **sharp** absorption peak (absorbs a very narrow range of frequencies), and OH, when it is in the NH region, usually gives a **broad** absorption peak. Primary amines give *two* absorptions in this region, whereas alcohols give only one.

Therefore, while you are studying the sample spectra in the pages that follow, you should also notice shapes and intensities. They are as important as the frequency at which an absorption occurs, and you must train your eye to recognize these features. In the literature of organic chemistry, you will often find absorptions referred to as strong (s), medium (m), weak (w), broad, or sharp. The author is trying to convey some idea of what the peak looks like without actually drawing the spectrum. Although the intensity of an absorption often provides useful information about the identity of a peak, be aware that the relative intensities of all the peaks in the spectrum are dependent on the amount of sample that is used and the sensitivity setting of the instrument. Therefore, the *actual* intensity of a particular peak may vary from spectrum to spectrum, and you must pay attention to *relative* intensities.

25.12 CORRELATION CHARTS AND TABLES

To extract structural information from infrared spectra, you must know the frequencies or wavelengths at which various functional groups absorb. Infrared **correlation tables** present as much information as is known about where the various functional groups absorb. The books listed at the end of this chapter present extensive lists of correlation tables. Sometimes the absorption information is given in a chart, called a **correlation chart.** A simplified correlation table is given in Table 25.1.

Although you may think assimilating the mass of data in Table 25.1 will be difficult, it is not if you make a modest start and then gradually increase your familiarity with the data. An ability to interpret the fine details of an infrared spectrum will follow. This is most easily accomplished by first establishing the broad visual patterns of Figure 25.13 firmly in mind. Then, as a second step, a "typical absorption value" can be memorized for each of the functional groups in this pattern. This value will be a single number that can be used as a pivot value for the memory. For instance, start with a simple aliphatic

Table 25.1 A Simplified Correlation Table

	Type of Vibration		Frequency (cm^{-1})	Intensity[a]
C—H	Alkanes	(stretch)	3000–2850	s
	—CH$_3$	(bend)	1450 and 1375	m
	—CH$_2$—	(bend)	1465	m
	Alkenes	(stretch)	3100–3000	m
		(bend)	1700–1000	s
	Aromatics	(stretch)	3150–3050	s
		(out-of-plane bend)	1000–700	s
	Alkyne	(stretch)	ca. 3300	s
	Aldehyde		2900–2800	w
			2800–2700	w
C—C	Alkane	Not interpretatively useful		
C=C	Alkene		1680–1600	m–w
	Aromatic		1600–1400	m–w
C≡C	Alkyne		2250–2100	m–w
C=O	Aldehyde		1740–1720	s
	Ketone (acyclic)		1725–1705	s
	Carboxylic acid		1725–1700	s
	Ester		1750–1730	s
	Amide		1700–1640	s
	Anhydride		ca. 1810	s
			ca. 1760	s
C—O	Alcohols, ethers, esters, carboxylic acids		1300–1000	s
O—H	Alcohol, phenols			
	Free		3650–3600	m
	H-Bonded		3400–3200	m
	Carboxylic acids		3300–2500	m
N—H	Primary and secondary amines		ca. 3500	m
C≡N	Nitriles		2260–2240	m
N=O	Nitro (R—NO$_2$)		1600–1500	s
			1400–1300	s
C—X	Fluoride		1400–1000	s
	Chloride		800–600	s
	Bromide, iodide		<600	s

[a] s, strong; m, medium; w, weak.

ketone as a model for all typical carbonyl compounds. The typical aliphatic ketone has carbonyl absorption of 1715 ± 10 cm^{-1}. Without worrying about the variation, memorize 1715 cm^{-1} as the base value for carbonyl absorption. Then, learn the extent of the carbonyl range and the visual pattern of how the different kinds of carbonyl groups are arranged throughout this region. See, for instance, Figure 25.27, which gives typical values for carbonyl compounds. Also learn how factors such as ring size (when the functional group is contained in a ring) and conjugation affect the base values (that is, in which direction the values are shifted). Learn the trends— always remembering the base value (1715 cm^{-1}). It might prove useful as a beginning to memorize the base values in Table 25.2 for this approach. Notice that there are only eight values.

Table 25.2 **Base Values for Absorptions of Bonds**

O—H	3400 cm^{-1}	C≡C	2150 cm^{-1}
N—H	3500 cm^{-1}	C=O	1715 cm^{-1}
C—H	3000 cm^{-1}	C=C	1650 cm^{-1}
C≡N	2250 cm^{-1}	C—O	1100 cm^{-1}

25.13 ANALYZING A SPECTRUM (OR WHAT YOU CAN TELL AT A GLANCE)

In analyzing the spectrum of an unknown, concentrate first on establishing the presence (or absence) of a few major functional groups. The most conspicuous peaks are C=O, O—H, N—H, C—O, C=C, C≡C, C≡N, and NO_2. If they are present, they give immediate structural information. Do not try to analyze in detail the CH absorptions near 3000 cm^{-1}; almost all compounds have these absorptions. Do not worry about subtleties of the exact type of environment in which the functional group is found. A checklist of the important gross features follows:

1. Is a carbonyl group present?
   The C=O group gives rise to a strong absorption in the region $1820–1600 \text{ cm}^{-1}$. The peak is often the strongest in the spectrum and of medium width. You can't miss it.
2. If C=O is present, check the following types. (If it is absent, go to item 3.)

Acids	Is OH also present?
	Broad absorption near $3300–2500 \text{ cm}^{-1}$ (usually overlaps C—H).
Amides	Is NH also present?
	Medium absorption near 3500 cm^{-1}, sometimes a double peak, equivalent halves.
Esters	Is C—O also present?
	Medium intensity absorptions near $1300–1000 \text{ cm}^{-1}$.
Anhydrides	Have *two* C=O absorptions near 1810 and 1760 cm^{-1}.
Aldehydes	Is aldehyde CH present?
	Two weak absorptions near 2850 and 2750 cm^{-1} on the right side of CH absorptions.
Ketones	The preceding five choices have been eliminated.

3. If C=O is absent

Alcohols or Phenols	Check for OH.
	Broad absorption near $3600–3300 \text{ cm}^{-1}$.
	Confirm this by finding C—O near $1300–1000 \text{ cm}^{-1}$.
Amines	Check for NH.
	Medium absorption(s) near 3500 cm^{-1}.
Ethers	Check for C—O (and absence of OH) near $1300–1000 \text{ cm}^{-1}$.

4. Double bonds or aromatic rings or both

C=C is a **weak** absorption near 1650 cm^{-1}.

Medium to strong absorptions in the region 1650–1450 cm^{-1} often imply an aromatic ring.

Confirm the above by consulting the CH region.

Aromatic and vinyl CH occur to the left of 3000 cm^{-1} (aliphatic CH occurs to the right of this value).

5. Triple bonds

C≡N is a medium, sharp absorption near 2250 cm^{-1}.

C≡C is a weak but sharp absorption near 2150 cm^{-1}.

Check also for acetylenic CH near 3300 cm^{-1}.

6. Nitro groups   *Two* strong absorptions near 1600–1500 cm^{-1} and 1390–1300 cm^{-1}.

7. Hydrocarbons

None of the above is found.

Main absorptions are in the CH region near 3000 cm^{-1}.

Very simple spectrum, only other absorptions are near 1450 cm^{-1} and 1375 cm^{-1}.

The beginning student should resist the idea of trying to assign or interpret *every* peak in the spectrum. You simply will not be able to do this. Concentrate first on learning the principal peaks and recognizing their presence or absence. This is best done by carefully studying the illustrative spectra in the section that follows.

> **NOTE:** In describing the shifts of absorption peaks or their relative positions, we have used the phrases "to the left" and "to the right." This was done to simplify descriptions of peak positions. The meaning is clear, because all spectra are conventionally presented left to right from 4000 to 600 cm^{-1}.

25.14 SURVEY OF THE IMPORTANT FUNCTIONAL GROUPS

A. Alkanes

The spectrum is usually simple, with a few peaks.

C–H   Stretch occurs around 3000 cm^{-1}.

    1. In alkanes (except strained ring compounds), absorption always occurs to the right of 3000 cm^{-1}.

    2. If a compound has vinylic, aromatic, acetylenic, or cyclopropyl hydrogens, the CH absorption is to the left of 3000 cm^{-1}.

CH$_2$ Methylene groups have a characteristic absorption at approximately 1450 cm^{-1}.

CH$_3$ Methyl groups have a characteristic absorption at approximately 1375 cm^{-1}.

C–C   Stretch—not interpretatively useful—has many peaks.

The spectrum of decane is shown in Figure 25.15.

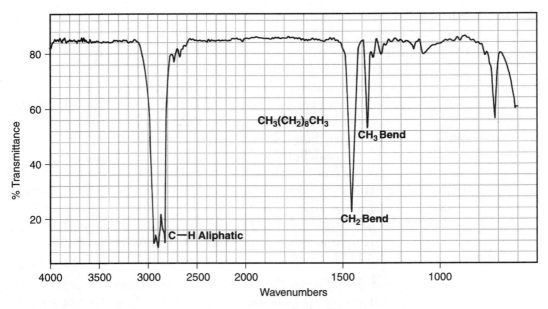

Figure 25.15 Infrared spectrum of decane (neat liquid, salt plates).

B. Alkenes

=C–H Stretch occurs to the left of 3000 cm^{-1}.

=C–H Out-of-plane ("oop") bending occurs at 1000–650 cm^{-1}.

> The C–H out-of-plane absorptions often allow you to determine the type of substitution pattern on the double bond, according to the number of absorptions and their positions. The correlation chart in Figure 25.16 shows the positions of these bands.

C=C Stretch 1675–1600 cm^{-1}, often weak.

> Conjugation moves C=C stretch to the right.
>
> Symmetrically substituted bonds, as in 2,3-dimethyl-2-butene, do not absorb in the infrared region (no dipole change). Highly substituted double bonds are often vanishingly weak in absorption.

The spectra of 4-methylcyclohexene and styrene are shown in Figures 25.17 and 25.18.

C. Aromatic Rings

=C–H Stretch is always to the left of 3000 cm^{-1}.

=C–H Out-of-plane (oop) bending occurs at 900 to 690 cm^{-1}.

> The CH out-of-plane absorptions often allow you to determine the type of ring substitution by their numbers, intensities, and positions. The correlation chart in Figure 25.19A indicates the positions of these bands.
>
> The patterns are generally reliable—they are most reliable for rings with alkyl substituents and least reliable for polar substituents.

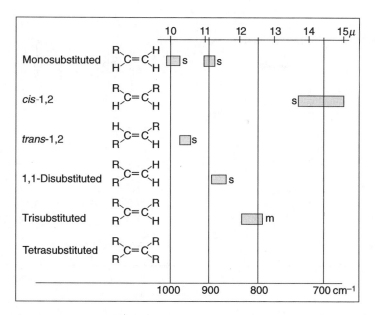

Figure 25.16 The C—H out-of-plane bending vibrations for substituted alkenes.

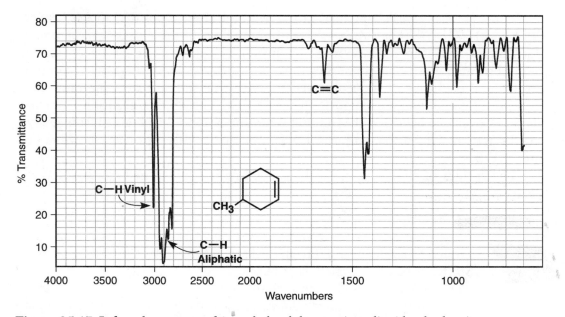

Figure 25.17 Infrared spectrum of 4-methylcyclohexene (neat liquid, salt plates).

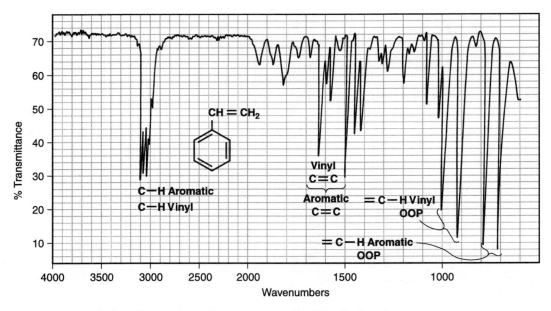

Figure 25.18 Infrared spectrum of styrene (neat liquid, salt plates).

Ring Absorptions (C=C).   There are often four sharp absorptions that occur in pairs at 1600 and 1450 cm^{-1} and are characteristic of an aromatic ring. See, for example, the spectra of anisole (Fig. 25.23), benzonitrile (Fig. 25.26), and methyl benzoate (Fig. 25.35).

There are many weak combination and overtone absorptions that appear between 2000 and 1667 cm^{-1}. The relative shapes and numbers of these peaks can be used to determine whether an aromatic ring is monosubstituted or di-, tri-, tetra-, penta-, or hexasubstituted. Positional isomers can also be distinguished. Because the absorptions are weak, these bands are best observed by using neat liquids or concentrated solutions. If the compound has a high-frequency carbonyl group, this absorption overlaps the weak overtone bands, so that no useful information can be obtained from analyzing this region. The various patterns that are obtained in this region are shown in Figure 25.19B.

The spectra of styrene and *o*-dichlorobenzene are shown in Figures 25.18 and 25.20.

D. Alkynes

≡C–H   Stretch is usually near 3300 cm^{-1}, sharp peak.
C≡C   Stretch is near 2150 cm^{-1}, sharp peak.
         Conjugation moves C≡C stretch to the right.
         Disubstituted or symmetrically substituted triple bonds give either no absorption or weak absorption.

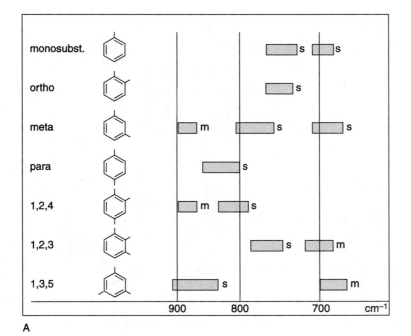

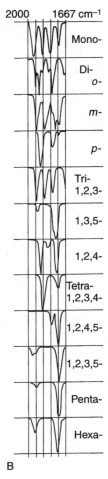

Figure 25.19 (A) The C—H out-of-plane bending vibrations for substituted benzenoid compounds. **(B)** The 2000–1667 cm⁻¹ region for substituted benzenoid compounds. (*From John R. Dyer, Applications of Absorption Spectroscopy of Organic Compounds. Englewood Cliffs, NJ: Prentice Hall, 1965.*)

E. Alcohols and Phenols

O—H Stretch is a sharp peak at 3650–3600 cm⁻¹ if no hydrogen bonding takes place. (This is usually observed only in dilute solutions.)

If there is hydrogen bonding (usual in neat or concentrated solutions), the absorption is *broad* and occurs more to the right at 3500–3200 cm⁻¹, sometimes overlapping C—H stretch absorptions.

C—O Stretch is usually in the range of 1300–1000 cm⁻¹.

Phenols are like alcohols. The 2-naphthol shown in Figure 25.21 has some molecules hydrogen bonded and some free. The spectrum of

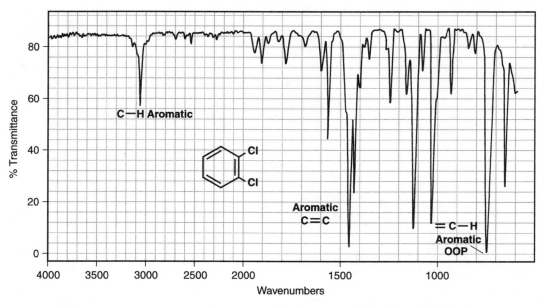

Figure 25.20 Infrared spectrum of *o*-dichlorobenzene (neat liquid, salt plates).

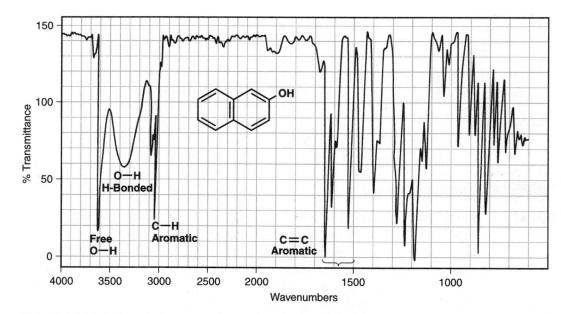

Figure 25.21 Infrared spectrum of 2-naphthol, showing both free and hydrogen-bonded OH (CHCl$_3$ solution).

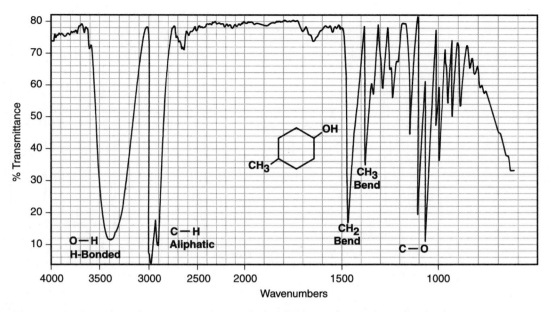

Figure 25.22 Infrared spectrum of 4-methylcyclohexanol (neat liquid, salt plates).

4-methylcyclohexanol is shown in Figure 25.22. This alcohol, which was determined neat, would also have had a free OH spike to the left of this hydrogen-bonded band if it had been determined in dilute solution.

F. Ethers

C—O   The most prominent band is due to C—O stretch at 1300–1000 cm^{-1}. Absence of C=O and O—H bands is required to be sure C—O stretch is not due to an alcohol or ester. Phenyl and vinyl ethers are found in the left portion of the range, aliphatic ethers in the right. (Conjugation with the oxygen moves the absorption to the left.)

The spectrum of anisole is shown in Figure 25.23.

G. Amines

N—H   Stretch occurs in the range of 3500–3300 cm^{-1}.
Primary amines have *two* bands typically 30 cm^{-1} apart.
Secondary amines have one band, often vanishingly weak.
Tertiary amines have no NH stretch.

C—N   Stretch is weak and occurs in the range of 1350–1000 cm^{-1}.

N—H   Scissoring bending mode occurs in the range of 1640–1560 cm^{-1} (broad).
An out-of-plane bending absorption can sometimes be observed at about 800 cm^{-1}.

The spectrum of *n*-butylamine is shown in Figure 25.24.

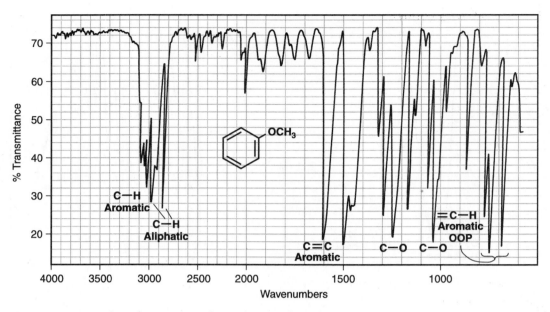

Figure 25.23 Infrared spectrum of anisole (neat liquid, salt plates).

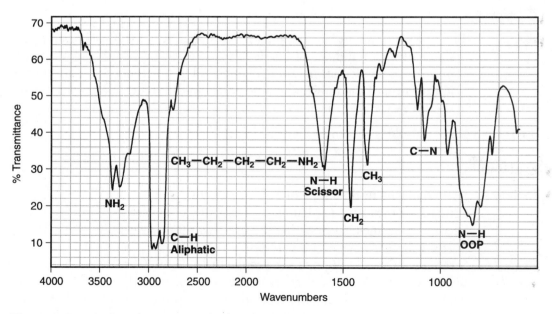

Figure 25.24 Infrared spectrum of *n*-butylamine (neat liquid, salt plates).

H. Nitro Compounds

N=O   Stretch is usually two strong bands at 1600–1500 cm^{-1} and 1390–1300 cm^{-1}.

The spectrum of nitrobenzene is shown in Figure 25.25.

I. Nitriles

C≡N   Stretch is a sharp absorption near 2250 cm^{-1}.
Conjugation with double bonds or aromatic rings moves the absorption to the right.

The spectrum of benzonitrile is shown in Figure 25.26.

J. Carbonyl Compounds

The carbonyl group is one of the most strongly absorbing groups in the infrared region of the spectrum. This is mainly due to its large dipole moment. It absorbs in a variety of compounds (aldehydes, ketones, acids, esters, amides, anhydrides, and so on) in the range of 1850–1650 cm^{-1}. In Figure 25.27 the normal values for the various types of carbonyl groups are compared. In the sections that follow, each type is examined separately.

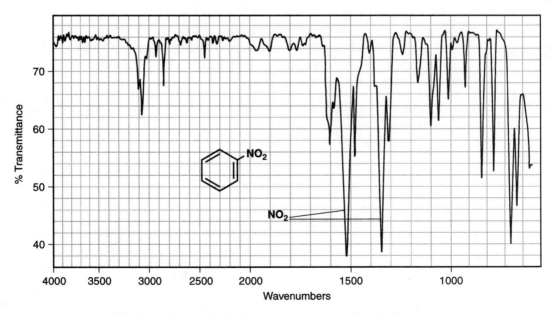

Figure 25.25 Infrared spectrum of nitrobenzene (neat liquid, salt plates).

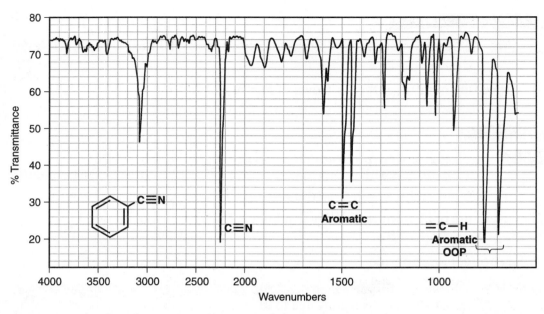

Figure 25.26 Infrared spectrum of benzonitrile (neat liquid, salt plates).

K. Aldehydes

C=O   Stretch at approximately 1725 cm^{-1} is normal.
Aldehydes *seldom* absorb to the left of this value.
Conjugation moves the absorption to the right.

C–H   Stretch, aldehyde hydrogen (–CHO), consists of *weak* bands at about
2750 and 2850 cm^{-1}. Note that the CH stretch in alkyl chains does not
usually extend this far to the right.

The spectrum of an unconjugated aldehyde, nonanal, is shown in Figure 25.28, while
the conjugated aldehyde, benzaldehyde, is shown in Figure 25.29.

1810	1760	1735	1725	1715	1710	1690	cm^{-1}
Anhydride (Band 1)		Esters		Ketones		Amides	
	Anhydride (Band 2)		Aldehydes		Carboxylic acids		

Figure 25.27 Normal values (±10 cm^{-1}) for various types of carbonyl groups.

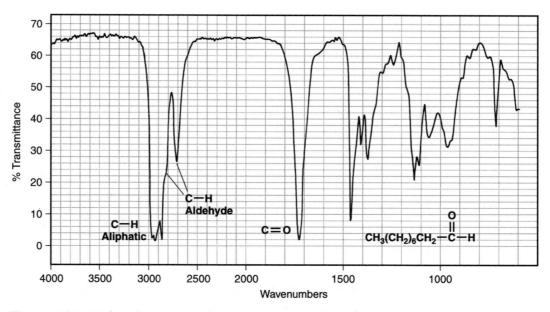

Figure 25.28 Infrared spectrum of nonanal (neat liquid, salt plates).

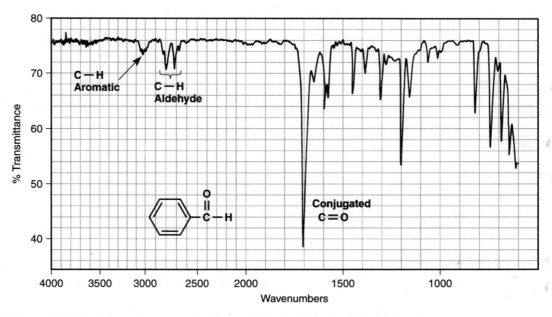

Figure 25.29 Infrared spectrum of benzaldehyde (neat liquid, salt plates).

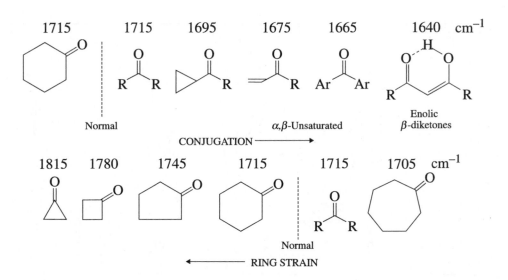

Figure 25.30 Effects of conjugation and ring strain on carbonyl frequencies in ketones.

L. Ketones

C=O   Stretch at approximately at 1715 cm^{-1} is normal.
Conjugation moves the absorption to the right.
Ring strain moves the absorption to the left in cyclic ketones. (See Fig. 25.30).

The spectra of methyl isopropyl ketone and mesityl oxide are shown in Figures 25.14 and 25.31. The spectrum of camphor, shown in Figure 25.32, has a carbonyl group that has been shifted to a higher frequency because of ring strain (1745 cm^{-1}).

M. Acids

O—H   Stretch, usually *very broad* (strongly hydrogen bonded) at 3300–2500 cm^{-1}, often interferes with C—H absorptions.

C=O   Stretch, broad, 1730–1700 cm^{-1}.
Conjugation moves the absorption to the right.

C—O   Stretch, in the range of 1320–1210 cm^{-1}, is strong.

The spectrum of benzoic acid is shown in Figure 25.33.

N. Esters (R—C(=O)—OR′)

C=O   Stretch occurs at about 1735 cm^{-1} in normal esters.

    1. Conjugation in the R part moves the absorption to the right.
    2. Conjugation with the O in the R′ part moves the absorption to the left.
    3. Ring strain (lactones) moves the absorption to the left.

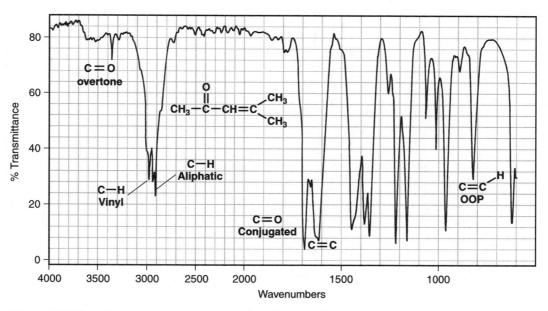

Figure 25.31 Infrared spectrum of mesityl oxide (neat liquid, salt plates).

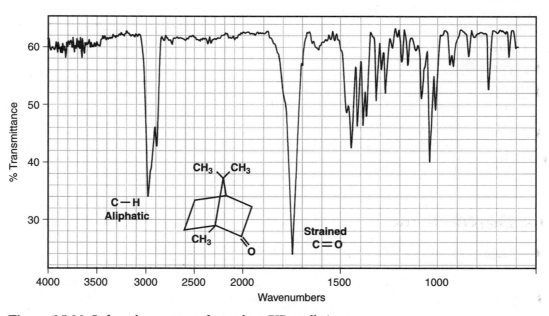

Figure 25.32 Infrared spectrum of camphor (KBr pellet).

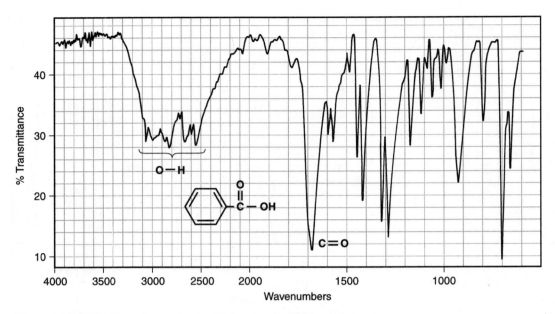

Figure 25.33 Infrared spectrum of benzoic acid (KBr pellet).

C—O   Stretch, two bands or more, one stronger than the others, is in the range of 1300–1000 cm^{-1}.

The spectrum of an unconjugated ester, isopentyl acetate, is shown in Figure 25.34 (C=O appears at 1740 cm^{-1}). A conjugated ester, methyl benzoate, is shown in Figure 25.35 (C=O appears at 1720 cm^{-1}).

O. Amides

C=O   Stretch is at approximately 1670–1640 cm^{-1}.
        Conjugation and ring size (lactams) have the usual effects.

N–H   Stretch (if monosubstituted or unsubstituted) is at 3500–3100 cm^{-1}.
        Unsubstituted amides have two bands (−NH$_2$) in this region.

N–H   Bending around 1640–1550 cm^{-1}.

The spectrum of benzamide is shown in Figure 25.36.

P. Anhydrides

C=O   Stretch always has *two* bands: 1830–1800 cm^{-1} and 1775–1740 cm^{-1}.
        Unsaturation moves the absorptions to the right.
        Ring strain (cyclic anhydrides) moves the absorptions to the left.

C–O   Stretch is at 1300–900 cm^{-1}. The spectrum of *cis*-norbornane-5,6-*endo*-dicarboxylic anhydride is shown in Figure 25.37.

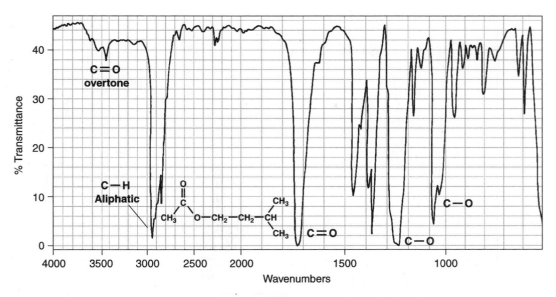

Figure 25.34 Infrared spectrum of isopentyl acetate (neat liquid, salt plates).

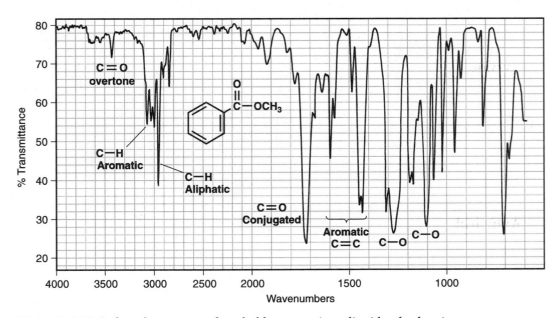

Figure 25.35 Infrared spectrum of methyl benzoate (neat liquid, salt plates).

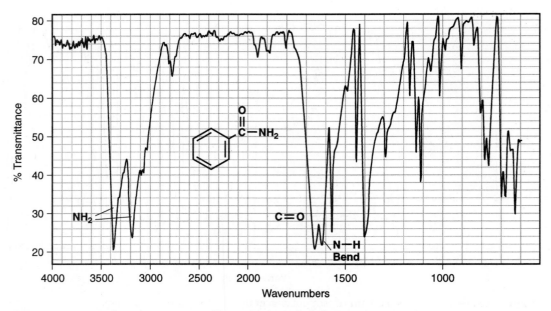

Figure 25.36 Infrared spectrum of benzamide (solid phase, KBr).

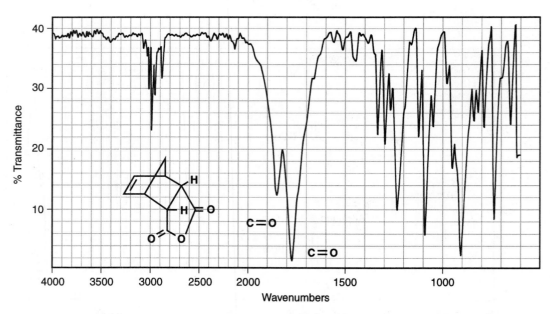

Figure 25.37 Infrared spectrum of *cis*-norbornene-5,6-*endo*-dicarboxylic anhydride (KBr pellet).

Q. Halides

It is often difficult to determine either the presence or the absence of a halide in a compound by infrared spectroscopy. The absorption bands cannot be relied on, especially if the spectrum is being determined with the compound dissolved in CCl_4 or $CHCl_3$ solution.

C—F   Stretch, 1350–960 cm^{-1}.
C—Cl   Stretch, 850–500 cm^{-1}.
C—Br   Stretch, to the right of 667 cm^{-1}.
C—I   Stretch, to the right of 667 cm^{-1}.

The spectra of the solvents, carbon tetrachloride and chloroform, are shown in Figures 25.7 and 25.8, respectively.

REFERENCES

Bellamy, L. J. *The Infra-red Spectra of Complex Molecules*, 3rd ed. New York: Methuen, 1975.
Colthup, N. B., Daly, L. H., and Wiberly, S. E. *Introduction to Infrared and Raman Spectroscopy*, 3rd ed. San Diego, CA: Academic Press, 1990.
Dyer, J. R. *Applications of Absorption Spectroscopy of Organic Compounds*. Englewood Cliffs, NJ: Prentice-Hall, 1965.
Lin-Vien, D., Colthup, N. B., Fateley, W. G., and Grasselli, J. G. *Infrared and Raman Characteristic Frequencies of Organic Molecules*. San Diego, CA: Academic Press, 1991.
Nakanishi, K., and Soloman, P. H. *Infrared Absorption Spectroscopy*, 2nd ed. San Francisco: Holden-Day, 1977.
Pavia, D. L., Lampman, G. M., and Kriz, G. S. *Introduction to Spectroscopy: A Guide for Students of Organic Chemistry*, 3rd ed. Philadelphia: Saunders, 2001.
Silverstein, R. M., and Webster, F. X. *Spectrometric Identification of Organic Compounds*, 6th ed., New York: John Wiley & Sons, 1998.

PROBLEMS

1. Comment on the suitability of running the infrared spectrum under each of the following conditions. If there is a problem with the conditions given, provide a suitable alternative method.
   (a) A neat spectrum of liquid with a boiling point of 150°C is determined using salt plates.
   (b) A neat spectrum of a liquid with a boiling point of 35°C is determined using salt plates.
   (c) A KBr pellet is prepared with a compound that melts at 200°C.
   (d) A KBr pellet is prepared with a compound that melts at 30°C.
   (e) A solid aliphatic hydrocarbon compound is determined as a Nujol mull.
   (f) Silver chloride plates are used to determine the spectrum of aniline.
   (g) Sodium chloride plates are selected to run the spectrum of a compound that contains some water.

2. Indicate how you could distinguish between the following pairs of compounds by using infrared spectroscopy.

(a) $CH_3CH_2CH_2\overset{\overset{\displaystyle O}{\|}}{C}-H$ $CH_3CH_2\overset{\overset{\displaystyle O}{\|}}{C}CH_3$

(b)

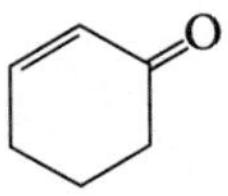

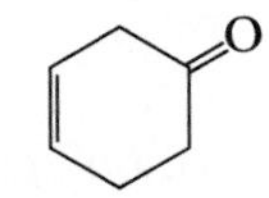

(c) $CH_3CH_2\overset{\overset{\displaystyle H}{|}}{N}CH_2CH_3$ $CH_3CH_2CH_2CH_2NH_2$

(d) $CH_3CH_2\overset{\overset{\displaystyle O}{\|}}{C}OCH_2CH_3$ $CH_3CH_2\overset{\overset{\displaystyle O}{\|}}{C}CH_2OCH_3$

(e) $CH_3CH_2\overset{\overset{\displaystyle O}{\|}}{C}OH$ $CH_3CH_2CH_2OH$

(f)

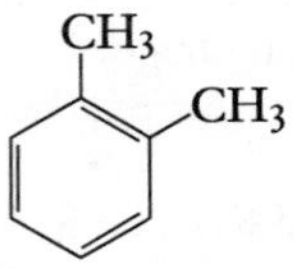

(g) $CH_3CH_2CH=CH_2$ $CH_3CH=CHCH_3$ (trans)

(h) $CH_3CH_2CH_2C\equiv CH$ $CH_3CH_2CH_2CH=CH_2$

(i)

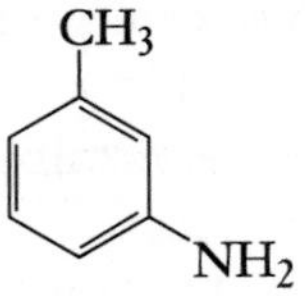

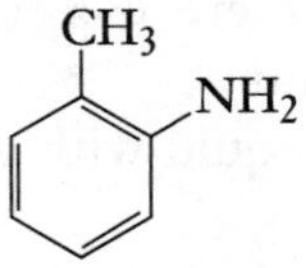

(j) $CH_3CH_2CH_2CH_2\overset{\overset{\displaystyle O}{\|}}{C}-OH$ $CH_3CH_2CH_2\overset{\overset{\displaystyle O}{\|}}{C}OCH_3$

(k) $CH_3CH_2CH_2CH_2CH_3$ $CH_2=CHCH_2CH_2CH_3$

(l) $CH_3CH_2CH_2CH_2C\equiv CH$ $CH_3CH_2CH_2C\equiv CCH_3$

CHAPTER 26

Nuclear Magnetic Resonance Spectroscopy (Proton NMR)

Nuclear magnetic resonance (NMR) spectroscopy is an instrumental technique that allows the number, type, and relative positions of certain atoms in a molecule to be determined. This type of spectroscopy applies only to those atoms that have nuclear magnetic moments because of their nuclear spin properties. Although many atoms meet this requirement, hydrogen atoms ($_1^1H$) are of the greatest interest to the organic chemist. Atoms of the ordinary isotopes of carbon ($_6^{12}C$) and oxygen ($_8^{16}O$) do not have nuclear magnetic moments, and ordinary nitrogen atoms ($_7^{14}N$), although they do have magnetic moments, generally fail to show typical NMR behavior for other reasons. The same is true of the halogen atoms, except for fluorine ($_9^{19}F$), which does show active NMR behavior. Of the atoms mentioned here, the hydrogen nucleus ($_1^1H$) and carbon-13 nucleus ($_6^{13}C$) are the most important to organic chemists. Proton (^1H) NMR is discussed in this chapter and carbon (^{13}C) NMR is described in Chapter 27.

Nuclei of NMR-active atoms placed in a magnetic field can be thought of as tiny bar magnets. In hydrogen, which has two allowed nuclear spin states ($+\frac{1}{2}$ and $-\frac{1}{2}$), either the nuclear magnets of individual atoms can be aligned with the magnetic field (spin $+\frac{1}{2}$), or they can be opposed to it (spin $-\frac{1}{2}$). A slight majority of the nuclei are aligned with the field, as this spin orientation constitutes a slightly lower-energy spin state. If radio-frequency waves of the appropriate energy are supplied, nuclei aligned with the field can absorb this radiation and reverse their direction of spin or become reoriented so that the nuclear magnet opposes the applied magnetic field (Fig. 26.1).

The frequency of radiation required to induce spin conversion is a direct function of the strength of the applied magnetic field. When a spinning hydrogen nucleus is placed in a magnetic field, the nucleus begins to precess with angular frequency ω,

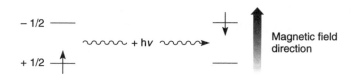

Figure 26.1 The NMR absorption process.

409

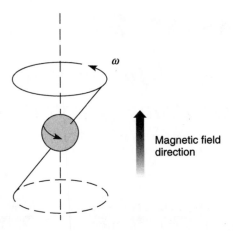

Figure 26.2 Precessional motion of a spinning nucleus in an applied magnetic field.

much like a child's toy top. This precessional motion is depicted in Figure 26.2. The angular frequency of nuclear precession ω increases as the strength of the applied magnetic field is increased. The radiation that must be supplied to induce spin conversion in a hydrogen nucleus of spin +½ must have a frequency that just matches the angular precessional frequency ω. This is called the resonance condition, and spin conversion is said to be a resonance process.

For the average proton (hydrogen atom), if a magnetic field of approximately 1.4 Tesla is applied, radiofrequency radiation of 60 MHz is required to induce a spin transition.[1] Fortunately, the magnetic field strength required to induce the various protons in a molecule to absorb 60-MHz radiation varies from proton to proton within the molecule and is a sensitive function of the immediate *electronic* environment of each proton. The proton nuclear magnetic resonance spectrometer supplies a basic radiofrequency radiation of 60 MHz to the sample being measured and *increases* the strength of the applied magnetic field over a range of several parts per million from the basic field strength. As the field increases, various protons come into resonance (absorb 60-MHz energy), and a resonance signal is generated for each proton. An NMR spectrum is a plot of the strength of the magnetic field versus the intensity of the absorptions. A typical 60-MHz NMR spectrum is shown in Figure 26.3.

Modern FT-NMR instruments produce the same type of NMR spectrum just described even though they do it by a different method. See your lecture textbook for a discussion of the differences between classical CW instruments and modern FT-NMR instruments. Fourier transform spectrometers operating at magnetic field strengths of at least 7.1 Tesla and at spectrometer frequencies of 300 MHz and above allow chemists to obtain both the proton and carbon NMR spectra on the same sample.

---

[1] Most modern instruments (FT-NMR instruments) use higher fields than described here and operate differently. The classical 60 MHz continuous wave (CW) instrument is used here as a simple example.

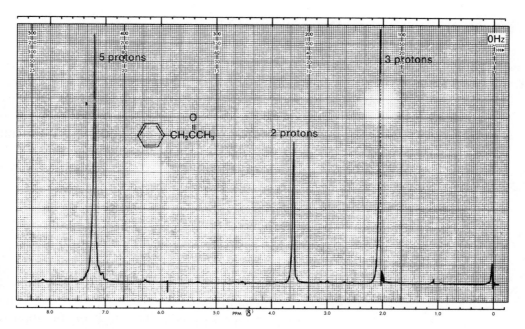

Figure 26.3 Nuclear magnetic resonance spectrum of phenylacetone (the absorption peak at the far right is caused by the added reference substance tetramethylsilane).

Part A. Preparing a Sample for NMR Spectroscopy

The NMR sample tubes used in most instruments are approximately 0.5 cm × 18 cm in overall dimension and are fabricated of uniformly thin glass tubing. These tubes are very fragile and expensive, so care must be taken to avoid breaking the tubes.

CAUTION NMR tubes are made out of very thin glass and break easily. Never place the cap on tightly and take special care when removing it.

To prepare the solution you must first choose the appropriate solvent. The solvent should not have NMR absorption peaks of its own, that is, it should contain no protons. Carbon tetrachloride (CCl_4) fits this requirement and can be used in some instruments. However, because FT-NMR spectrometers require deuterium to stabilize (lock) the field, organic chemists usually use deuterated chloroform ($CDCl_3$) as a solvent. This solvent dissolves most organic compounds and is relatively inexpensive. You can use this solvent with any NMR instrument. You should not use normal chloroform $CHCl_3$, because the solvent contains a proton. Deuterium ^2H does not absorb in the proton region and is thus "invisible," or not seen, in the proton NMR spectrum. Use deuterated chloroform to dissolve your sample unless you are instructed to use another solvent, such as deuterated derivatives of water, acetone, or dimethylsulfoxide.

26.1 ROUTINE SAMPLE PREPARATION USING DEUTERATED CHLOROFORM

1. Most organic liquids and low-melting solids will dissolve in deuterated chloroform. However, you should first determine if your sample will dissolve in ordinary $CHCl_3$ before using the deuterated solvent. If your sample does not dissolve in chloroform, consult your instructor about a possible alternative solvent, or consult Section 26.2.

   ▨▨▨ CAUTION Chloroform, deuterated chloroform, and carbon tetrachloride are all toxic solvents. In addition, they may be carcinogenic substances.

2. If you are using an FT-NMR spectrometer, add 30 mg (0.030 g) of your liquid or solid sample to a tared conical vial or test tube. Use a Pasteur pipet to transfer a liquid or a spatula to transfer a solid. Non-FT instruments usually require a more concentrated solution in order to obtain an adequate spectrum. Typically, a 10–30% sample concentration (weight/weight) is used.
3. Transfer about 0.5 mL of the deuterated chloroform with a *clean and dry Pasteur pipet* to your sample. Swirl the test tube or conical vial to help dissolve the sample. At this point, the sample should have completely dissolved. Add a little more solvent, if necessary, to dissolve the sample fully.
4. Transfer the solution to the NMR tube using a clean and dry Pasteur pipet. Be careful when transferring the solution so as to avoid breaking the edge of the fragile NMR tube. It is best to hold the NMR tube and the container with the solution in the same hand when making the transfer.
5. Once the solution has been transferred to the NMR tube, use a clean pipet to add enough deuterated chloroform to bring the total solution height to about 35 mm (Fig. 26.4). In some cases, you will need to add a small amount of tetramethylsilane (TMS) as a reference substance (Section 26.3). Check with your instructor to see if you need to add TMS to your sample. Deuterated chloroform has a small amount of $CHCl_3$ impurity, which gives rise to a low-intensity peak in the NMR spectrum at 7.27 parts per million (ppm). This impurity may also help you to "reference" your spectrum.
6. Cap the NMR tube. Do this firmly, but not too tightly. If you jam the cap on, you may have trouble removing it later without breaking the end off of the very thin glass tube. Make sure that the cap is on straight. Invert the NMR tube several times to mix the contents.
7. You are now ready to record the NMR spectrum of your sample. Insert the NMR tube into its holder and adjust its depth by using the gauge provided to you.

Cleaning the NMR Tube

1. Carefully uncap the tube so that you do not break it. Turn the tube upside-down, and hold it vertically over a beaker. Shake the tube up and down gently so that the contents of the tube empties into the beaker.

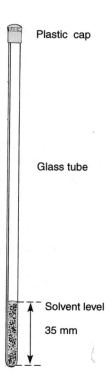

Plastic cap

Glass tube

Solvent level

35 mm

Figure 26.4 An NMR sample tube.

2. Partially refill the NMR tube with acetone using a Pasteur pipet. Carefully replace the cap and invert the tube several times to rinse it.
3. Remove the cap and drain the tube as before. Place the open tube upside-down in a beaker with a Kimwipe or paper towel placed in the bottom of the beaker. Leave the tube standing in this position for at least one laboratory period so that the acetone completely evaporates. Alternatively, you may place the beaker and NMR tube in an oven for at least 2 hours. If you need to use the NMR tube before the acetone has fully evaporated, attach a piece of pressure tubing to the tube, and pull a vacuum with an aspirator. After several minutes, the acetone should have fully evaporated. Because acetone contains protons, you must not use the NMR tube until the acetone has evaporated completely.[2]
4. Once the acetone is evaporated, place the clean tube and its cap (do not cap the tube) in its storage container and place it in your desk. The storage container will prevent the tube from being crushed.

---

[2] If you can't wait to be sure all of the acetone has evaporated, you may rinse the tube once or twice with a *very small* amount of $CDCl_3$ before using it.

Health Hazards Associated with NMR Solvents

Carbon tetrachloride, chloroform (and chloroform-d), and benzene (and benzene-d_6) are hazardous solvents. Besides being highly toxic, they are also suspected carcinogens. In spite of these health problems, these solvents are commonly used in NMR spectroscopy, because there are no suitable alternatives. These solvents are used because they contain no protons and because they are excellent solvents for most organic compounds. Therefore, you must learn to handle these solvents with great care to minimize the hazard. These solvents should be stored either under a hood or in septum-capped bottles. If the bottles have screw caps, a pipet should be attached to each bottle. A recommended way of attaching the pipet is to store it in a test tube taped to the side of the bottle. Septum-capped bottles can be used only by withdrawing the solvent with a hypodermic syringe that has been designated solely for this use. All samples should be prepared under a hood, and solutions should be disposed of in an appropriately designated waste container that is stored under the hood. Wear rubber or plastic gloves when preparing or discarding samples.

26.2 NONROUTINE SAMPLE PREPARATION

Some compounds do not dissolve readily in $CDCl_3$. A commercial solvent called **Unisol** will often dissolve the difficult cases. Unisol is a mixture of $CDCl_3$ and DMSO-d_6.

With highly polar substances you may find that your sample will not dissolve in deuterated chloroform or Unisol. If this is the case, you may be able to dissolve the sample in deuterium oxide D_2O. Spectra determined in D_2O often show a small peak at about 5 ppm because of OH impurity. If the sample compound has acidic hydrogens, they may *exchange* with D_2O, leading to the appearance of an OH peak in the spectrum and the *loss* of the original absorption from the acidic proton, owing to the exchanged hydrogen. In many cases, this will also alter the splitting patterns of a compound.

$$
\underset{\sim 12.0 \text{ ppm}}{R-\overset{\displaystyle O}{\overset{\|}{C}}-O-H} + D_2O \rightleftharpoons R-\overset{\displaystyle O}{\overset{\|}{C}}-\underset{\text{becomes invisible}}{O-D} + \underset{\text{OH peak appears}}{D-OH}
$$

$$
\underset{CH_3 \quad CH_3}{\overset{\displaystyle O}{\overset{\|}{C}}} + D_2O \rightleftharpoons \underset{D-CH_2 \quad CH_3}{\overset{\displaystyle O}{\overset{\|}{C}}} + D-OH
$$

$$
CH_3CH_2OH + D_2O \rightleftharpoons CH_3CH_2OD + D-OH
$$

Many solid carboxylic acids do not dissolve in $CDCl_3$ or even D_2O. In such cases, add a small piece of sodium metal to about 1 mL of D_2O. The acid is then dissolved in this solution. The resulting basic solution enhances the solubility of the carboxylic acid. In such a case, the hydroxyl proton of the carboxylic acid cannot be observed in the NMR spectrum, because it exchanges with the solvent. A large DOH peak is observed, however, due to the exchange and the H_2O impurity in the D_2O solvent.

When the above solvents fail, other special solvents can be used. Acetone, acetonitrile, dimethylsulfoxide, pyridine, benzene, and dimethylformamide can be used if you are not interested in the region or regions of the NMR spectrum in which they give rise to absorption. The deuterated (but expensive) analogues of these compounds are also used in special instances (for example, acetone-d_6, dimethylsulfoxide-d_6, dimethylformamide-d_7, and benzene-d_6). If the sample is not sensitive to acid, trifluoroacetic acid (which has no protons with $\delta < 12$) can be used. You must be aware that these solvents often lead to chemical shift values different from those determined in CCl_4 or $CDCl_3$. Variations of as much as 0.5–1.0 ppm have been observed. In fact, it is sometimes possible, by switching to pyridine, benzene, acetone, or dimethylsulfoxide as solvents, to separate peaks that overlap when CCl_4 or $CDCl_3$ solutions are used.

26.3 REFERENCE SUBSTANCES

To provide the internal reference standard, tetramethylsilane (TMS) must be added to the sample solution. This substance has the formula $(CH_3)_4Si$. By universal convention, the chemical shifts of the protons in this substance are defined as 0.00 ppm. The spectrum should be shifted so that the TMS signal appears at this position on precalibrated paper.

The concentration of TMS in the sample should range from 1 to 3%. Some people prefer to add one to two drops of TMS to the sample just before determining the spectrum. Because TMS has 12 equivalent protons, not much of it needs to be added. A Pasteur pipet or a syringe may be used for the addition. It is far easier to have available in the laboratory a prepared solvent that already contains TMS. Deuterated chloroform and carbon tetrachloride often have TMS added to them. Because TMS is highly volatile (bp 26.5°C), such solutions should be stored, tightly stoppered, in a refrigerator. Tetramethylsilane itself is best stored in a refrigerator as well.

Tetramethylsilane does not dissolve in D_2O. For spectra determined in D_2O, a different internal standard, sodium 2,2-dimethyl-2-silapentane-5-sulfonate, must be used. This standard is water soluble and gives a resonance peak at 0.00 ppm.

$$CH_3-\underset{\underset{CH_3}{|}}{\overset{\overset{CH_3}{|}}{Si}}-CH_2-CH_2-CH_2-SO_3^-Na^+$$

Sodium 2,2-dimethyl-2-silapentane-5-sulfonate (DSS)

Part B. Nuclear Magnetic Resonance (^{1}H NMR)

26.4 THE CHEMICAL SHIFT

The differences in the applied field strengths at which the various protons in a molecule absorb 60-MHz radiation are extremely small. The different absorption positions amount to a difference of only a few parts per million (ppm) in the magnetic field strength. Because it is experimentally difficult to measure the precise field strength at which each proton absorbs to less than one part in a million, a technique has been developed whereby the *difference* between two absorption positions is measured directly. To achieve this measurement, a standard reference substance is used, and the positions of the absorptions of all other protons are measured relative to the values for the reference substance. The reference substance that has been universally accepted is **tetramethylsilane** $(CH_3)_4Si$, which is also called **TMS.** The proton resonances in this molecule appear at a higher field strength than the proton resonances in most other molecules, and all the protons of TMS have resonance at the same field strength.

To give the position of absorption of a proton, a quantitative measurement, a parameter called the **chemical shift** (δ), has been defined. One δ unit corresponds to a one-ppm change in the magnetic field strength. To determine the chemical shift value for the various protons in a molecule, the operator determines an NMR spectrum of the molecule with a small quantity of TMS added directly to the sample. That is, both spectra are determined *simultaneously*. The TMS absorption is adjusted to correspond to the $\delta = 0$ ppm position on the recording chart, which is calibrated in δ units, and the δ values of the absorption peaks for all other protons can be read directly from the chart.

Because the NMR spectrometer increases the magnetic field as the pen moves from left to right on the chart, the TMS absorption appears at the extreme right edge of the spectrum ($\delta = 0$ ppm) or at the *upfield* end of the spectrum. The chart is calibrated in δ units (or ppm), and most other protons absorb at a lower field strength (or *downfield*) from TMS.

The shift from TMS for a given proton depends on the strength of the applied magnetic field. In an applied field of 1.41 Tesla the resonance of a proton is approximately 60 MHz, whereas in an applied field of 2.35 Tesla (23,500 Gauss) the resonance appears at approximately 100 MHz. The ratio of the resonance frequencies is the same as the ratio of the two field strengths:

$$\frac{100 \text{ MHz}}{60 \text{ MHz}} = \frac{2.35 \text{ Tesla}}{1.41 \text{ Tesla}} = \frac{23,500 \text{ Gauss}}{14,100 \text{ Gauss}} = \frac{5}{3}$$

Hence, for a given proton, the shift (in Hertz) from TMS is five-thirds larger in the 100-MHz range than in the 60-MHz range. This can be confusing for workers trying to compare data if they have spectrometers that differ in the strength of the applied magnetic field. The confusion is easily overcome by defining a new parameter that is independent of field strength—for instance, by dividing the shift in Hertz of a given proton by the frequency in megahertz of the spectrometer with which the shift value was obtained. In this manner, a field-independent measure called the **chemical shift** (δ) is obtained

$$\delta = \frac{\text{(shift in Hz)}}{\text{(spectrometer frequency in MHz)}} \qquad (26.1)$$

The chemical shift in δ units expresses the amount by which a proton resonance is shifted from TMS, in parts per million (ppm), of the spectrometer's basic operating frequency. Values of δ for a given proton are always the same irrespective of whether the measurement was made at 60 MHz, 100 MHz, or 300 MHz. For instance, at 60 MHz the shift of the protons in CH_3Br is 162 Hz from TMS, while at 100 MHz the shift is 270 Hz, and at 300 MHz the shift is 810 Hz. However, all three correspond to the same value of $\delta = 2.70$ ppm:

$$\delta = \frac{162 \text{ Hz}}{60 \text{ MHz}} = \frac{270 \text{ Hz}}{100 \text{ MHz}} = \frac{810 \text{ Hz}}{300 \text{ MHz}} = 2.70 \text{ ppm}$$

26.5 CHEMICAL EQUIVALENCE—INTEGRALS

All the protons in a molecule that are in chemically identical environments often exhibit the same chemical shift. Thus, all the protons in TMS or all the protons in benzene, cyclopentane, or acetone have their own respective resonance values all at the same δ value. Each compound gives rise to a single absorption peak in its NMR spectrum. The protons are said to be **chemically equivalent.** On the other hand, molecules that have sets of protons that are chemically distinct from one another may give rise to an absorption peak from each set.

Molecules giving rise to one NMR absorption peak—all protons chemically equivalent

Molecules giving rise to two NMR absorption peaks—two different sets of chemically equivalent protons

The NMR spectrum given in Figure 26.3 is that of phenylacetone, a compound having *three* chemically distinct types of protons:

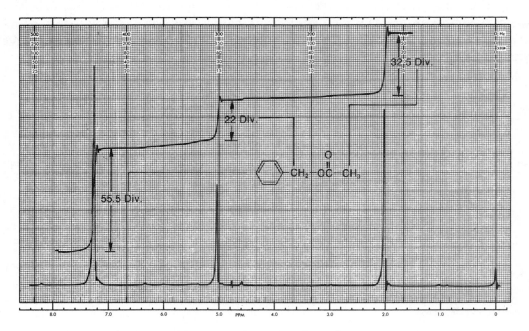

You can immediately see that the NMR spectrum furnishes valuable information on this basis alone. In fact, the NMR spectrum cannot only distinguish how many types of protons a molecule has but also can reveal *how many* of each type are contained within the molecule.

In the NMR spectrum, the area under each peak is proportional to the number of hydrogens generating that peak. Hence, in the case of phenylacetone, the area ratio of the three peaks is 5:2:3, the same as the ratio of the numbers of each type of hydrogen. The NMR spectrometer can electronically "integrate" the area under each peak. It does this by tracing over each peak a vertically rising line, which rises in height by an amount proportional to the area under the peak. Shown in Figure 26.5 is an NMR spectrum of benzyl acetate, with each of the peaks integrated in this way.

Figure 26.5 Determination of the integral ratios for benzyl acetate.

It is important to note that the height of the integral line does not give the absolute number of hydrogens; it gives the *relative* numbers of each type of hydrogen. For a given integral to be of any use, there must be a second integral to which it is referred. The benzyl acetate case provides a good example of this. The first integral rises for 55.5 divisions on the chart paper, the second for 22.0 divisions, and the third for 32.5 divisions. These numbers are relative and give the *ratios* of the various types of protons. You can find these ratios by dividing each of the larger numbers by the smallest number:

$$\frac{55.5 \text{ div}}{22.0 \text{ div}} = 2.52 \qquad \frac{22.0 \text{ div}}{22.0 \text{ div}} = 1.00 \qquad \frac{32.5 \text{ div}}{22.0 \text{ div}} = 1.48$$

Thus, the number ratio of the protons of each type is 2.52:1.00:1.48. If you assume that the peak at 5.1 ppm is really caused by two hydrogens, and if you assume that the integrals are slightly in error (this can be as much as 10%), then you can arrive at the true ratios by multiplying each figure by 2 and rounding off; we then get 5:2:3. Clearly the peak at 7.3 ppm, which integrates for 5, arises from the resonance of the aromatic ring protons, and the peak at 2.0 ppm, which integrates for 3, is caused by the methyl protons. The two-proton resonance at 5.1 ppm arises from the benzyl protons. Notice then that the integrals give the simplest ratios, but not necessarily the true ratios, of the number of protons in each type.

In addition to the rising integral line, modern instruments usually give digitized numerical values for the integrals. Like the heights of the integral lines, these digitized integral values are not absolute but relative, and they should be treated as explained in the preceding paragraph. These digital values are also not exact; like the integral lines, they have the potential for a small degree of error (up to 10%). Figure 26.6 is an example of an integrated spectrum of benzyl acetate determined on a 300-MHz pulsed FT-NMR instrument. The digitized values of the integrals appear under the peaks.

26.6 CHEMICAL ENVIRONMENT AND CHEMICAL SHIFT

If the resonance frequencies of all protons in a molecule were the same, NMR would be of little use to the organic chemist. However, not only do different types of protons have different chemical shifts, they also have a value of chemical shift that characterizes the type of proton they represent. Every type of proton has only a limited range of δ values over which it gives resonance. Hence, the numerical value of the chemical shift for a proton indicates the *type of proton* originating the signal, just as the infrared frequency suggests the type of bond or functional group. Notice, for instance, that the aromatic protons of both phenylacetone (Fig. 26.3) and benzyl acetate (Fig. 26.5) have resonance near 7.3 ppm and that both methyl groups attached directly to a carbonyl group have a resonance of approximately 2.1 ppm. Aromatic protons characteristically have resonance near 7–8 ppm, and acetyl groups (the methyl protons) have their resonance near 2 ppm. These values of chemical shift are

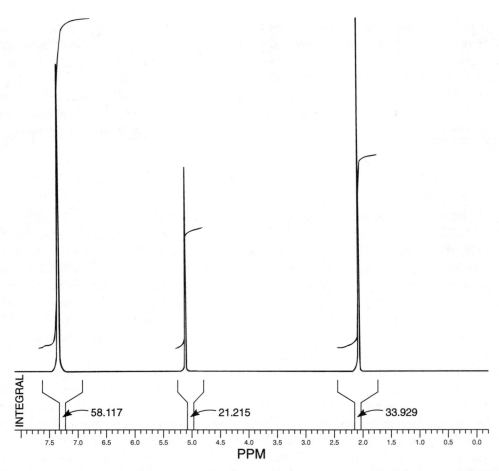

Figure 26.6 An integrated spectrum of benzyl acetate determined on a 300-MHz FT-NMR.

diagnostic. Notice also how the resonance of the benzyl (—CH_2—) protons comes at a higher value of chemical shift (5.1 ppm) in benzyl acetate than in phenylacetone (3.6 ppm). Being attached to the electronegative element, oxygen, these protons are more deshielded (see Section 26.7) than the protons in phenylacetone. A trained chemist would have readily recognized the probable presence of the oxygen by the chemical shift shown by these protons.

It is important to learn the ranges of chemical shifts over which the most common types of protons have resonance. Figure 26.7 is a correlation chart that contains the most essential and frequently encountered types of protons. Table 26.1 lists the chemical shift ranges for selected types of protons. For the beginner it is often difficult to memorize a large body of numbers relating to chemical shifts and proton types. However, this needs to be done only crudely. It is more important to "get a feel" for the regions and the types of protons than to know a string of actual numbers. To do this, study Figure 26.7 carefully.

Table 26.1 Approximate Chemical Shift Ranges (ppm) for Selected Types of Protons[a]

Left structure	Range		Right structure	Range
R–CH$_3$	0.7 – 1.3		R–N–C–H	2.2 – 2.9
R–CH$_2$–R	1.2 – 1.4			
R$_3$CH	1.4 – 1.7		R–S–C–H	2.0 – 3.0
R–C=C–C–H	1.6 – 2.6		I–C–H	2.0 – 4.0
R–C(=O)–C–H, H–C(=O)–C–H	2.1 – 2.4		Br–C–H	2.7 – 4.1
RO–C(=O)–C–H, HO–C(=O)–C–H	2.1 – 2.5		Cl–C–H	3.1 – 4.1
N≡C–C–H	2.1 – 3.0		R–S(=O)(=O)–O–C–H	ca. 3.0
C$_6$H$_5$–C–H	2.3 – 2.7		RO–C–H, HO–C–H	3.2 – 3.8
R–C≡C–H	1.7 – 2.7		R–C(=O)–O–C–H	3.5 – 4.8
R–S–H var	1.0 – 4.0[b]		O$_2$N–C–H	4.1 – 4.3
R–N–H var	0.5 – 4.0[b]		F–C–H	4.2 – 4.8
R–O–H var	0.5 – 5.0[b]			
C$_6$H$_5$–O–H var	4.0 – 7.0[b]		R–C=C–H	4.5 – 6.5
			C$_6$H$_5$–H	6.5 – 8.0
C$_6$H$_5$–N–H var	3.0 – 5.0[b]		R–C(=O)–H	9.0 – 10.0
R–C(=O)–N–H var	5.0 – 9.0[b]		R–C(=O)–OH	11.0 – 12.0

[a]For those hydrogens shown as –C–H, if that hydrogen is part of a methyl group (CH$_3$) the shift is generally at the low end of the range given, if the hydrogen is in a methylene group (–CH$_2$–) the shift is intermediate, and if the hydrogen is in a methine group (–CH–), the shift is typically at the high end of the range given.

[b]The chemical shift of these groups is variable, depending not only on the chemical environment in the molecule, but also on concentration, temperature, and solvent.

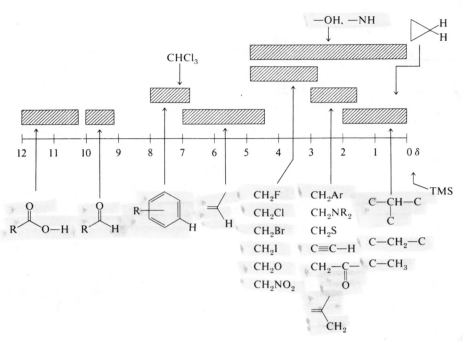

Figure 26.7 A simplified correlation chart for proton chemical shift values.

The values of chemical shift given in Figure 26.7 and in Table 26.1 can be easily understood in terms of two factors: local diamagnetic shielding and anisotropy. These two factors are discussed in Sections 26.7 and 26.8.

26.7 LOCAL DIAMAGNETIC SHIELDING

The trend of chemical shifts that is easiest to explain is that involving electronegative elements substituted on the same carbon to which the protons of interest are attached. The chemical shift simply increases as the electronegativity of the attached element increases. This is illustrated in Table 26.2 for several compounds of the type CH_3X.

Multiple substituents have a stronger effect than a single substituent. The influence of the substituent drops off rapidly with distance. An electronegative element has little effect on protons that are more than three carbons away from it. These effects are illustrated in Table 26.3.

Table 26.2 Dependence of Chemical Shift of CH_3X on the Element X

Compound CH_3X	CH_3F	CH_3OH	CH_3Cl	CH_3Br	CH_3I	CH_4	$(CH_3)_4Si$
Element X	F	O	Cl	Br	I	H	Si
Electronegativity of X	4.0	3.5	3.1	2.8	2.5	2.1	1.8
Chemical shift (ppm)	4.26	3.40	3.05	2.68	2.16	0.23	0

Table 26.3 Substitution Effects[a]

	CHCl₃	CH₂Cl₂	CH₃Cl	—CH₂Br	—CH₂—CH₂Br	—CH₂—CH₂CH₂Br
δ (ppm)	7.27	5.30	3.05	3.3	1.69	1.25

[a]Values apply to underlined hydrogens.

Electronegative substituents attached to a carbon atom, because of their electron-withdrawing effects, reduce the valence electron density around the protons attached to that carbon. These electrons *shield* the proton from the applied magnetic field. This effect, called **local diamagnetic shielding,** occurs because the applied magnetic field induces the valence electrons to circulate. This circulation generates an induced magnetic field, which *opposes* the applied field. This is illustrated in Figure 26.8. Electronegative substituents on carbon reduce the local diamagnetic shielding in the vicinity of the attached protons because they reduce the electron density around those protons. Substituents that produce this effect are said to *deshield* the proton. The greater the electronegativity of the substituent, the more the deshielding of the protons and, hence, the greater the chemical shift of those protons.

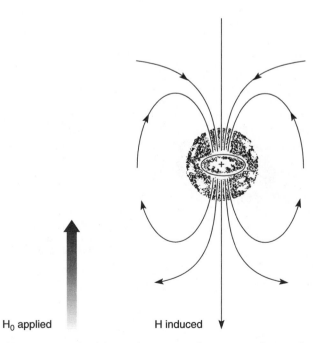

H₀ applied   H induced

Figure 26.8 Local diamagnetic shielding of a proton due to its valence electrons.

26.8 ANISOTROPY

Figure 26.7 clearly shows that several types of protons have chemical shifts not easily explained by a simple consideration of the electronegativity of the attached groups. Consider, for instance, the protons of benzene or other aromatic systems. Aryl protons generally have a chemical shift that is as large as that for the proton of chloroform. Alkenes, alkynes, and aldehydes also have protons whose resonance values are not in line with the expected magnitude of any electron-withdrawing effects. In each of these cases, the effect is due to the presence of an unsaturated system (π electrons) in the vicinity of the proton in question. In benzene, for example, when the π electrons in the aromatic ring system are placed in a magnetic field, they are induced to circulate around the ring. This circulation is called a **ring current.** Moving electrons (the ring current) generate a magnetic field much like that generated in a loop of wire through which a current is induced to flow. The magnetic field covers a spatial volume large enough to influence the shielding of the benzene hydrogens. This is illustrated in Figure 26.9. The benzene hydrogens are deshielded by the **diamagnetic anisotropy** of the ring. An applied magnetic field is nonuniform (anisotropic) in the vicinity of a benzene molecule because of the labile electrons in the ring that interact with the applied field. Thus, a proton attached to a benzene ring is influenced by *three* magnetic fields: the strong magnetic field applied by the magnets of the NMR spectrometer and two weaker fields, one due to the usual shielding by the valence electrons around the proton and the other due to the anisotropy generated by the ring system electrons. It is this anisotropic effect that gives the benzene protons a greater chemical shift than is expected. These protons just hap-

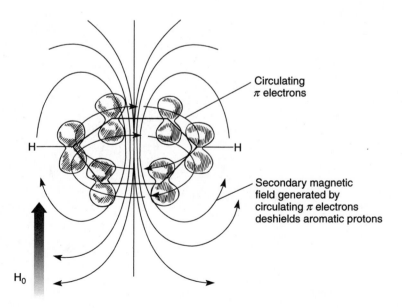

Figure 26.9 Diamagnetic anisotropy in benzene.

pen to lie in a **deshielding** region of this anisotropic field. If a proton were placed in the center of the ring rather than on its periphery, the proton would be shielded, because the field lines would have the opposite direction.

All groups in a molecule that have π electrons generate secondary anisotropic fields. In acetylene, the magnetic field generated by induced circulation of π electrons has a geometry such that the acetylene hydrogens are **shielded.** Hence, acetylenic hydrogens come at a higher field than expected. The shielding and deshielding regions due to the various π electron functional groups have characteristic shapes and directions; they are illustrated in Figure 26.10. Protons falling within the cones are shielded, and those falling outside the conical areas are deshielded. Because the magnitude of the anisotropic field diminishes with distance, beyond a certain distance anisotropy has essentially no effect.

26.9 SPIN–SPIN SPLITTING (N + 1 RULE)

We have already considered how the chemical shift and the integral (peak area) can give information about the numbers and types of hydrogens contained in a molecule. A third type of information available from the NMR spectrum is derived from spin–spin splitting. Even in simple molecules, each type of proton rarely gives a single resonance peak. For instance, in 1,1,2-trichloroethane there are two chemically distinct types of hydrogen:

$$Cl-\overset{\overset{\textstyle \textcircled{H}}{|}}{\underset{\underset{\textstyle Cl}{|}}{C}}-\boxed{CH_2}-Cl$$

From information given thus far, you would predict *two* resonance peaks in the NMR spectrum of 1,1,2-trichloroethane with an area ratio (integral ratio) of 2:1. In fact, the NMR spectrum of this compound has *five* peaks. A group of three peaks (called a

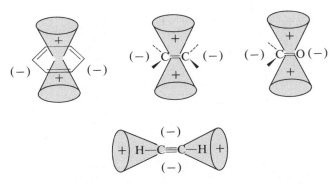

Figure 26.10 Anisotropy caused by the presence of π electrons in some common multiple-bond systems.

triplet) exists at 5.77 ppm and a group of two peaks (called a **doublet**) is found at 3.95 ppm. The spectrum is shown in Figure 26.11. The methine (CH) resonance (5.77 ppm) is split into a triplet, and the methylene resonance (3.95 ppm) is split into a doublet. The area under the three triplet peaks is *one*, relative to an area of *two* under the two doublet peaks.

This phenomenon is called **spin–spin splitting**. Empirically, spin–spin splitting can be explained by the "$n + 1$ rule." Each type of proton "senses" the number of equivalent protons (n) on the carbon atom or atoms next to the one to which it is bonded, and its resonance peak is split into $n + 1$ components.

Let's examine the case at hand, 1,1,2-trichloroethane, using the $n + 1$ rule. First, the lone methine hydrogen is situated next to a carbon bearing two methylene protons. According to the rule, it has two equivalent neighbors ($n = 2$) and is split into $n + 1 = 3$ peaks (a triplet). The methylene protons are situated next to a carbon bearing only one methine hydrogen. According to the rule, they have one neighbor ($n = 1$) and are split into $n + 1 = 2$ peaks (a doublet).

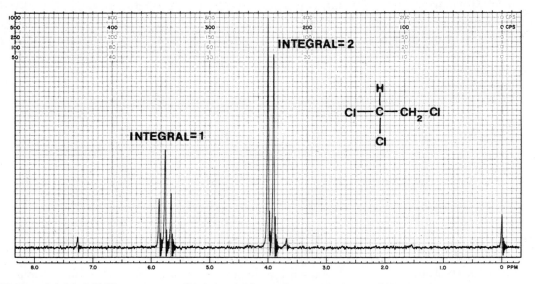

Figure 26.11 NMR spectrum of 1,1,2-trichloroethane. (*Courtesy of Varian Associates.*)

The spectrum of 1,1,2-trichloroethane can be explained easily by the interaction, or coupling, of the spins of protons on adjacent carbon atoms. The position of absorption of proton H_a is affected by the spins of protons H_b and H_c attached to the neighboring (adjacent) carbon atom. If the spins of these protons are aligned with the applied magnetic field, the small magnetic field generated by their nuclear spin properties will augment the strength of the field experienced by the first-mentioned proton H_a. The proton H_a will thus be *deshielded*. If the spins of H_b and H_c are opposed to the applied field, they will decrease the field experienced by proton H_a. It will then be *shielded*. In each of these situations, the absorption position of H_a will be altered. Among the many molecules in the solution, you will find all the various possible spin combinations for H_b and H_c; hence, the NMR spectrum of the molecular solution will give *three* absorption peaks (a triplet) for H_a because H_b and H_c have three different possible spin combinations (Fig. 26.12). By a similar analysis, it can be seen that protons H_b and H_c should appear as a doublet.

Some common splitting patterns that can be predicted by the $n + 1$ rule and that are frequently observed in a number of molecules are shown in Figure 26.13. Notice particularly the last entry, where *both* methyl groups (six protons in all) function as a unit and split the methine proton into a septet $(6 + 1 = 7)$.

26.10 THE COUPLING CONSTANT

The quantitative amount of spin–spin interaction between two protons can be defined by the **coupling constant**. The spacing between the component peaks in a single multiplet is called the coupling constant J. This distance is measured on the same scale as the chemical shift and is expressed in Hertz (Hz).

Coupling constants for protons on adjacent carbon atoms have magnitudes of from about 6 to 8 Hz (see Table 26.4). You should expect to see a coupling constant in this range for compounds where there is free rotation about a single bond. Since three bonds separate protons from each other on adjacent carbon atoms, we label these coupling constants as ^3J. For example, the coupling constant for the compound

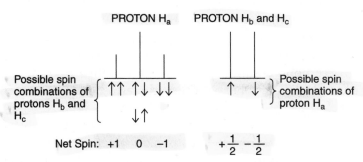

Figure 26.12 Analysis of spin–spin splitting pattern for 1,1,2-trichloroethane.

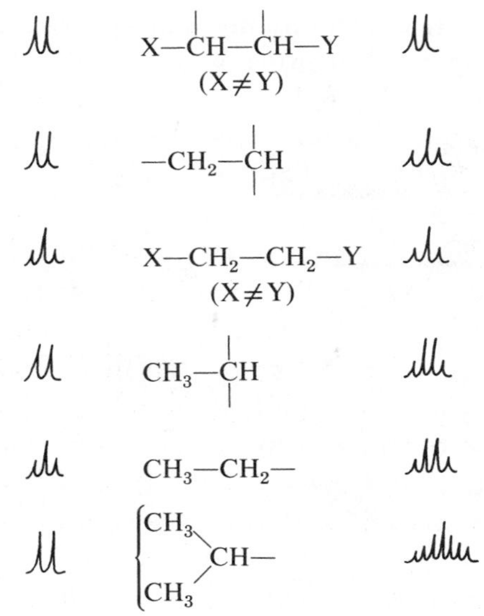

Figure 26.13 Some common splitting patterns.

shown in Figure 26.11 would be written as $^3J = 6$ Hz. The bold faced lines in the following diagram show how the protons on adjacent carbon atoms are three bonds away from each other.

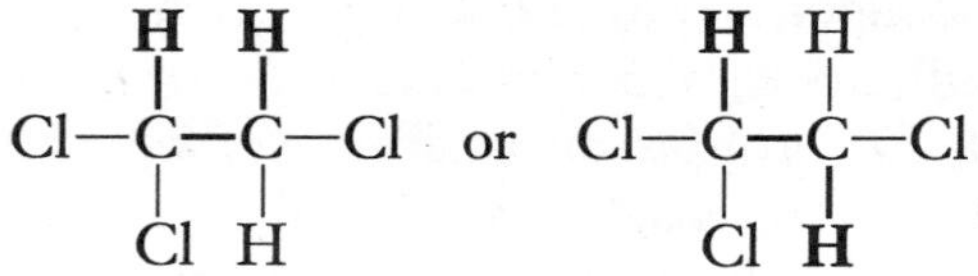

In compounds where there is a C=C double bond, free rotation is restricted. In compounds of this kind, we often find two types of ^3J coupling constants; $^3J_{trans}$ and $^3J_{cis}$. These coupling constants vary in value as shown in Table 26.4, but $^3J_{trans}$ is almost always larger than $^3J_{cis}$. The magnitudes of these ^3Js often provide important structural clues. You can distinguish, for example, between a *cis* alkene and a *trans* alkene on the basis of the observed coupling constants for the two vinyl protons on disubstituted alkenes. Most of the coupling constants shown in the first column of Table 26.4 are three bond couplings, but you will notice that there is a two bond (^2J) coupling constant listed. These protons that are bonded to a common carbon atom are often referred to as *geminal* protons and can be labeled as $^2J_{gem}$. Notice that the coupling constants for *geminal* protons are quite small for alkenes. The ^2J couplings are observed only when the protons on a methylene group are in a different environment (see Section 26.11). The following structure shows the various

Table 26.4 **Representative Coupling Constants and Approximate Values (Hz)**

types of couplings that you observe for protons on a C=C double bond in a typical alkene, vinyl acetate. The spectrum for this compound is described in detail in Section 26.11.

Longer range couplings that occur over four or more bonds are observed in some alkenes and also in aromatic compounds. Thus, in Table 26.4, we see that it is possible to observe a small H—H coupling (^4J = 0–3 Hz) occurring over four bonds in an alkene. In an aromatic compound, you often observe a small but measurable coupling between *meta* protons that are four bonds away from each other (^4J = 1–4 Hz). Couplings over five bonds are usually quite small, with values close to 0 Hz. The long range couplings are usually observed only in *unsaturated* compounds. The spectra of saturated compounds are often more easily interpreted because they usually have only three bond couplings. Aromatic compounds are discussed in detail in Section 26.13.

26.11 MAGNETIC EQUIVALENCE

In the example of spin–spin splitting in 1,1,2-trichloroethane (Fig. 26.11), notice that the two protons H_b and H_c, which are attached to the same carbon atom, do not split one another. They behave as an integral group. Actually the two protons H_b and H_c *are* coupled to one another; however, for reasons we cannot explain fully here, protons that are attached to the same carbon and both of which have the *same chemical shift* do not show spin–spin splitting. Another way of stating this is that protons coupled to the same extent to *all* other protons in a molecule do not show spin–spin splitting. Protons that have the same chemical shift and are coupled equivalently to all other protons are *magnetically equivalent* and do not show spin–spin splitting. Thus, in 1,1,2-trichloroethane, protons H_b and H_c have the same value of δ and are coupled by the same value of J to proton H_a. They are magnetically equivalent and $^2J_{gem} = 0$.

It is important to differentiate magnetic equivalence and chemical equivalence. Note the following two compounds.

In the cycloproprane compound, the two geminal hydrogens H_A and H_B are chemically equivalent; however, they are not magnetically equivalent. Proton H_A is on the same side of the ring as the two halogens. Proton H_B is on the same side of the ring as the two methyl groups. Protons H_A and H_B will have different chemical shifts, will couple to one another, and will show spin–spin splitting. Two doublets will be seen for H_A and H_B. For cyclopropane rings, $^2J_{gem}$ is usually around 5 Hz.

The general vinyl structure (alkene) shown above and the specific example of vinyl acetate shown in Figure 26.14 are examples of cases in which the methylene protons H_A and H_B are nonequivalent. They appear at different chemical shift values and will split each other. This coupling constant, $^2J_{gem}$, is usually small with vinyl compounds (about 2 Hz).

The spectrum of vinyl acetate is shown in Figure 26.14. H_C appears downfield at about 7.3 ppm because of the electronegativity of the attached oxygen atom. This proton is split by H_B into a doublet ($^3J_{trans} = {}^3J_{BC} = 15$ Hz) and then each leg of the doublet is split by H_A into a doublet ($^3J_{cis} = {}^3J_{AC} = 7$ Hz). Notice that the $n + 1$ rule is applied individually to each adjacent proton. The pattern that results is usually referred to as a doublet of doublets (dd). The graphic analysis shown in Figure 26.15 should help you understand the pattern obtained for proton H_C.

Now, look at the pattern shown in Figure 26.14 for proton H_B at 4.85 ppm. It is also a doublet of doublets (dd). Proton H_B is split by proton H_C into a doublet ($^3J_{trans} = {}^3J_{BC} = 15$ Hz) and then each leg of the doublet is split by the geminal proton H_A into doublets ($^2J_{gem} = {}^2J_{AB} = 2$ Hz).

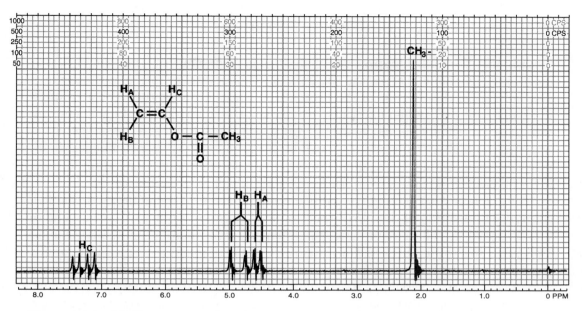

Figure 26.14 NMR spectrum of vinyl acetate. *(Courtesy of Varian Associates.)*

Proton H_A shown in Figure 26.14 appears at 4.55 ppm. This pattern is also a doublet of doublets (dd). Proton H_A is split by proton H_C into a doublet ($^3J_{cis} = ^3J_{AC} = 7$ Hz) and then each leg of the doublet is split by the geminal proton H_B into doublets ($^2J_{gem} = ^2J_{AB} = 2$ Hz). For each proton shown in Figure 26.14, the NMR spectrum must be analyzed graphically, splitting by splitting. This complete graphic analysis is shown in Figure 26.15.

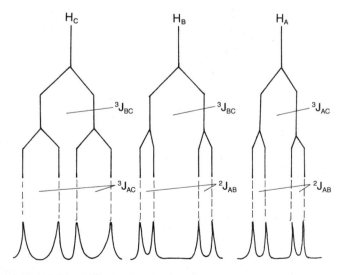

Figure 26.15 Analysis of the splittings in vinyl acetate.

26.12 SPECTRA AT HIGHER FIELD STRENGTH

Occasionally the 60-MHz spectrum of an organic compound, or a portion of it, is almost undecipherable because the chemical shifts of several groups of protons are all very similar. In these cases, all the proton resonances occur in the same area of the spectrum, and peaks often overlap so extensively that individual peaks and splittings cannot be extracted. One way to simplify such a situation is to use a spectrometer that operates at a higher frequency. Although both 60- and 100-MHz instruments are still in use, it is becoming increasingly common to find instruments operating at much higher fields, and with spectrometer frequencies 300, 400, or 500 MHz.

Although NMR coupling constants do not depend on the frequency or the field strength of operation of the NMR spectrometer, chemical shifts in Hertz depend on these parameters. This circumstance can often be used to simplify an otherwise undecipherable spectrum. Suppose, for instance, that a compound contained three multiplets derived from groups of protons with very similar chemical shifts. At 60 MHz these peaks might overlap, as illustrated in Figure 26.16, and simply give an unresolved envelope of absorption. It turns out that the $n + 1$ rule fails to make the proper predictions when chemical shifts are similar for the protons in a molecule. The spectral patterns that result are said to be **second order** and what you end up seeing is an amorphous blob of unrecognizable patterns!

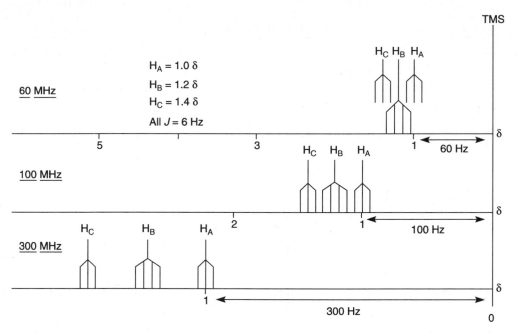

Figure 26.16 A comparison of the spectrum of a compound with overlapping multiplets at 60 MHz, with spectra of the same compound also determined at 100 MHz and 300 MHz.

Figure 26.16 also shows the spectrum of the same compound at two higher frequencies (100 MHz and 300 MHz). In redetermining the spectrum at a higher frequency, the coupling constants (*J*) do not change, but the chemical shifts in *Hertz* (not ppm) of the proton groups (H_A, H_B, H_C) responsible for the multiplets do increase. It is important to realize, however, that the chemical shift in *ppm* is a constant and it will not change when the frequency of the spectrometer is increased (see Eq. 26.1 on page 417).

Notice that at 300 MHz, the individual multiplets are cleanly separated and resolved. At high frequency the chemical shift differences of each proton increase, resulting in more clearly recognizable patterns (i.e., triplets, quartets, etc.) and less overlap of proton patterns in the spectrum. At high frequency the chemical shift differences are large and the *n* + 1 rule will more likely correctly predict the patterns. Thus, it is a clear advantage to use NMR spectrometers operating at high frequency (300 MHz or above) because the resulting spectra are more likely to provide nonoverlapped and well-resolved peaks. When the protons in a spectrum follow the *n* + 1 rule, the spectrum is said to be **first order.** The result is that you will obtain a spectrum with much more recognizable patterns, as shown in Figure 26.16.

26.13 AROMATIC COMPOUNDS—SUBSTITUTED BENZENE RINGS

Phenyl rings are so common in organic compounds that it is important to know a few facts about NMR absorptions in compounds that contain them. In general, the ring protons of a benzenoid system have resonance near 7.3 ppm; however, electron-withdrawing ring substituents (e.g., nitro, cyano, carboxyl, or carbonyl) move the resonance of these protons downfield (larger ppm values), and electron-donating ring substituents (e.g., methoxy or amino) move the resonance of these protons upfield (smaller ppm values). Table 26.5 shows these trends for a series of symmetrically *p*-disubstituted benzene compounds. The *p*-disubstituted compounds were chosen because their two planes of symmetry render all of the hydrogens equivalent. Each compound gives only one aromatic

Table 26.5 Proton Chemical Shifts in *p*-Disubstituted Benzene Compounds

Substituent X	δ (ppm)	
—OCH_3	6.80	Electron donating (shielding)
—OH	6.60	
—NH_2	6.36	
—CH_3	7.05	
—H	7.32	
—COOH	8.20	Electron withdrawing (deshielding)
—NO_2	8.48	

peak (a singlet) in the proton NMR spectrum. Later you will see that some positions are affected more strongly than others in systems with substitution patterns different from this one.

In the sections that follow, we will attempt to cover some of the most important types of benzene ring substitution. In some cases it will be necessary to examine sample spectra taken at both 60 MHz and 300 MHz. Many benzenoid rings show second-order splittings at 60 MHz but are essentially first order at 300 MHz.

A. Monosubstituted Rings

Alkylbenzenes. In monosubstituted benzenes in which the substituent is neither a strongly electron-withdrawing nor a strongly electron-donating group, all the ring protons give rise to what appears to be a *single resonance* when the spectrum is determined at 60 MHz. This is a particularly common occurrence in alkyl-substituted benzenes. Although the protons *ortho*, *meta*, and *para* to the substituent are not chemically equivalent, they generally give rise to a single unresolved absorption peak. A possible explanation is that the chemical shift differences, which should be small in any event, are somehow eliminated by the presence of the ring current, which tends to equalize them. All of the protons are nearly equivalent under these conditions. The NMR spectra of the aromatic portions of alkylbenzene compounds are good examples of this type of circumstance. Figure 26.17A is the 60-MHz ^{1}H spectrum of ethylbenzene.

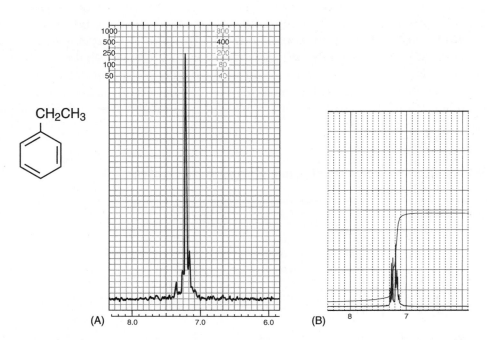

Figure 26.17 The aromatic ring portions of the ^{1}H NMR spectra of ethylbenzene at **(A)** 60 MHz and **(B)** 300 MHz.

The 300-MHz spectrum of ethylbenzene, shown in Figure 26.17B, presents quite a different picture. With the increased frequency shifts at 300 MHz, the nearly equivalent (at 60 MHz) protons are neatly separated into two groups. The *ortho* and *para* protons appear upfield from the *meta* protons. The splitting pattern is clearly second order.

Electron-Donating Groups.   When electron-donating groups are attached to the ring, the ring protons are not equivalent, even at 60 MHz. A highly activating substituent such as methoxy clearly increases the electron density at the *ortho* and *para* positions of the ring (by resonance) and helps to give these protons greater shielding than those in the *meta* positions and, thus, a substantially different chemical shift.

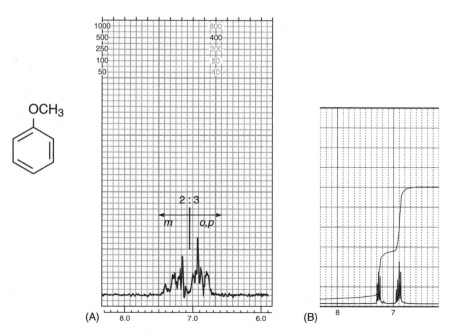

At 60 MHz, this chemical shift difference results in a complicated second-order splitting pattern for anisole (methoxybenzene), but the protons do fall clearly into two groups, the *ortho/para* protons and the *meta* protons. The 60-MHz NMR spectrum of the aromatic portion of anisole (Fig. 26.18A) has a complex multiplet for the *o,p,*

Figure 26.18 The aromatic ring portions of the ^{1}H NMR spectra of anisole at **(A)** 60 MHz and **(B)** 300 MHz.

protons (integrating for three protons) that is upfield from the *meta* protons (integrating for two protons), with a clear distinction (gap) between the two types. Aniline (aminobenzene) provides a similar spectrum, also with a 3:2 split, owing to the electron-releasing effect of the amino group.

The 300-MHz spectrum of anisole (Fig. 26.18B) shows the same separation between the *ortho/para* hydrogens (upfield) and the *meta* hydrogens (downfield). However, because the actual shift in Hertz between the two types of hydrogens is greater, there is less second-order interaction, and the lines in the pattern are sharper at 300 MHz. In fact, it might be tempting to try to interpret the observed pattern as if it were first order, a triplet at 7.25 ppm (*meta*, 2H) and an overlapping triplet (*para*, 1 H) with a doublet (*ortho*, 2 H) at about 6.9 ppm.

Anisotropy—Electron-Withdrawing Groups. A carbonyl or a nitro group would be expected to show (aside from anisotropy effects) a reverse effect, since these groups are electron withdrawing. It would be expected that the group would act to decrease the electron density around the *ortho* and *para* positions, thus deshielding the *ortho* and *para* hydrogens and providing a pattern exactly the reverse of the one shown for anisole (3:2 ratio, downfield:upfield). Convince yourself of this by drawing resonance structures. Nevertheless, the actual NMR spectra of nitrobenzene and benzaldehyde do not have the appearances that would be predicted on the basis of resonance structures. Instead, the *ortho* protons are much more deshielded than the *meta* and *para* protons, due to the magnetic anisotropy of the π bonds in these groups.

Anisotropy is observed when a substituent group bonds a carbonyl group directly to the benzene ring (Fig. 26.19). Once again, the ring protons fall into two groups, with the *ortho* protons downfield from the *meta/para* protons. Benzaldehyde (Fig. 26.20) and acetophenone both show this effect in their NMR spectra. A similar effect is sometimes observed when a carbon–carbon double bond is attached to the ring. The 300-MHz spectrum of benzaldehyde (Fig. 26.20b) is a nearly first-order spectrum and shows a doublet (H_C, 2 H), a triplet (H_B, 1 H), and a triplet (H_A, 2 H). It can be analyzed by the $n + 1$ rule.

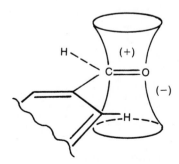

Figure 26.19 Anisotropic deshielding of the *ortho* protons of benzaldehyde.

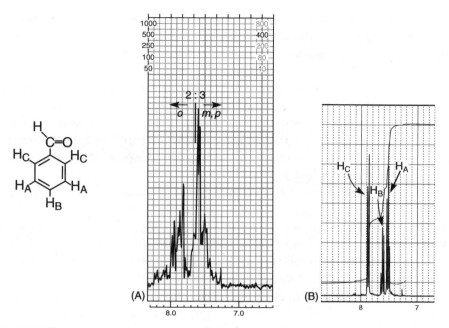

Figure 26.20 The aromatic ring portions of the ^{1}H NMR spectra of benzaldehyde at **(A)** 60 MHz and **(B)** 300 MHz.

B. *para*-Disubstituted Rings

Of the possible substitution patterns of a benzene ring, some are easily recognized. One of these is the *para*-disubstituted benzene ring. Examine anethole (Fig. 26.21) as a first example.

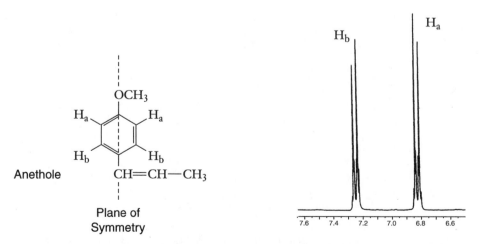

Figure 26.21 The aromatic ring protons of the 300-MHz ^{1}H NMR spectrum of anethole showing a *para*-disubstituted pattern.

On one side of the anethole ring shown in Figure 26.21, proton H_a is coupled to H_b, $^3J = 8$ Hz, resulting in a doublet at about 6.80 ppm in the spectrum. Proton H_a appears upfield (smaller ppm value) relative to H_b because of shielding by the electron-releasing effect of the methoxy group (see page 435). Likewise, H_b is coupled to H_a, $^3J = 8$ Hz, producing another doublet at 7.25 ppm for this proton. Because of the plane of symmetry, both halves of the ring are equivalent. Thus, H_a and H_b on the other side of the ring also appear at 6.80 and 7.25 ppm, respectively. Each doublet, therefore, integrates for two protons each. A *para*-disubstituted ring, with two different substituents attached, is easily recognized by the appearance of two doublets, each integrating for two protons each.

As the chemical shifts of H_a and H_b approach each other in value, the *para*-disubstituted pattern becomes similar to that of 4-allyloxyanisole (Fig. 26.22). The inner peaks move closer together, and the outer ones become smaller or even disappear. Ultimately, when H_a and H_b approach each other closely enough in chemical shift, the outer peaks disappear and the two inner peaks merge into a *singlet*; 1,4-dimethylbenzene (*para*-xylene), for instance, gives a singlet at 7.05 ppm. Hence, a single aromatic resonance integrating for four protons could easily represent a *para*-disubstituted ring, but the substituents would obviously be either identical or very similar.

C. Other Substitution

Figure 26.23 shows the 300-MHz ^{1}H spectra of the aromatic ring portions of 2-, 3-, and 4-nitroaniline (the *ortho*, *meta*, and *para* isomers). The characteristic pattern of a *para*-disubstituted ring, with its pair of doublets, makes it easy to recognize 4-nitroaniline.

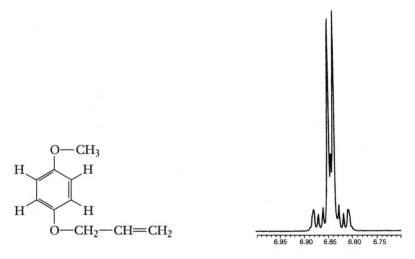

Figure 26.22 The aromatic ring protons of the 300-MHz ^{1}H NMR spectrum of 4-allyloxyanisole.

The splitting patterns for 2- and 3-nitroaniline are first order, and they can be analyzed by the $n + 1$ rule. As an exercise, see if you can analyze these patterns, assigning the multiplets to specific protons on the ring. Use the indicated multiplicities (s, d, t) and expected chemical shifts to help your assignments. Remember that the amino group releases electrons by resonance while the nitro group shows a significant anisotropy toward *ortho* protons. You may ignore any *meta* and *para* couplings, remembering that these long-range couplings will be too small in magnitude to be observed on the scale that these figures are presented. If the spectra were expanded, you would be able to observe ^4J couplings.

The spectrum shown in Figure 26.24 is of 2-nitrophenol. It is helpful to look also at the coupling constants for the benzene ring found in Table 26.4. Because the spectrum is expanded, it is now possible to see ^3J couplings (about 8 Hz) as well as ^4J couplings (about 1.5 Hz). ^5J couplings are not observed ($^5J \approx 0$). Each of the protons on this compound is assigned on the spectrum. Proton H_d appears downfield at 8.11 ppm as a doublet of doublets ($^3J_{ad} = 8$ Hz and $^4J_{cd} = 1.5$ Hz), H_c appears at 7.6 ppm as a triplet of doublets ($^3J_{ac} = {}^3J_{bc} = 8$ Hz and $^4J_{cd} = 1.5$ Hz), H_b appears at 7.17 ppm as a doublet of doublets ($^3J_{bc} = 8$ Hz and $^4J_{ab} = 1.5$ Hz), and H_a appears at 7.0 ppm as a triplet of doublets ($^3J_{ac} = {}^3J_{ad} = 8$ Hz and $^4J_{ab} = 1.5$ Hz). H_d appears the furthest downfield because of the anisotropy of the nitro group. H_a and H_b are relatively

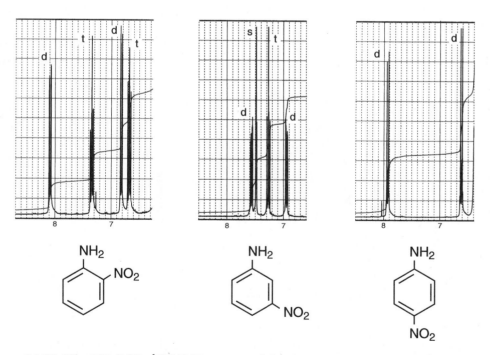

Figure 26.23 The 300-MHz ^1H NMR spectra of the aromatic ring portions of 2-, 3-, and 4-nitroaniline. s, singlet; d, doublet; t, triplet.

shielded because of the resonance releasing effect of the hydroxyl group, which shields these two protons. H_c is assigned by a process of elimination in the absence of these two effects.

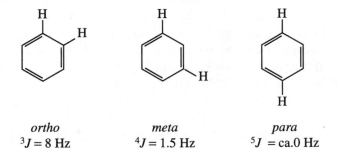

ortho	meta	para
$^3J = 8$ Hz	$^4J = 1.5$ Hz	$^5J = $ ca.0 Hz

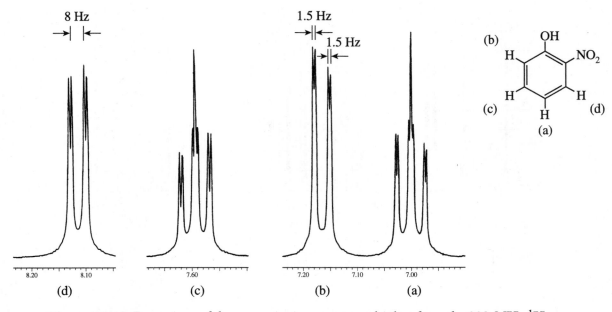

Figure 26.24 Expansions of the aromatic ring proton multiplets from the 300-MHz ^{1}H spectrum of 2-nitrophenol. The accompanying hydroxyl absorption (OH) is not shown. Coupling constants are indicated on some of the peaks of the spectrum to give an idea of scale.

26.14 PROTONS ATTACHED TO ATOMS OTHER THAN CARBON

Protons attached to atoms other than carbon often have a widely variable range of absorptions. Several of these groups are tabulated in Table 26.6. In addition, under the usual conditions of determining an NMR spectrum, protons on heteroelements normally do not couple with protons on adjacent carbon atoms to give spin–spin splitting. This is primarily because such protons often exchange very rapidly with those of the solvent medium. The absorption position is variable because these groups also undergo varying degrees of hydrogen bonding in solutions of different concentrations. The amount of hydrogen bonding that occurs with a proton radically affects the valence electron density around that proton and produces correspondingly large changes in the chemical shift. The absorption peaks for protons that have hydrogen bonding or are undergoing exchange are frequently broad relative to other singlets and can often be recognized on that basis. For a different reason, called **quadrupole broadening,** protons attached to nitrogen atoms often show an extremely broad resonance peak, often almost indistinguishable from the baseline.

Table 26.6 **Typical Ranges for Groups with Variable Chemical Shift**

Acids	RCOOH	10.5–12.0 ppm
Phenols	ArOH	4.0–7.0
Alcohols	ROH	0.5–5.0
Amines	RNH_2	0.5–5.0
Amides	$RCONH_2$	5.0–8.0
Enols	CH=CH—OH	≥15

26.15 CHEMICAL SHIFT REAGENTS

Researchers have known for some time that interactions between molecules and solvents, such as those due to hydrogen bonding, can cause large changes in the resonance positions of certain types of protons (e.g., hydroxyl and amino). They have also known that the resonance positions of some groups of protons can be greatly affected by changing from the usual NMR solvents such as CCl_4 and $CDCl_3$ to solvents such as benzene, which impose local anisotropic effects on surrounding molecules. In many cases, it is possible to resolve partially overlapping multiplets by such a solvent change. The use of **chemical shift**

reagents for this purpose dates from about 1969. Most of these chemical shift reagents are organic complexes of paramagnetic rare earth metals from the lanthanide series of elements. When these metal complexes are added to the compound whose spectrum is being determined, profound shifts in the resonance positions of the various groups of protons are observed. The direction of the shift (upfield or downfield) depends primarily on which metal is being used. Complexes of europium, erbium, thulium, and ytterbium shift resonances to lower field; complexes of cerium, praseodymium, neodymium, samarium, terbium, and holmium generally shift resonances to higher field. The advantage of using such reagents is that shifts similar to those observed at higher field can be induced without the purchase of an expensive higher field instrument.

Of the lanthanides, europium is probably the most commonly used metal. Two of its widely used complexes are *tris*-(dipivalomethanato)europium and *tris*-(6,6,7,7,8,8,8,-heptafluoro-2,2-dimethyl-3,5-octanedionato)europium. These are frequently abbreviated $Eu(dpm)_3$ and $Eu(fod)_3$, respectively.

$Eu(dpm)_3$
or $Eu(thd)_3$

$Eu(fod)_3$

These lanthanide complexes produce spectral simplifications in the NMR spectrum of any compound that has a relatively basic pair of electrons (unshared pair) that can coordinate with Eu^{3+}. Typically, aldehydes, ketones, alcohols, thiols, ethers, and amines will all interact:

$$2B: + Eu(dpm)_3 \longrightarrow \begin{array}{c} B: \\ \\ B: \end{array} \begin{array}{c} dpm \\ Eu-dpm \\ dpm \end{array}$$

The amount of shift that a given group of protons will experience depends (1) on the distance separating the metal (Eu^{3+}) and that group of protons, and (2) on the concentration of the shift reagent in the solution. Because of the latter dependence, it is necessary when reporting a lanthanide-shifted spectrum to report the number of mole equivalents of shift reagent used or its molar concentration.

The distance factor is illustrated in the spectra of hexanol, which are given in Figures 26.25 and 26.26. In the absence of shift reagent, the normal spectrum is obtained (Fig. 26.25). Only the triplet of the terminal methyl group and the triplet of

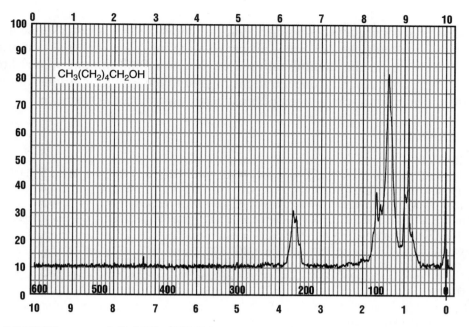

Figure 26.25 The normal 60-MHz ^{1}H NMR spectrum of hexanol. *(Courtesy of Aldrich Chemical Co.)*

the methylene group next to the hydroxyl are resolved in the spectrum. The other protons (aside from OH) are found together in a broad unresolved group. With shift reagent added (Fig. 26.26), each of the methylene groups is clearly separated and resolved into the proper multiplet structure. The spectrum is first order and simplified; all the splittings are explained by the $n + 1$ rule.

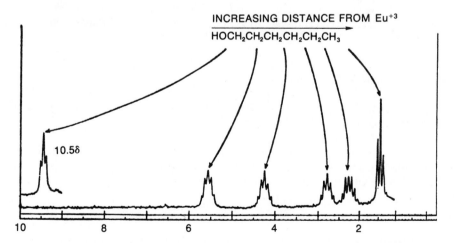

Figure 26.26 The 100-MHz ^{1}H NMR spectrum of hexanol with 0.29 mole equivalents of Eu(dpm)$_3$ added. *(Reprinted with permission from J. K. M. Sanders and D. H. Williams, Chemical Communications, [1970]: 422.)*

One final consequence of using a shift reagent should be noted. Notice in Figure 26.26 that the multiplets are not as nicely resolved into sharp peaks as you might expect. This is due to the fact that shift reagents cause a small amount of peak broadening. At high shift reagent concentrations, this problem becomes serious, but at most useful concentrations the amount of broadening experienced is tolerable.

REFERENCES

Textbooks

Friebolin, H. *Basic One- and Two-Dimensional NMR Spectroscopy*, 2nd ed. New York: VCH Publishers, 1993.

Gunther, H. *NMR Spectroscopy*, 2nd ed. New York: John Wiley & Sons, 1995.

Jackman, L. M., and Sternhell, S. *Nuclear Magnetic Resonance Spectroscopy in Organic Chemistry*, 2d ed. New York: Pergamon Press, 1969.

Macomber, R. S. *NMR Spectroscopy: Essential Theory and Practice*. New York: College Outline Series, Harcourt, Brace Jovanovich, 1988.

Macomber, R. S. *A Complete Introduction to Modern NMR Spectroscopy*. New York: John Wiley & Sons, 1997.

Pavia, D. L., Lampman, G. M., and Kriz, G. S. *Introduction to Spectroscopy*, 3rd ed. Philadelphia: Harcourt College Publishers, 2001.

Sanders, J. K. M., and Hunter, B. K. *Modern NMR Spectroscopy—A Guide for Chemists*, 2nd ed. Oxford: Oxford University Press, 1993.

Silverstein, R. M., and Webster, F. X. *Spectrometric Identification of Organic Compounds*, 6th ed. New York: John Wiley & Sons, 1998.

Williams, D. H., and Fleming, I. *Spectroscopic Methods in Organic Chemistry*, 4th ed. London–New York: McGraw-Hill Book Co. Ltd., 1987.

Compilations of Spectra

Pouchert, C. J. *The Aldrich Library of NMR Spectra, 60 MHz*, 2d ed. Milwaukee, WI: Aldrich Chemical Company, 1983.

Pouchert, C. J., and J. Behnke, *The Aldrich Library of ^{13}C and ^1H FT-NMR Spectra, 300 MHz*, Aldrich Chemical Company, Milwaukee, WI, 1993.

Pretsch, E., Clerc, T., Seibl, J., and Simon, W. *Tables of Spectral Data for Structure Determination of Organic Compounds*, 2d ed. Berlin and New York: Springer-Verlag, 1989. Translated from the German by K. Biemann.

Web sites

http://www.aist.go.jp/RIODB/SDBS/menu-e.html

   Integrated Spectral DataBase System for Organic Compounds, National Institute of Materials and Chemical Research, Tsukuba, Ibaraki 305-8565, Japan. This database includes infrared, mass spectra, and NMR data (proton and carbon-13) for a large number of compounds.

http://www.chem.ucla.edu/~webnmr

   UCLA Department of Chemistry and Biochemistry in connection with Cambridge University Isotope Laboratories, maintains a web site, WebSpectra, that provides NMR and IR spectroscopy problems for students to interpret. They provide links to other sites with problems for students to solve.

PROBLEMS

1. Describe the method that you should use to determine the proton NMR spectrum of a carboxylic acid, which is insoluble in *all* the common organic solvents that your instructor is likely to make available.

2. To save money, a student uses chloroform instead of deuterated chloroform to run a carbon-13 NMR spectrum. Is this a good idea?

3. Look up the solubilities for the following compounds and decide whether you would select deuterated chloroform or deuterated water to dissolve the substances for NMR spectroscopy.
   (a) Glycerol (1,2,3-propanetriol)
   (b) 1,4-Diethoxybenzene
   (c) Propyl pentanoate (propyl ester of pentanoic acid)

4. Assign each of the proton patterns in the spectra of 2-, 3-, and 4-nitroaniline as shown in Figure 26.23.

5. The following compounds are isomeric esters derived from acetic acid, each with formula $C_5H_{10}O_2$. The peaks of the spectrum have been labeled to indicate the degrees of splitting. With the first spectrum as an example, use the integral curve traced on the spectrum to calculate the number of hydrogens represented in each multiplet. The multiplets appear both on the spectrum and in the first column of the following table. The second column is obtained by dividing through by the lowest number (1.7 div). The third column is obtained by multiplying by 2 and rounding off the values. Notice that the sum of the numbers in the third column equals the number of hydrogen atoms (10) present in the formula. Often one can inspect the spectrum and visually approximate the relative numbers of hydrogen atoms, thus avoiding the more mathematical approach demonstrated in the following table. Using either method, the second spectrum yields a ratio of 1:3:6. What are the structures of the two esters?

1.7 div	1.0	2 H
2.5 div	1.47	3 H
1.7 div	1.0	2 H
2.5 div	1.47	3 H

(a)

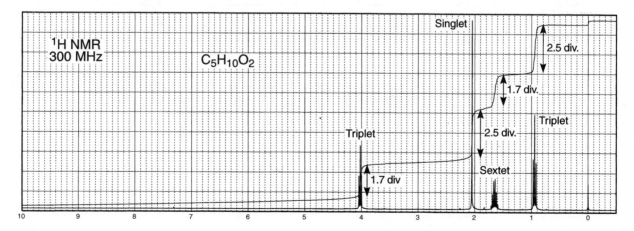

(b)

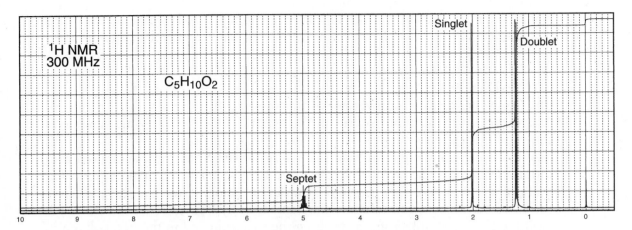

6. The compound that gives the following NMR spectrum has the formula $C_3H_6Br_2$. Draw the structure.

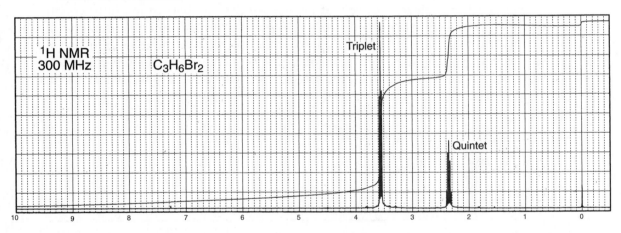

7. Draw the structure of an ether with formula $C_5H_{12}O_2$ that fits the following NMR spectrum.

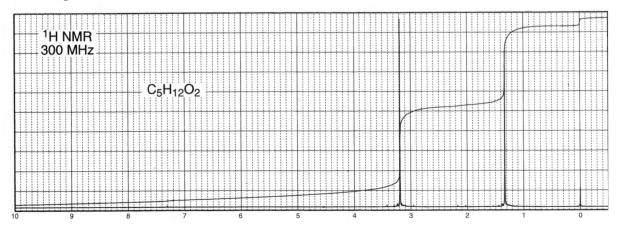

8. Following are the NMR spectra of three isomeric esters with the formula $C_7H_{14}O_2$, all derived from propanoic acid. Provide a structure for each.

(a)

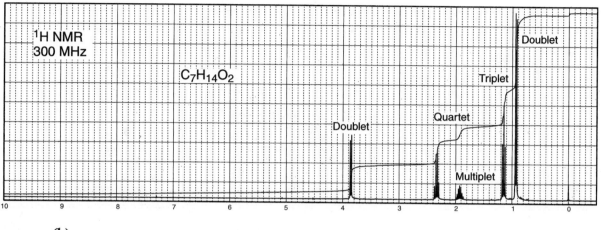

(b)

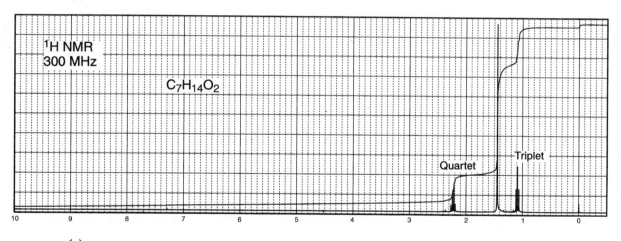

(c)

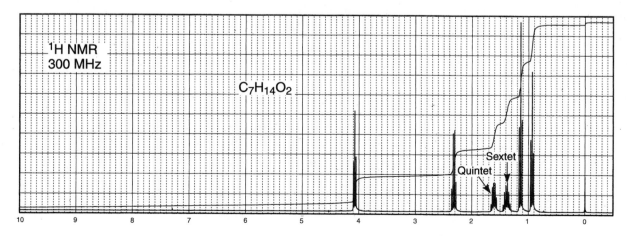

9. The two isomeric carboxylic acids that give the following NMR spectra both have the formula $C_3H_5ClO_2$. Draw their structures.

(a)

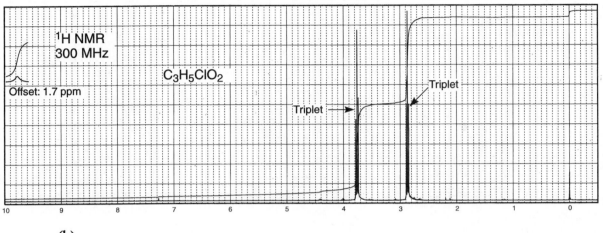

(b)

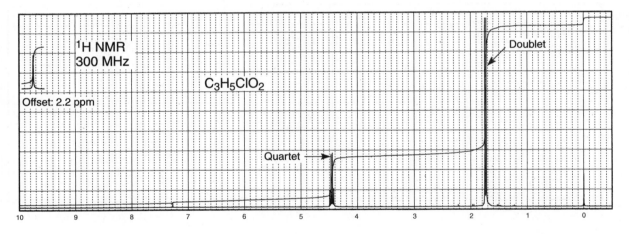

10. The following compounds are isomers with formula $C_{10}H_{12}O$. Their infrared spectra show strong bands near 1715 cm^{-1} and in the range from 1600 to 1450 cm^{-1}. Draw their structures.

(a)

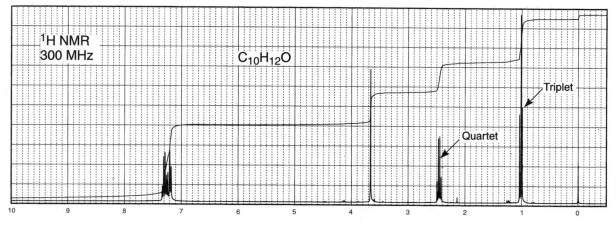

(b)

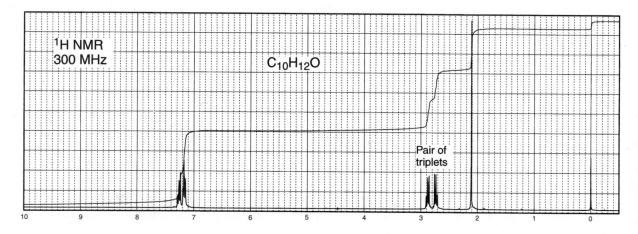

CHAPTER 27

Carbon-13 Nuclear Magnetic Resonance Spectroscopy

Carbon-12, the most abundant isotope of carbon, does not possess spin ($I = 0$); it has both an even atomic number and an even atomic weight. The second principal isotope of carbon, ^{13}C, however, does have the nuclear spin property ($I = ½$). ^{13}C atom resonances are not easy to observe, due to a combination of two factors. First, the natural abundance of ^{13}C is low; only 1.08% of all carbon atoms are ^{13}C. Second, the magnetic moment μ of ^{13}C is low. For these two reasons, the resonances of ^{13}C are about *6000 times weaker* than those of hydrogen. With special Fourier transform instrumental techniques, which are not discussed here, it is possible to observe ^{13}C nuclear magnetic resonance (carbon-13) spectra on samples that contain only the natural abundance of ^{13}C.

The most useful parameter derived from carbon-13 spectra is the chemical shift. Integrals are unreliable and are not necessarily related to the relative numbers of ^{13}C atoms present in the sample. Hydrogens that are attached to ^{13}C atoms cause spin–spin splitting, but spin–spin interaction between adjacent carbon atoms is rare. With its low natural abundance (0.0108), the probability of finding two ^{13}C atoms adjacent to one another is extremely low.

Carbon spectra can be used to determine the number of nonequivalent carbons and to identify the types of carbon atoms (methyl, methylene, aromatic, carbonyl, and so on) that may be present in a compound. Thus, carbon NMR provides direct information about the carbon skeleton of a molecule. Because of the low natural abundance of carbon-13 in a sample, it is often necessary to acquire multiple scans over what is needed for proton NMR.

For a given magnetic field strength, the resonance frequency of a ^{13}C nucleus is about one-fourth the frequency required to observe proton resonances. For example, in a 7.05-Tesla applied magnetic field, protons are observed at 300 MHz while ^{13}C nuclei are observed at about 75 MHz.

27.1 PREPARING A SAMPLE FOR CARBON-13 NMR

Chapter 26, Section 26.1 describes the technique for preparing samples for proton NMR. Much of what is described there also applies to carbon NMR. There are some differences, however, in determining a carbon spectrum. Fourier transform instruments require a deuterium signal to stabilize (lock) the field. Therefore, the solvents must contain deuterium. Deuterated chloroform $CDCl_3$ is used most commonly for this purpose because of its relatively low cost. Other deuterated solvents may also be used.

Modern FT-NMR spectrometers allow chemists to obtain both the proton and carbon NMR spectra of the same sample in the same NMR tube. After changing several parameters in the program operating the spectrometer, you can obtain both spectra without removing the sample from the probe. The only real difference is that a proton spectrum may be obtained after a few scans, whereas the carbon spectrum may require 10 to 100 times more scans.

TMS may be added as an internal reference standard, where the chemical shift of the methyl carbon is defined as 0.00 ppm. Alternatively, you may use the center peak of the $CDCl_3$ pattern, which is found at 77.0 ppm. This pattern can be observed as a small "triplet" near 77.0 ppm in a number of the spectra given in this chapter.

27.2 CARBON-13 CHEMICAL SHIFTS

An important parameter derived from carbon-13 spectra is the chemical shift. The correlation chart in Figure 27.1 shows typical ^{13}C chemical shifts, listed in parts per million (ppm) from tetramethylsilane (TMS), where the carbons of the methyl groups of TMS (not the hydrogens) are used for reference. Notice that the chemical shifts appear over a range (0 to 220 ppm) much larger than that observed for protons (0 to 12 ppm). Because of the very large range of values, nearly every nonequivalent carbon atom in an organic molecule gives rise to a peak with a different chemical shift. Peaks rarely overlap as they often do in proton NMR.

The correlation chart is divided into four sections. Saturated carbon atoms appear at the highest field, nearest to TMS (8 to 60 ppm). The next section of the correlation chart demonstrates the effect of electronegative atoms (40 to 80 ppm). The third section of the chart includes alkene and aromatic-ring carbon atoms (100 to 175 ppm). Finally, the fourth section of the chart contains carbonyl carbons, which appear at the lowest field values (155 to 220 ppm).

Electronegativity, hybridization, and anisotropy all affect ^{13}C chemical shifts in nearly the same fashion as they affect ^1H chemical shifts; however ^{13}C chemical shifts are about 20 times larger. Electronegativity (Section 26.7) produces the same deshielding effect in carbon NMR as in proton NMR—the electronegative element produces a large downfield shift. The shift is greater for a ^{13}C atom than for a proton since the electronegative atom is directly attached to the ^{13}C atom, and the effect occurs through only a single bond, C—X. With protons, the electronegative atoms are attached to carbon, not hydrogen; the effect occurs through two bonds, H—C—X, rather than one.

Analogous with ^1H shifts, changes in hybridization also produce larger shifts for the carbon-13 that is *directly involved* (no bonds) than they do for the hydrogens attached to that carbon (one bond). In ^{13}C NMR, the carbons of carbonyl groups have the largest chemical shifts, due both to sp^2 hybridization and to the fact that an

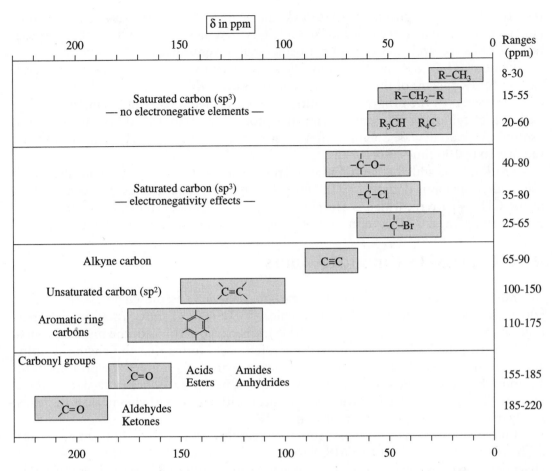

Figure 27.1 A correlation chart for ^{13}C chemical shifts (chemical shifts are listed in parts per million from tetramethylsilane).

electronegative oxygen is directly attached to the carbonyl carbon, deshielding it even further. Anisotropy (Section 26.8) is responsible for the large chemical shifts of the carbons in aromatic rings and alkenes.

Notice that the range of chemical shifts is larger for carbon atoms than for hydrogen atoms. Because the factors affecting carbon shifts operate either through one bond or directly on carbon, they are greater than those for hydrogen, which operate through more bonds. As a result, the entire range of chemical shifts becomes larger for ^{13}C (0 to 220 ppm) than for ^{1}H (0 to 12 ppm).

Many of the important functional groups of organic chemistry contain a carbonyl group. In determining the structure of a compound containing a carbonyl group, it is frequently helpful to have some idea of the type of carbonyl group in the unknown. Figure 27.2 illustrates the typical ranges of ^{13}C chemical shifts for some carbonyl-containing functional groups. Although there is some overlap in the ranges, ketones

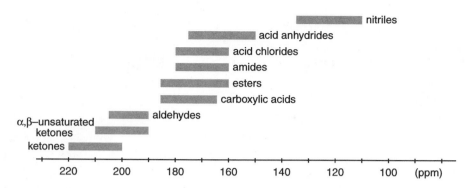

Figure 27.2 A ^{13}C correlation chart for carbonyl and nitrile functional groups.

and aldehydes are easy to distinguish from the other types. Chemical shift data for carbonyl carbons are particularly powerful when combined with data from an infrared spectrum.

27.3 PROTON-COUPLED ^{13}C SPECTRA—SPIN–SPIN SPLITTING OF CARBON-13 SIGNALS

Unless a molecule is artificially enriched by synthesis, the probability of finding two ^{13}C atoms in the same molecule is low. The probability of finding two ^{13}C atoms adjacent to each other in the same molecule is even lower. Therefore, we rarely observe **homonuclear** (carbon–carbon) spin–spin splitting patterns where the interaction occurs between two ^{13}C atoms. However, the spins of protons attached directly to ^{13}C atoms do interact with the spin of carbon and cause the carbon signal to be split according to the $n + 1$ rule. This is **heteronuclear** (carbon–hydrogen) coupling involving two different types of atoms. With ^{13}C NMR, we generally examine splitting that arises from the protons *directly attached* to the carbon atom being studied. This is a one-bond coupling. In proton NMR, the most common splittings are *homonuclear* (hydrogen–hydrogen) that occur between protons attached to *adjacent* carbon atoms. In these cases, the interaction is a three-bond coupling, H−C−C−H.

Figure 27.3 illustrates the effect of protons directly attached to a ^{13}C atom. The $n + 1$ rule predicts the degree of splitting in each case. The resonance of a ^{13}C atom with three attached protons, for instance, is split into a quartet ($n + 1 = 3 + 1 = 4$). Since the hydrogens are directly attached to the carbon-13 (one-bond couplings), the coupling constants for this interaction are quite large, with J values of about 100 to 250 Hz. Compare the typical three-bond H−C−C−H couplings that are common in NMR spectra, which have J values of about 4 to 18 Hz.

It is important to note while examining Figure 27.3 that you are not "seeing" protons directly when looking at a ^{13}C spectrum (proton resonances occur at frequencies outside the range used to obtain ^{13}C spectra); you are observing only the effect of the protons on ^{13}C atoms. Also, remember that we cannot observe ^{12}C, because it is NMR inactive.

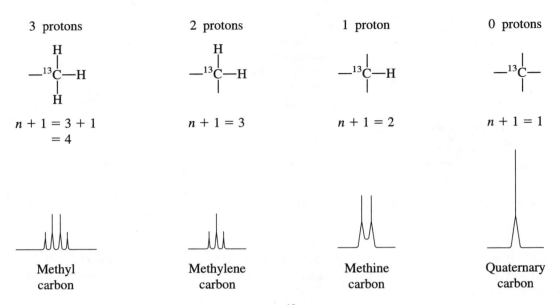

Figure 27.3 The effect of attached protons on ^{13}C resonances.

Spectra that show the spin–spin splitting, or coupling, between carbon-13 and the protons directly attached to it are called **proton-coupled spectra.** Figure 27.4A is the proton-coupled ^{13}C NMR spectrum of ethyl phenylacetate. In this spectrum, the first quartet downfield from TMS (14.2 ppm) corresponds to the carbon of the methyl group. It is split into a quartet ($J = 127$ Hz) by the three attached hydrogen atoms (^{13}C—H, one-bond couplings). In addition, although it cannot be seen on the scale of this spectrum (an expansion must be used), each of the quartet lines is split into a closely spaced triplet ($J =$ ca. 1 Hz). This additional fine splitting is caused by the two protons on the adjacent —CH$_2$— group. These are two-bond couplings (H—C—^{13}C) of a type that occurs commonly in ^{13}C spectra, with coupling constants that are generally quite small ($J = 0$–2 Hz) for systems with carbon atoms in an aliphatic chain. Because of their small size, these couplings are frequently ignored in the routine analysis of spectra, with greater attention being given to the larger one-bond splittings seen in the quartet itself.

There are two —CH$_2$— groups in ethyl phenylacetate. The one corresponding to the ethyl —CH$_2$— group is found farther downfield (60.6 ppm), as this carbon is deshielded by the attached oxygen. It is a triplet because of the two attached hydrogens (one-bond couplings). Again, although it is not seen in this unexpanded spectrum, the three hydrogens on the adjacent methyl group finely split each of the triplet peaks into a quartet. The benzyl —CH$_2$— carbon is the intermediate triplet (41.4 ppm). Farthest downfield is the carbonyl-group carbon (171.1 ppm). On the scale of this presentation it is a singlet (no directly attached hydrogens), but because of the adjacent benzyl —CH$_2$— group, it is actually split finely into a triplet. The aromatic-ring carbons also appear in the spectrum, and they have resonances in the range from 127 to 136 ppm. Section 27.7 will discuss aromatic-ring ^{13}C resonances.

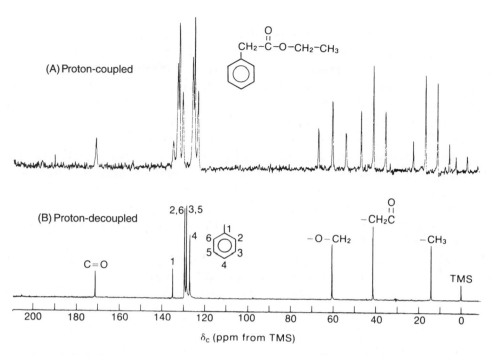

Figure 27.4 Ethyl phenylacetate. **(A)** The proton-coupled ^{13}C NMR spectrum (20 MHz). **(B)** The proton-decoupled ^{13}C spectrum (20 MHz). *(From Moore, J. A., Dalrymple, D. L., and Rodig, O. R. Experimental Methods in Organic Chemistry, 3rd ed. Philadelphia: W. B. Saunders, 1982.)*

Proton-coupled spectra for large molecules are often difficult to interpret. The multiplets from different carbons commonly overlap because the ^{13}C—H coupling constants are frequently larger than the chemical shift differences of the carbons in the spectrum. Sometimes, even simple molecules such as ethyl phenylacetate (Fig. 27.4A) are difficult to interpret. Proton decoupling, which is discussed in the next section, avoids this problem.

27.4 PROTON-DECOUPLED ^{13}C SPECTRA

By far the great majority of ^{13}C NMR spectra are obtained as **proton-decoupled spectra.** The decoupling technique obliterates all interactions between protons and ^{13}C nuclei; therefore, only **singlets** are observed in a decoupled ^{13}C NMR spectrum. Although this technique simplifies the spectrum and avoids overlapping multiplets, it has the disadvantage that the information on attached hydrogens is lost.

Proton **decoupling** is accomplished in the process of determining a ^{13}C NMR spectrum by simultaneously irradiating all of the protons in the molecule with a broad spectrum of frequencies in the proper range for protons. Modern NMR spectrometers provide a second, tunable radiofrequency generator, the **decoupler,** for this purpose. Irradiation causes the protons to become saturated, and they undergo rapid upward and downward transitions, among all their possible spin states. These rapid transitions decouple any spin–spin interactions between the hydrogens and the ^{13}C nuclei being observed. In effect, all spin interactions are averaged to zero by the rapid changes. The carbon nucleus "senses" only one average spin state for the attached hydrogens rather than two or more distinct spin states.

Figure 27.4B is a proton-decoupled spectrum of ethyl phenylacetate. The proton-coupled spectrum (Fig. 27.4A) was discussed in Section 27.3. It is interesting to compare the two spectra to see how the proton-decoupling technique simplifies the spectrum. Every chemically and magnetically distinct carbon gives only a single peak. Notice, however, that the two *ortho* ring carbons (carbons 2 and 6) and the two *meta* ring carbons (carbons 3 and 5) are equivalent by symmetry and that each pair gives only a single peak.

Figure 27.5 is a second example of a proton-decoupled spectrum. Notice that the spectrum shows three peaks, corresponding to the exact number of carbon atoms in 1-propanol. If there are no equivalent carbon atoms in a molecule, a ^{13}C peak will be

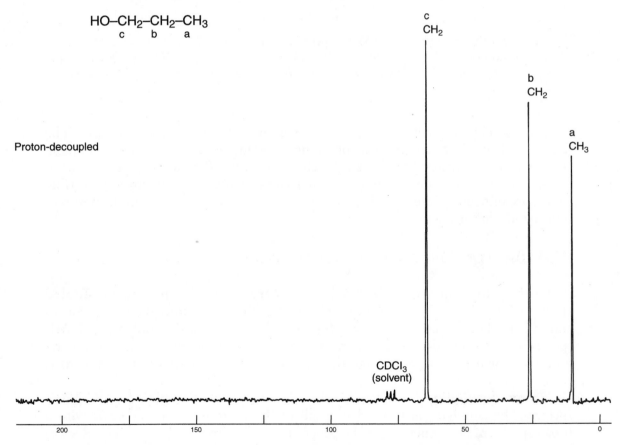

Figure 27.5 The proton-decoupled ^{13}C NMR spectrum of 1-propanol (22.5 MHz).

observed for *each* carbon. Notice also that the assignments given in Figure 27.5 are consistent with the values in the chemical shift table (Fig. 27.1). The carbon atom closest to the electronegative oxygen is farthest downfield, and the methyl carbon is at highest field.

The three-peak pattern centered at $\delta = 77$ ppm is due to the solvent $CDCl_3$. This pattern results from the coupling of a deuterium (^2H) nucleus to the ^{13}C nucleus. Often the $CDCl_3$ pattern is used as an internal reference, in place of TMS.

27.5 SOME SAMPLE SPECTRA—EQUIVALENT CARBONS

Equivalent ^{13}C atoms appear at the same chemical shift value. Figure 27.6 shows the proton-decoupled carbon spectrum for 2,2-dimethylbutane. The three methyl groups at the left side of the molecule are equivalent by symmetry.

$$CH_3-\overset{\overset{\textstyle CH_3}{|}}{\underset{\underset{\textstyle CH_3}{|}}{C}}-CH_2-CH_3$$

Although this compound has a total of six carbons, there are only four peaks in the ^{13}C NMR spectrum. The ^{13}C atoms that are equivalent appear at the same chemical shift. The single methyl carbon, **a**, appears at highest field (9 ppm), while the three equivalent methyl carbons, **b**, appear at 29 ppm. The quaternary carbon **c** gives rise to the small peak at 30 ppm, and the methylene carbon **d** appears at 37 ppm. The relative sizes of the peaks are related, in part, to the number of each type

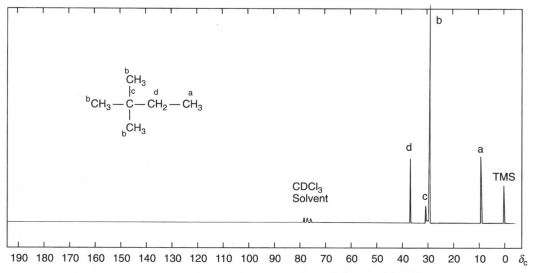

Figure 27.6 The proton-decoupled ^{13}C NMR spectrum of 2,2-dimethylbutane.

of carbon atom present in the molecule. For example, notice in Figure 27.6 that the peak at 29 ppm **(b)** is much larger than the others. This peak is generated by three carbons. The quaternary carbon at 30 ppm **(c)** is very weak. Since no hydrogens are attached to this carbon, there is very little nuclear Overhauser enhancement (NOE) (Section 27.6). Without attached hydrogen atoms, relaxation times are also longer than for other carbon atoms. Quaternary carbons, those with no hydrogens attached, frequently appear as weak peaks in proton-decoupled ^{13}C NMR spectra (see Section 27.6).

Figure 27.7 is a proton-decoupled ^{13}C spectrum of cyclohexanol. This compound has a plane of symmetry passing through its hydroxyl group, and it shows only four carbon resonances. Carbons **a** and **c** are doubled due to symmetry and give rise to larger peaks than carbons **b** and **d**. Carbon **d**, bearing the hydroxyl group, is deshielded by oxygen and has its peak at 70.0 ppm. Notice that this peak has the lowest intensity of all of the peaks. Its intensity is lower than that of carbon **b** in part because the carbon **d** peak receives the least amount of NOE; there is only one hydrogen attached to the hydroxyl carbon, whereas each of the other carbons has two hydrogens.

A carbon attached to a double bond is deshielded due to its sp^2 hybridization and some diamagnetic anisotropy. This effect can be seen in the ^{13}C NMR spectrum of cyclohexene (Fig. 27.8). Cyclohexene has a plane of symmetry that runs perpendicular to the double bond. As a result, we observe only three absorption peaks. There are two of each type of sp^3 carbon. Each of the double-bond carbons **c** has only one hydrogen, whereas each of the remaining carbons has two. As a result of a reduced NOE, the double-bond carbons (127 ppm) have a lower-intensity peak in the spectrum.

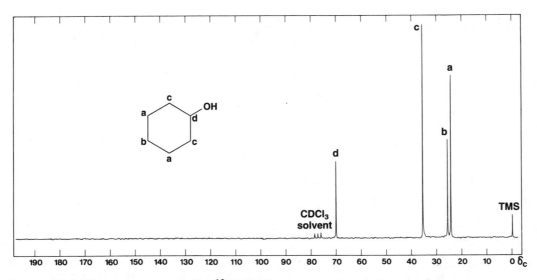

Figure 27.7 The proton-decoupled ^{13}C NMR spectrum of cyclohexanol.

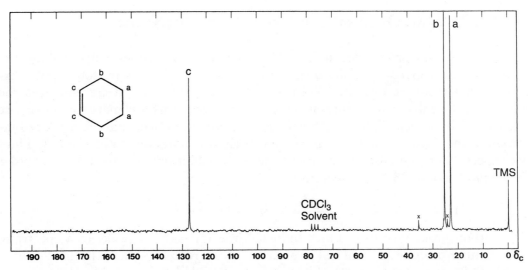

Figure 27.8 The proton-decoupled ^{13}C NMR spectrum of cyclohexene. (The peaks marked with an x are impurities.)

In Figure 27.9, the spectrum of cyclohexanone, the carbonyl carbon has the lowest intensity. This is due not only to reduced NOE (no hydrogen attached) but also to the long relaxation time of the carbonyl carbon (Section 27.6). Notice also that Figure 27.2 predicts the large chemical shift for this carbonyl carbon (211 ppm).

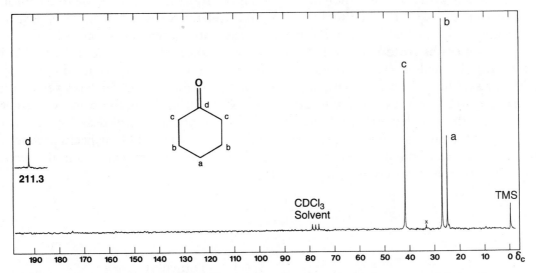

Figure 27.9 The proton-decoupled ^{13}C NMR spectrum of cyclohexanone. (The peak marked with an x is an impurity.)

27.6 NUCLEAR OVERHAUSER ENHANCEMENT (NOE)

When we obtain a proton-decoupled ^{13}C spectrum, the intensities of many of the carbon resonances increase significantly above those observed in a proton-coupled experiment. Carbon atoms with hydrogen atoms directly attached are enhanced the most, and the enhancement increases (but not always linearly) as more hydrogens are attached. This effect is known as the nuclear Overhauser effect, and the degree of increase in the signal is called the **nuclear Overhauser enhancement (NOE).** Thus, we expect that the intensity of the carbon peaks should increase in the following order in a typical carbon-13 NMR spectrum:

$$CH_3 > CH_2 > CH > C$$

Carbon atom relaxation times influence the intensity of peaks in a spectrum. When more protons are attached to a carbon atom, relaxation times become shorter resulting in more intense peaks. Thus, we expect methyl and methylene groups to be relatively more intense than that observed for quaternary carbon atoms where there are no attached protons. Thus, a weak intensity peak is observed for the quaternary carbon atom at 30 ppm in 2,2-dimethylbutane (Fig. 27.6). In addition, weak carbonyl carbon peaks are observed at 171 ppm in ethyl phenylacetate (Fig. 27.4) and at 211 ppm in cyclohexanone (Fig. 27.9).

27.7 COMPOUNDS WITH AROMATIC RINGS

Compounds with carbon–carbon double bonds or aromatic rings give rise to chemical shifts in the range from 100 to 175 ppm. Since relatively few other peaks appear in this range, a great deal of useful information is available when peaks appear here.

A **monosubstituted** benzene ring shows *four* peaks in the aromatic carbon area of a proton-decoupled ^{13}C spectrum, since the *ortho* and *meta* carbons are doubled by symmetry. Often the carbon with no protons attached, the *ipso* carbon, has a very weak peak due to a long relaxation time and a weak NOE. In addition, there are two larger peaks for the doubled *ortho* and *meta* carbons and a medium-sized peak for the *para* carbon. In many cases, it is not important to be able to assign all of the peaks precisely. In the example of toluene, shown in Figure 27.10, notice that carbons **c** and **d** are not easy to assign by inspection of the spectrum.

Toluene — Difficult to assign

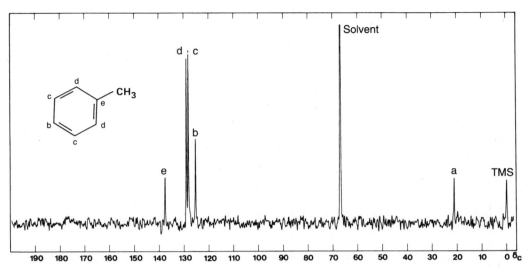

Figure 27.10 The proton-decoupled ^{13}C NMR spectrum of toluene.

In a proton-coupled ^{13}C spectrum, a monosubstituted benzene ring shows three doublets and one singlet. The singlet arises from the *ipso* carbon, which has no attached hydrogen. Each of the other carbons in the ring (*ortho*, *meta*, and *para*) has one attached hydrogen and yields a doublet.

Figure 27.4B is the proton-decoupled spectrum of ethyl phenylacetate, with the assignments noted next to the peaks. Notice that the aromatic-ring region shows four peaks between 125 and 135 ppm, consistent with a monosubstituted ring. There is one peak for the methyl carbon (13 ppm) and two peaks for the methylene carbons. One of the methylene carbons is directly attached to an electronegative oxygen atom and appears at 61 ppm, while the other is more shielded (41 ppm). The carbonyl carbon (an ester) has resonance at 171 ppm. All of the carbon chemical shifts agree with the values in the correlation chart (Fig. 27.1).

Depending on the mode of substitution, a symmetrically **disubstituted** benzene ring can show two, three, or four peaks in the proton-decoupled ^{13}C spectrum. The following drawings illustrate this for the isomers of dichlorobenzene.

Three unique carbon atoms   Four unique carbon atoms   Two unique carbon atoms

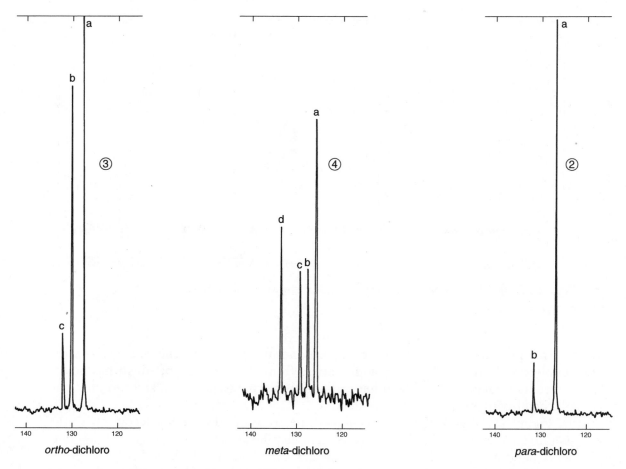

Figure 27.11 The proton-decoupled ^{13}C NMR spectra of the three isomers of dichlorobenzene (25 MHz).

Figure 27.11 shows the spectra of all three dichlorobenzenes, each of which has the number of peaks consistent with the analysis just given. You can see that ^{13}C NMR spectroscopy is very useful in the identification of isomers.

Most other polysubstitution patterns on a benzene ring yield six different peaks in the proton-decoupled ^{13}C NMR spectrum, one for each carbon. However, when identical substituents are present, watch carefully for planes of symmetry that may reduce the number of peaks.

REFERENCES

Textbooks

Friebolin, H. *Basic One- and Two-Dimensional NMR Spectroscopy*, 2nd ed. New York: VCH Publishers, 1993.

Gunther, H. *NMR Spectroscopy*, 2nd ed. New York: John Wiley & Sons, 1995.

Levy, G. C. *Topics in Carbon-13 Spectroscopy*. New York: John Wiley & Sons, 1984.

Levy, G. C., Lichter, R. L., and Nelson, G. L. *Carbon-13 Nuclear Magnetic Resonance Spectroscopy*, 2d ed. New York: John Wiley & Sons, 1980.

Macomber, R. S. *NMR Spectroscopy—Essential Theory and Practice*. New York: College Outline Series, Harcourt, Brace Jovanovich, 1988.

Macomber, R. S. *A Complete Introduction to Modern NMR Spectroscopy*. New York: John Wiley & Sons, 1997.

Pavia, D. L., Lampman, G. M., and Kriz, G. S. *Introduction to Spectroscopy*, 3rd ed. Philadelphia: Harcourt College Publishers, 2001.

Sanders, J. K. M., and Hunter, B. K. *Modern NMR Spectroscopy—A Guide for Chemists*, 2d ed. Oxford, England: Oxford University Press, 1993.

Silverstein, R. M., and Webster, F. X. *Spectrometric Identification of Organic Compounds*, 6th ed. New York: John Wiley & Sons, 1998.

Compilations of Spectra

Johnson, L. F. and Jankowski, W. C. *Carbon-13 NMR Spectra: A Collection of Assigned, Coded, and Indexed Spectra, 25 MHz*. New York: Wiley–Interscience, 1972.

Pouchert, C. J., and Behnke, J. *The Aldrich Library of ^{13}C and ^{1}H FT-NMR Spectra, 75 and 300 MHz*. Milwaukee, WI: Aldrich Chemical Company, 1993.

Pretsch, E., Clerc, T., Seibl, J., and Simon, W. *Tables of Spectral Data for Structure Determination of Organic Compounds*, 2nd ed. Berlin and New York: Springer-Verlag, 1989. Translated from the German by K. Biemann.

Web sites

http://www.aist.go.jp/RIODB/SDBS/menu-e.html

Integrated Spectral DataBase System for Organic Compounds, National Institute of Materials and Chemical Research, Tsukuba, Ibaraki 305-8565, Japan. This database includes infrared, mass spectra, and NMR data (proton and carbon-13) for a number of compounds.

http://www.chem.ucla.edu/~webnmr

UCLA Department of Chemistry and Biochemistry in connection with Cambridge University Isotope Laboratories, maintains a web site, WebSpectra, that provides NMR and IR spectroscopy problems for students to interpret. They provide links to other sites with problems for students to solve.

PROBLEMS

1. Predict the number of peaks that you would expect in the proton-decoupled ^{13}C spectrum of each of the following compounds. Problems 1a and 1b are provided as examples. Dots are used to show the nonequivalent carbon atoms in these two examples.

(a)

$$CH_3-C(=O)-O-CH_2-CH_3 \quad \text{Four peaks}$$

(b)

Five peaks

(c)

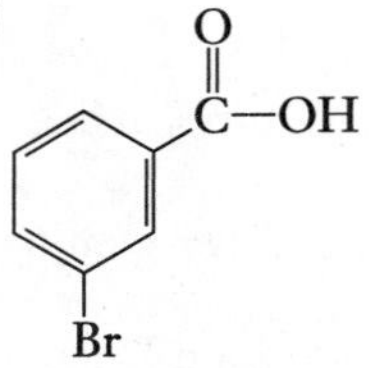

(d)

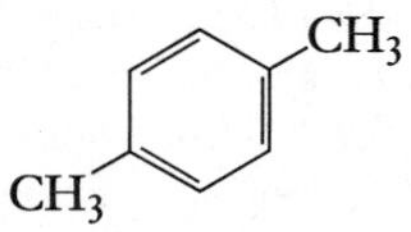

(e)

$$Br-CH_2-CH=CH-\overset{O}{\overset{\|}{C}}-O-CH_3$$

(f)

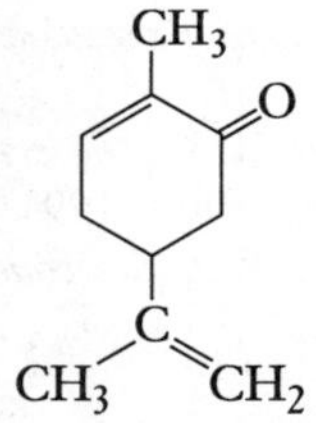

(g)

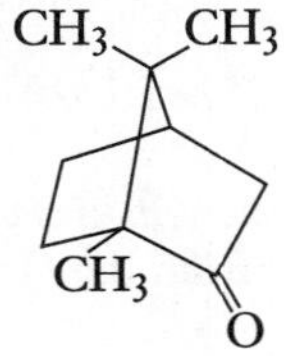

(h)

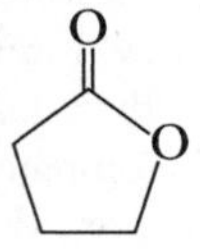

(i)

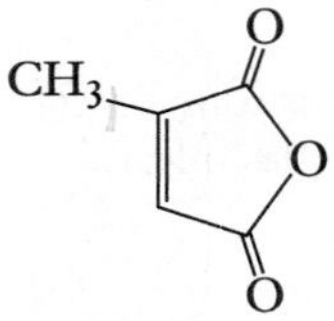

(j)

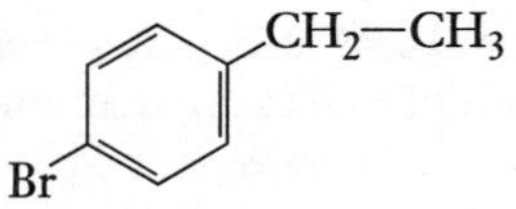

(k)

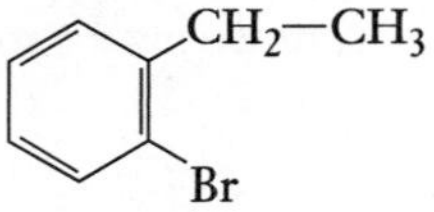

2. Following are the ^1H and ^{13}C spectra for each of four isomeric bromoalkanes with formula C_4H_9Br. Assign a structure to each pair of spectra.

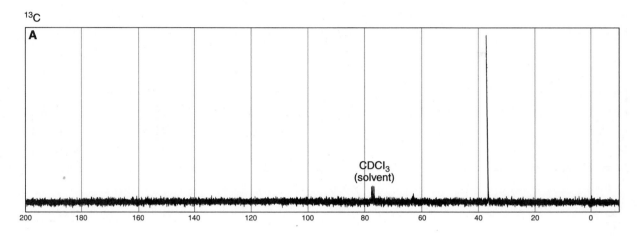

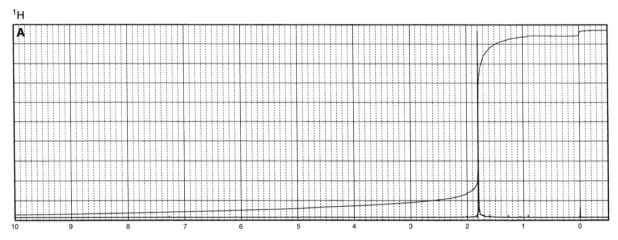

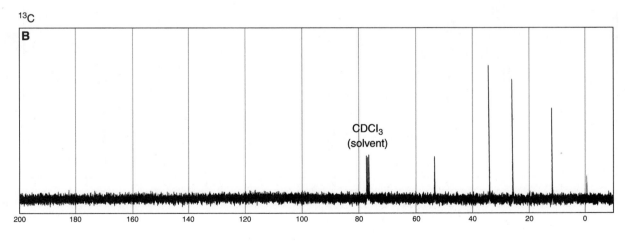

¹H

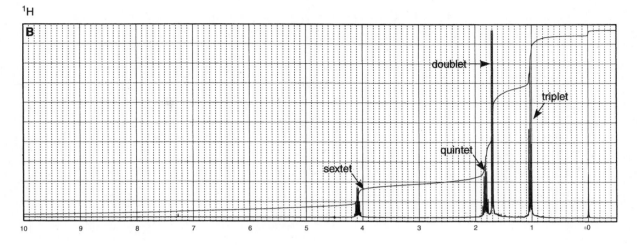

¹³C

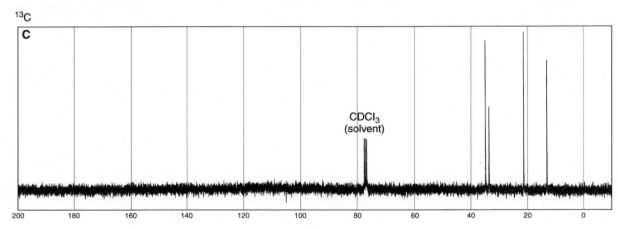

¹H

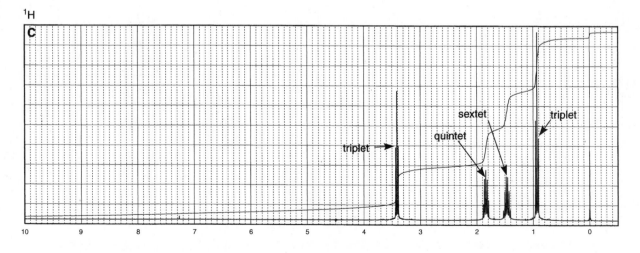

¹³C

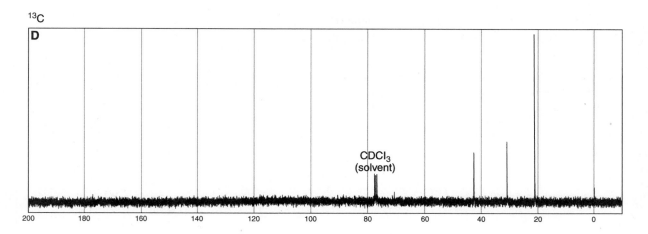

¹H

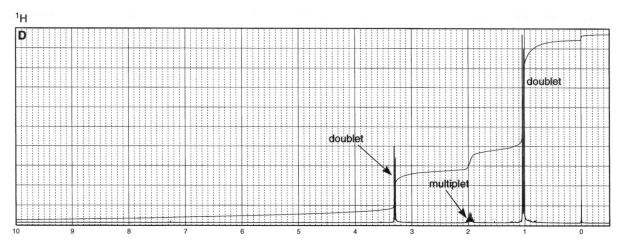

3. Following are the ^1H and ^{13}C spectra for each of three isomeric ketones with formula $C_7H_{14}O$. Assign a structure to each pair of spectra.

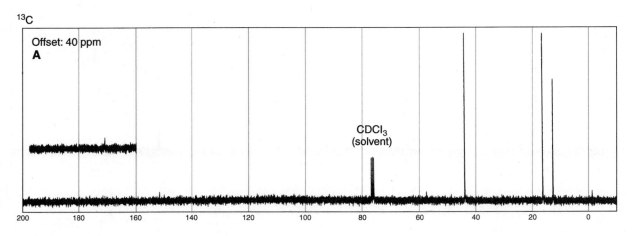

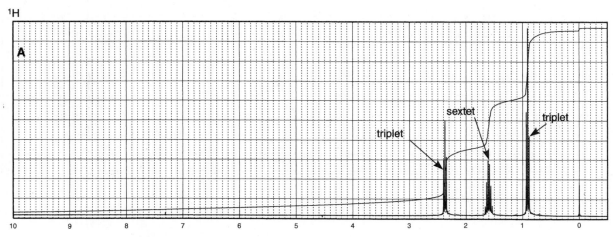

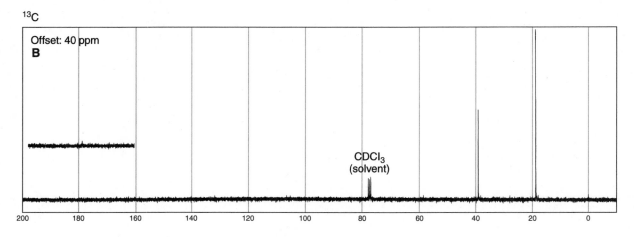

¹H

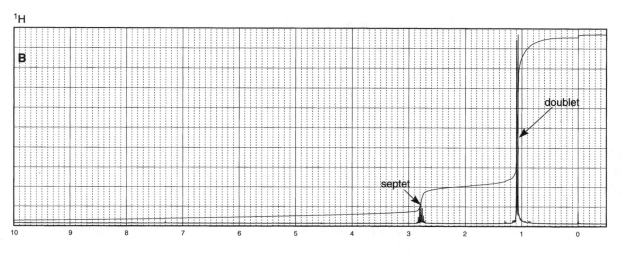

¹³C

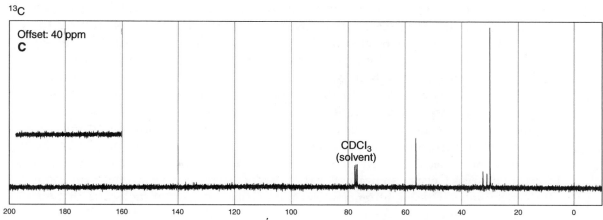

¹H

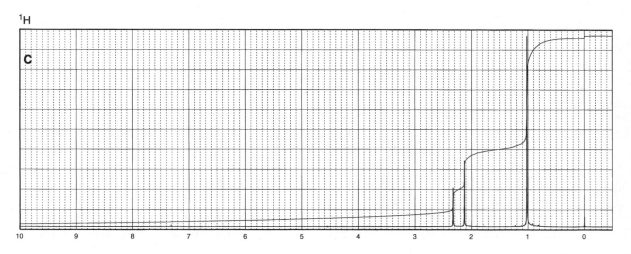

CHAPTER 28
Mass Spectrometry

In its simplest form, the mass spectrometer performs three essential functions. First, molecules are bombarded by a stream of high-energy electrons, converting some of the molecules to positive ions. Because of their high energy, some of these ions **fragment,** or break apart into smaller ions. All of these ions are accelerated in an electric field. Second, the accelerated ions are separated according to their mass-to-charge ratio in a magnetic or electric field. Finally, the ions with a particular mass-to-charge ratio are detected by a device that is able to count the number of ions that strike it. The output of the detector is amplified and fed to a recorder. The trace from the recorder is a **mass spectrum**—a graph of the number of particles detected as a function of mass-to-charge ratio.

Ions are formed in an **ionization chamber.** The sample is introduced into the ionization chamber using a sample inlet system. In the ionization chamber, a heated **filament** emits a beam of high-energy electrons. The filament is heated to several thousand degrees Celsius. In normal operation, the electrons have an energy of about 70 electron-volts. These high-energy electrons strike a stream of molecules that has been admitted from the sample system, and ionize the molecules in the sample stream by removing electrons from them. The molecules are thus converted into **radical-cations.**

$$e^- + M \rightarrow 2e^- + M^{+\bullet}$$

The energy required to remove an electron from an atom or molecule is its **ionization potential.** The ionized molecules are accelerated and focused into a beam of rapidly moving ions by means of charged plates.

From the ionization chamber, the beam of ions passes through a short field-free region. From there, the beam enters the **mass analyzer,** where the ions are separated according to their mass-to-charge ratio.

The detector of most instruments consists of a counter that produces a current proportional to the number of ions that strike it. Electron multiplier circuits allow accurate measurement of the current from even a single ion striking the detector. The signal from the detector is fed to a **recorder,** which produces the actual mass spectrum.

28.1 THE MASS SPECTRUM

The **mass spectrum** is a plot of ion abundance versus mass-to-charge (*m/e*) ratio. A typical mass spectrum is shown in Figure 28.1. The spectrum shown is that of dopamine, a substance that acts as a neurotransmitter in the central nervous system. The spectrum is displayed as a bar graph of percentage ion abundance (relative abundance) plotted against *m/e*.

$$HO-\underset{HO}{\bigcirc}-CH_2-CH_2-NH_2$$

Dopamine

The most abundant ion formed in the ionization chamber gives rise to the tallest peak in the mass spectrum, called the **base peak.** For dopamine, the base peak appears at *m/e* = 124. The relative abundances of all the other peaks in the spectrum are reported as percentages of the abundance of the base peak.

The beam of electrons in the ionization chamber converts some of the sample molecules into positive ions. Removal of a single electron from a molecule yields an ion whose weight is the actual molecular weight of the original molecule. This ion is the **molecular ion,** frequently symbolized as M^+. The value of *m/e* at which the molecular ion appears on the mass spectrum, assuming that the ion has only one electron removed, gives the molecular weight of the original molecule. In the mass spectrum of dopamine, the molecular ion appears at *m/e* = 153, the molecular weight of dopamine. If you can identify the molecular ion peak in the mass spectrum, you can use the spectrum to determine the molecular weight of an unknown substance. Ignoring the presence of heavy isotopes for the moment, the molecular ion peak corresponds to the heaviest particle observed in the mass spectrum.

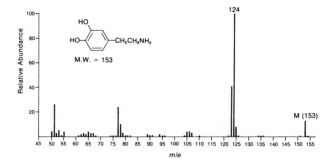

Figure 28.1 The mass spectrum of dopamine.

Molecules do not occur in nature as isotopically pure species. Virtually all atoms have heavier isotopes that occur in varying natural abundances. Hydrogen occurs largely as ^1H, but a few percent of hydrogen atoms occur as the isotope ^2H. Further, carbon normally occurs as ^{12}C, but a few percent of carbon atoms are the heavier isotope, ^{13}C. With the exception of fluorine, most other elements have a certain percentage of heavier isotopes that occur naturally. Peaks caused by ions bearing these heavier isotopes are also found in the mass spectrum. The relative abundances of these isotopic peaks are proportional to the abundances of the isotopes in nature. Most often, the isotopes occur at one or two mass units above the mass of the "normal" atom. Therefore, besides looking for the molecular ion (M^+) peak, you should also attempt to locate the M+1 and M+2 peaks. As will be demonstrated later, you can use the relative abundances of these M+1 and M+2 peaks to determine the molecular formula of the substance being studied.

The beam of electrons in the ionization chamber can produce the molecular ion. This beam also has sufficient energy to break some of the bonds in the molecule, producing a series of molecular fragments. Fragments that are positively charged are also accelerated in the ionization chamber, sent through the analyzer, detected, and recorded on the mass spectrum. These **fragment ion peaks** appear at *m/e* values corresponding to their individual masses. Very often a fragment ion, rather than the molecular ion, will be the most abundant ion produced in the mass spectrum (the base peak). A second means of producing fragment ions occurs with the molecular ion, which, once it is formed, is so unstable that it disintegrates before it can pass into the accelerating region of the ionization chamber. Lifetimes of less than 10^{-5} seconds are typical in this type of fragmentation. Those fragments that are charged then appear as fragment ions in the mass spectrum. As a result of these fragmentation processes, the typical mass spectrum can be quite complex, containing many more peaks than the molecular ion and M+1 and M+2 peaks. Structural information about a substance can be determined by examining the fragmentation pattern in the mass spectrum. Fragmentation patterns are discussed further in Section 28.3.

28.2 MOLECULAR FORMULA DETERMINATION

Mass spectrometry can be used to determine the molecular formulas of molecules that provide reasonably abundant molecular ions. Although there are at least two principal techniques of determining a molecular formula, only one will be described here.

The molecular formula of a substance can be determined through the use of **precise atomic masses.** High-resolution mass spectrometers are required for this method. Atoms are normally thought of as having integral atomic masses; for example, H = 1, C = 12, and O = 16. If you can determine atomic masses with sufficient precision, however, you find that the masses do not have values that are exactly integral. The mass of each atom actually differs from a whole mass number by a small fraction of a mass unit. The actual masses of some atoms are given in Table 28.1.

Table 28.1 Precise Masses of Some Common Elements

Element	Atomic Weight	Nuclide	Precise Mass
Hydrogen	1.00797	^1H	1.00783
		^2H	2.01410
Carbon	12.01115	^{12}C	12.0000
		^{13}C	13.00336
Nitrogen	14.0067	^{14}N	14.0031
		^{15}N	15.0001
Oxygen	15.9994	^{16}O	15.9949
		^{17}O	16.9991
		^{18}O	17.9992
Fluorine	18.9984	^{19}F	18.9984
Silicon	28.086	^{28}Si	27.9769
		^{29}Si	28.9765
		^{30}Si	29.9738
Phosphorus	30.974	^{31}P	30.9738
Sulfur	32.064	^{32}S	31.9721
		^{33}S	32.9715
		^{34}S	33.9679
Chlorine	35.453	^{35}Cl	34.9689
		^{37}Cl	36.9659
Bromine	79.909	^{79}Br	78.9183
		^{81}Br	80.9163
Iodine	126.904	^{127}I	126.9045

Depending on the atoms that are contained within a molecule, it is possible for particles of the same nominal mass to have slightly different measured masses when precise mass determinations are possible. To illustrate, a molecule whose molecular weight is 60 could be C_3H_8O, $C_2H_8N_2$, $C_2H_4O_2$, or CH_4N_2O. The species have the following precise masses:

$$C_3H_8O \quad 60.05754$$
$$C_2H_8N_2 \quad 60.06884$$
$$C_2H_4O_2 \quad 60.02112$$
$$CH_4N_2O \quad 60.03242$$

Observing a molecular ion with a mass of 60.058 would establish that the unknown molecule was C_3H_8O. Distinguishing among these possibilities is well within the capability of a modern high-resolution instrument.

In another method, these four compounds may also be distinguished by differences in the relative intensities of their M, M+1, and M+2 peaks. The predicted intensities are either calculated by formula or looked up in tables. Details of this method may be found in the references (page 491).

28.3 DETECTING HALOGENS

When chlorine or bromine is present in a molecule, the isotope peak that is two mass units heavier than the molecular ion (the M+2 peak) becomes very significant. The heavy isotope of each of these elements is two mass units heavier than the lighter isotope. The natural abundance of ^{37}Cl is 32.5% that of ^{35}Cl; the natural abundance of ^{81}Br is 98.0% that of ^{79}Br. When these elements are present, the M+2 peak becomes quite intense, and the pattern is characteristic of the particular halogen present. If a compound contains two chlorine or bromine atoms, a quite distinct M+4 peak should be observed, as well as an intense M+2 peak. In these cases, you should exercise caution in identifying the molecular ion peak in a mass spectrum, but the pattern of peaks is very characteristic of the nature of the halogen substitution in the molecule. Table 28.2 gives the relative intensities of isotope peaks for various combinations of bromine and chlorine atoms. The patterns of molecular ion and isotopic peaks observed with halogen substitution are shown in Figure 28.2. Examples of these patterns can be seen in the mass spectra of chloroethane (Fig. 28.3) and bromoethane (Fig. 28.4).

28.4 FRAGMENTATION PATTERNS

When the molecule has been bombarded by high-energy electrons in the ionization chamber of a mass spectrometer, besides losing one electron to form an ion, the molecule also absorbs some of the energy transferred in the collision between the molecule and the incident electrons. This extra energy puts the molecular ion in an excited vibrational state. The vibrationally excited molecular ion is often unstable, and may lose some of this extra energy by breaking apart into fragments. If the lifetime of an individual molecular ion is longer than 10^{-5} seconds, a peak corresponding to the molecular ion will be observed in the mass spectrum. Those molecular ions with lifetimes shorter than 10^{-5} seconds will break apart into fragments

Table 28.2 Relative Intensities of Isotope Peaks for Various Combinations of Bromine and Chlorine

Halogen	M	M+2	M+4	M+6
Br	100	97.7	—	—
Br_2	100	195.0	95.4	—
Br_3	100	293.0	286.0	93.4
Cl	100	32.6	—	—
Cl_2	100	65.3	10.6	—
Cl_3	100	97.8	31.9	3.47
BrCl	100	130.0	31.9	—
Br_2Cl	100	228.0	159.0	31.2
$BrCl_2$	100	163.0	74.4	10.4

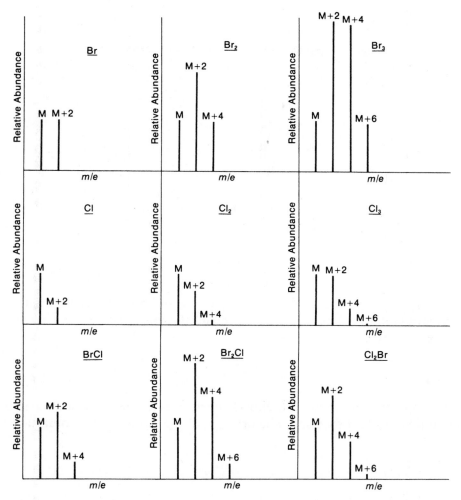

Figure 28.2 Mass spectra expected for various combinations of bromine and chlorine.

before they are accelerated within the ionization chamber. In such cases, peaks corresponding to the mass-to-charge ratios for these fragments will also appear in the mass spectrum. For a given compound, not all the molecular ions formed by ionization have precisely the same lifetime. The ions have a range of lifetimes; some individual ions may have shorter lifetimes than others. As a result, peaks are usually observed arising from both the molecular ion and the fragment ions in a typical mass spectrum.

For most classes of compounds, the mode of fragmentation is somewhat characteristic. In many cases, it is possible to predict how a molecule will fragment. Remember that the ionization of the sample molecule forms a molecular ion that not only carries a positive charge but that also has an unpaired electron. The molecular ion, then, is actually a **radical-cation,** and it contains an odd number of electrons. In the

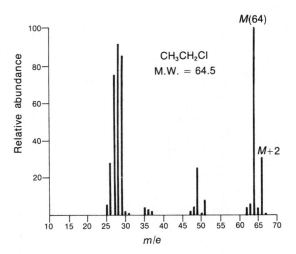

Figure 28.3 The mass spectrum of chloroethane.

structural formulas that follow, the radical-cation is indicated by enclosing the structure in square brackets. The positive charge and the unshared electron are shown as superscripts.

$$[R-CH_3]^{\cdot+}$$

When fragment ions form in the mass spectrometer, they almost always form by means of unimolecular processes. The pressure of the sample in the ionization cham-

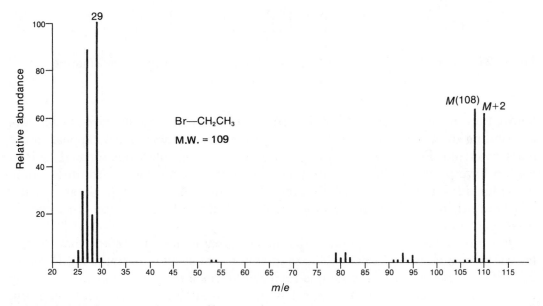

Figure 28.4 The mass spectrum of bromoethane.

ber is too low to permit a significant number of bimolecular collisions. Those unimolecular processes that require the least energy will give rise to the most abundant fragment ions.

Fragment ions are cations. Much of the chemistry of these fragment ions can be explained in terms of what is known about carbocations in solution. For example, alkyl substitution stabilizes fragment ions (and promotes their formation) in much the same way that it stabilizes carbocations. Those fragmentation processes that lead to more stable ions will be favored over processes that lead to the formation of less stable ions.

Fragmentation often involves the loss of an electrically neutral fragment. The neutral fragment does not appear in the mass spectrum, but you can deduce its existence by noting the difference in masses of the fragment ion and the original molecular ion. Again, processes that lead to the formation of a more stable neutral fragment will be favored over those that lead to the formation of a less stable neutral fragment. The loss of a stable neutral molecule, such as water, is commonly observed in the mass spectrometer.

A. Cleavage of One Bond

The most common mode of fragmentation involves the cleavage of one bond. In this process, the odd-electron molecular ion yields an odd-electron neutral fragment and an even-electron fragment ion. The neutral fragment that is lost is a **free radical,** whereas the ionic fragment is of the carbocation type. Cleavages that lead to the formation of more stable carbocations will be favored. Thus, the ease of fragmentation to form ions increases in the following order:

$$CH_3^+ < RCH_2^+ < R_2CH^+ < R_3C^+ < CH_2{=}CH{-}CH_2^+ < C_6H_5{-}CH_2^+$$
Increasing Ease of Formation $\rightarrow$

The following reactions show examples of fragmentation that take place with the cleavage of one bond:

$$\left[R{-}CH_3\right]^{+\cdot} \longrightarrow R^+ + {\cdot}CH_3$$

$$\left[R{-}\overset{\overset{\displaystyle O}{\|}}{C}{-}R'\right]^{+\cdot} \longrightarrow {}^+\overset{\overset{\displaystyle O}{\|}}{C}{-}R + {\cdot}R$$

$$\left[R{-}X\right]^{+\cdot} \longrightarrow R^+ + {\cdot}X$$

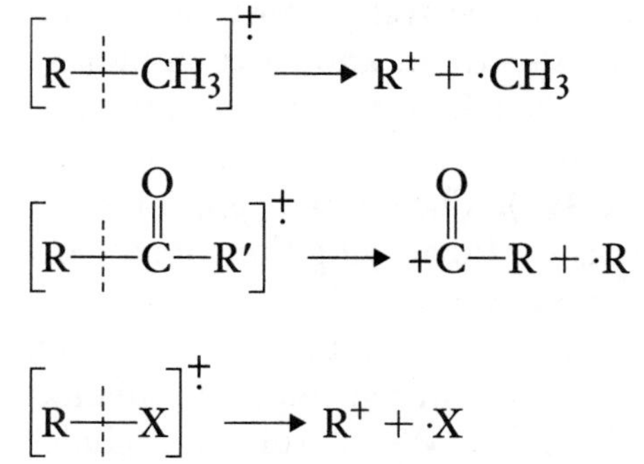

where X = halogen, OR,
SR, or NR$_2$, and where R =
H, alkyl, or aryl

B. Cleavage of Two Bonds

The next most important type of fragmentation involves the cleavage of two bonds. In this type of process, the odd-electron molecular ion yields an odd-electron fragment ion and an even-electron neutral fragment, usually a small, stable molecule. Examples of this type of cleavage are shown next:

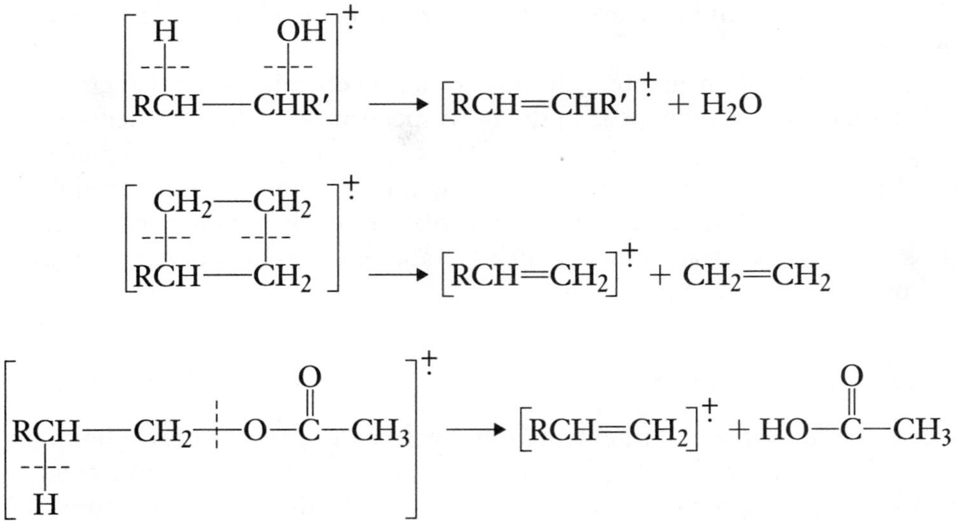

$$\left[\begin{array}{cc} \overset{H}{\underset{|}{\text{---}}} & \overset{OH}{\underset{|}{\text{---}}} \\ RCH & CHR' \end{array} \right]^{+\cdot} \longrightarrow \left[RCH{=}CHR' \right]^{+\cdot} + H_2O$$

$$\left[\begin{array}{cc} CH_2{-}CH_2 \\ \overset{|}{\text{---}} \;\; \overset{|}{\text{---}} \\ RCH{-}CH_2 \end{array} \right]^{+\cdot} \longrightarrow \left[RCH{=}CH_2 \right]^{+\cdot} + CH_2{=}CH_2$$

$$\left[\begin{array}{c} \overset{}{\underset{|}{RCH}}{-}CH_2{+}O{-}\overset{\overset{O}{\|}}{C}{-}CH_3 \\ \overset{|}{H} \end{array} \right]^{+\cdot} \longrightarrow \left[RCH{=}CH_2 \right]^{+\cdot} + HO{-}\overset{\overset{O}{\|}}{C}{-}CH_3$$

C. Other Cleavage Processes

In addition to the processes just mentioned, fragmentation reactions involving rearrangements, migrations of groups, and secondary fragmentations of fragment ions are also possible. These processes occur less often than the types of processes just described. Nevertheless, the pattern of molecular ion and fragment ion peaks observed in the typical mass spectrum is quite complex and unique for each particular molecule. As a result, the mass spectral pattern observed for a given substance can be compared with the mass spectra of known compounds as a means of identification. The mass spectrum is like a fingerprint. For a treatment of the specific modes of fragmentation characteristic of particular classes of compounds, refer to more advanced textbooks (see page 491). The unique appearance of the mass spectrum for a given compound is the basis for identifying the components of a mixture in the **gas chromatography–mass spectrometry (GC–MS)** technique (see Chapter 22, Section 22.13, page 352). The mass spectrum of every component in a mixture is compared with standard spectra stored in the computer memory of the instrument. The printed output produced by a GC-MS instrument includes an identification, based on the results of the computer matching of mass spectra.

28.5 INTERPRETED MASS SPECTRA

In this section, the mass spectra of some representative organic compounds are presented. The important fragment ion peaks in each mass spectrum are identified. In some of the examples, identification of the fragments is presented without explanation, although some interpretation is provided where an unusual or interesting process takes place. In the first example, that of butane, a more complete explanation of the symbolism used is offered.

Butane; C_4H_{10}, MW = 58 (Fig. 28.5)

$$CH_3\!+\!CH_2\!+\!CH_2\!+\!CH_3$$

$$\overset{\blacktriangleleft}{15} \qquad \overset{\blacktriangleleft}{29} \qquad \overset{\blacktriangleleft}{43}$$

In the structural formula of butane, the dashed lines represent the location of bond-breaking processes that occur during fragmentation. In each case, the fragmentation process involves the breaking of one bond to yield a neutral radical and a cation. The arrows point toward the fragment that bears the positive charge. This positive fragment is the ion that appears in the mass spectrum. The mass of the fragment ion is indicated beneath the arrow.

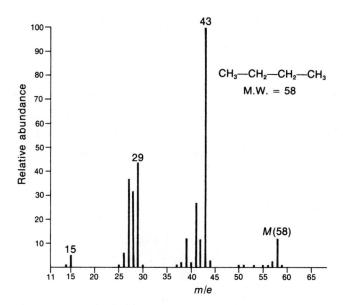

Figure 28.5 The mass spectrum of butane.

The mass spectrum shows the molecular ion at $m/e = 58$. Breaking of the C1–C2 bond yields a three-carbon fragment with a mass of 43.

$$CH_3-CH_2-CH_2 \!\mid\! CH_3 \longrightarrow CH_3-CH_2-CH_2^+ + \cdot CH_3$$
$$m/e = 43$$

Cleavage of the central bond yields an ethyl cation, with a mass of 29.

$$CH_3-CH_2 \!\mid\! CH_2-CH_3 \longrightarrow CH_3-CH_2^+ + \cdot CH_2-CH_3$$
$$m/e = 29$$

The terminal bond can also break to yield a methyl cation, which has a mass of 15.

$$CH_3 \!\mid\! CH_2-CH_2-CH_3 \longrightarrow CH_3^+ + \cdot CH_2-CH_2-CH_3$$
$$m/e = 15$$

Each of these fragments appears in the mass spectrum of butane and has been identified.

2,2,4-Trimethylpentane; C_8H_{18}, MW = 114 (Fig. 28.6)

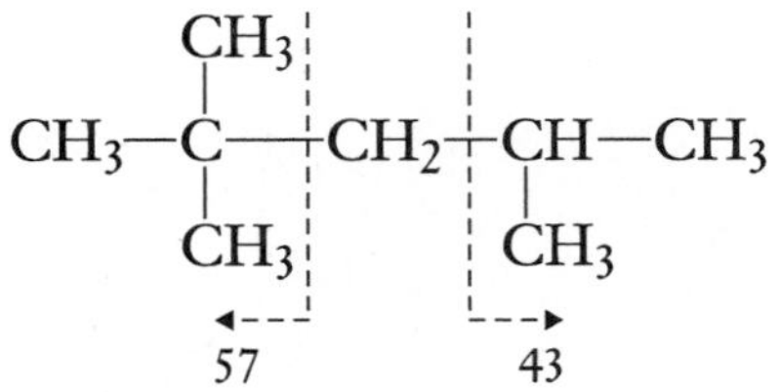

$$
\begin{array}{c}
\qquad\quad CH_3 \\
\qquad\quad | \\
CH_3-C \!\mid\! CH_2 \!\mid\! CH-CH_3 \\
\qquad\quad | \qquad\quad | \\
\qquad\quad CH_3 \qquad CH_3 \\
\qquad\;\; \underset{57}{\longleftarrow} \quad \underset{43}{\longrightarrow}
\end{array}
$$

Notice that in the case of 2,2,4-trimethylpentane, by far the most abundant fragment is the *tert*-butyl cation ($m/e = 57$). This result is not surprising, when one considers that the *tert*-butyl cation is a particularly stable carbocation.

Cyclopentane; C_5H_{10}, MW = 70 (Fig. 28.7)

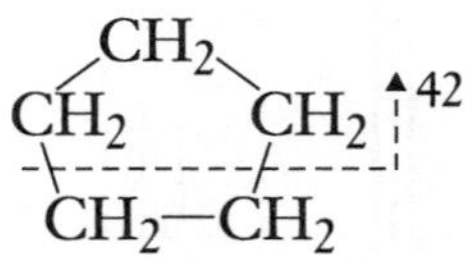

$$
\begin{array}{c}
\qquad CH_2 \\
CH_2 \qquad CH_2 \quad {}^{\blacktriangle 42} \\
CH_2-CH_2
\end{array}
$$

In the case of cyclopentane, the most abundant fragment results from the simultaneous cleavage of two bonds. This mode of fragmentation eliminates a neutral molecule of ethene (MW = 28), and results in the formation of a cation at $m/e = 42$.

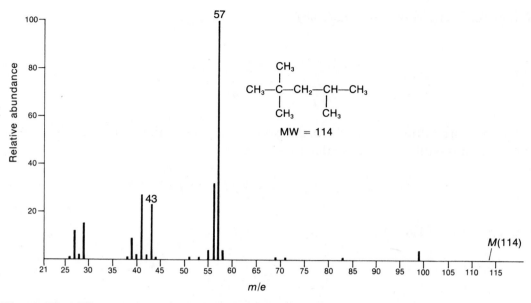

Figure 28.6 The mass spectrum of 2,2,4-trimethylpentane ("isooctane").

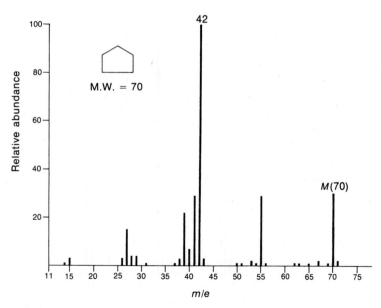

Figure 28.7 The mass spectrum of cyclopentane.

1-Butene; C_4H_8, MW = 56 (Fig. 28.8)

$$CH_2{=}CH{-}CH_2 {+} CH_3$$
41

An important fragment in the mass spectra of alkenes is the allyl cation (m/e = 41). This cation is particularly stable due to resonance.

$$\left[{}^{+}CH_2{-}CH{=}CH_2 \longleftrightarrow CH_2{=}CH{-}CH_2{}^{+} \right]$$

Toluene; C_7H_8; MW = 92 (Fig. 28.9)

$$\text{—}CH_2{+}H$$
91

When an alkyl group is attached to a benzene ring, preferential fragmentation occurs at a benzylic position to form a fragment ion of the formula $C_7H_7{}^{+}$ (m/e = 91). In the mass spectrum of toluene, loss of hydrogen from the molecule ion gives a strong peak at m/e = 91. Although it may be expected that this fragment ion peak is due to the benzyl carbocation, evidence suggests the benzyl carbocation actually rearranges to

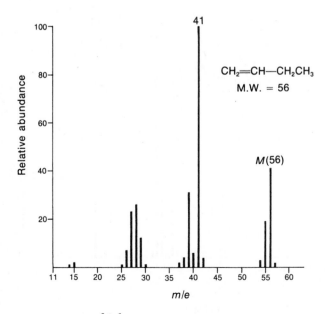

Figure 28.8 The mass spectrum of 1-butene.

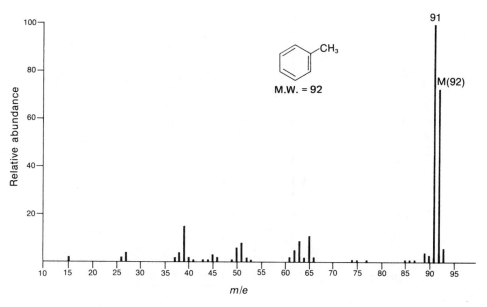

Figure 28.9 The mass spectrum of toluene.

form the **tropylium ion.** Isotope-labeling experiments tend to confirm the formation of the tropylium ion. The tropylium ion is a seven-carbon ring system that contains six electrons in π-molecular orbitals and hence is resonance stabilized in a manner similar to that observed in benzene.

Benzyl cation   Tropylium ion

1-Butanol; $C_4H_{10}O$, MW = 74 (Fig. 28.10)

$$CH_3-CH_2-CH_2\overset{\vdots}{|}CH_2-OH$$
$$31$$

The most important fragmentation reaction for alcohols is loss of an alkyl group:

$$\left[\begin{matrix} R' \\ | \\ R-C-OH \\ | \\ R'' \end{matrix}\right]^{+} \longrightarrow R\cdot + \begin{matrix} R' \\ \diagdown \\ \diagup \\ R'' \end{matrix} C=OH^{+}$$

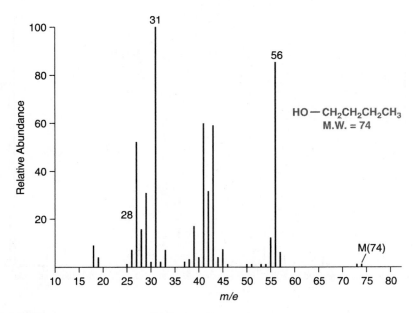

Figure 28.10 The mass spectrum of 1-butanol.

The largest alkyl group is the one that is lost most readily. In the spectrum of 1-butanol, the intense peak at $m/e = 31$ is due to the loss of a propyl group to form

$$\underset{H}{\overset{H}{\diagdown}}C{=}OH^{+}$$

A second common mode of fragmentation involves dehydration. Loss of a molecule of water from 1-butanol leaves a cation of mass 56.

$$CH_3{-}CH_2{-}\underset{H}{\overset{|}{CH}}{-}\underset{OH}{\overset{|}{CH_2}}{\uparrow}\,{}^{56}$$

Benzaldehyde; C_7H_6O, MW = 106 (Fig. 28.11)

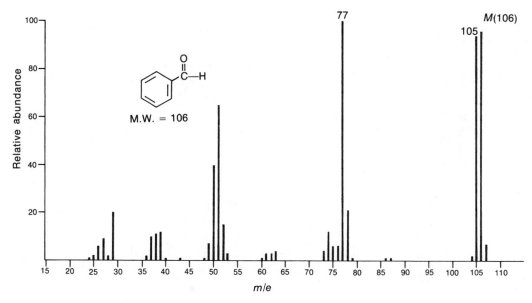

Figure 28.11 The mass spectrum of benzaldehyde.

The loss of a hydrogen atom from an aldehyde is a very favorable process. The resulting fragment ion is a benzoyl cation, a particular stable type of carbocation.

Loss of the entire aldehyde functional group leaves a phenyl cation. This ion can be seen in the spectrum of an *m/e* value of 77.

2-Butanone; C_4H_8O, MW = 72 (Fig. 28.12)

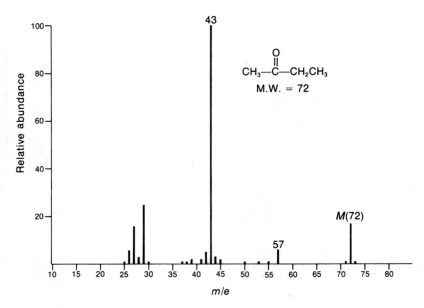

Figure 28.12 The mass spectrum of 2-butanone.

If the methyl group is lost as a neutral fragment, the resulting cation, an **acylium ion,** has an *m/e* value of 57. If the ethyl group is lost, the resulting acylium ion appears at an *m/e* value of 43.

$$CH_3-CH_2-\overset{\overset{\displaystyle O}{\|}}{C}\!\!\mid\!\!CH_3 \longrightarrow CH_3-CH_2-\overset{\overset{\displaystyle O}{\|}}{C}{}^+ + \cdot CH_3$$
$$m/e = 57$$

$$CH_3-CH_2\!\!\mid\!\!\overset{\overset{\displaystyle O}{\|}}{C}-CH_3 \longrightarrow CH_3-\overset{\overset{\displaystyle O}{\|}}{C}{}^+ + \cdot CH_2CH_3$$
$$m/e = 43$$

Acetophenone; C_8H_8O, MW = 120 (Fig. 28.13)

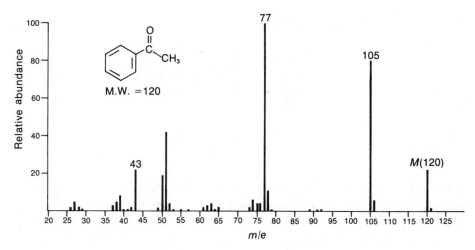

Figure 28.13 The mass spectrum of acetophenone.

Aromatic ketones undergo α-cleavage to lose the alkyl group and form the benzoyl cation (*m/e* = 105). This ion subsequently loses carbon monoxide to form the phenyl cation (*m/e* = 77). Aromatic ketones also undergo α-cleavage on the other side of the carbonyl group, forming an alkyl acylium ion. In the case of acetophenone, this ion appears at an *m/e* value of 43.

Propanoic acid; $C_3H_6O_2$, MW = 74 (Fig. 28.14)

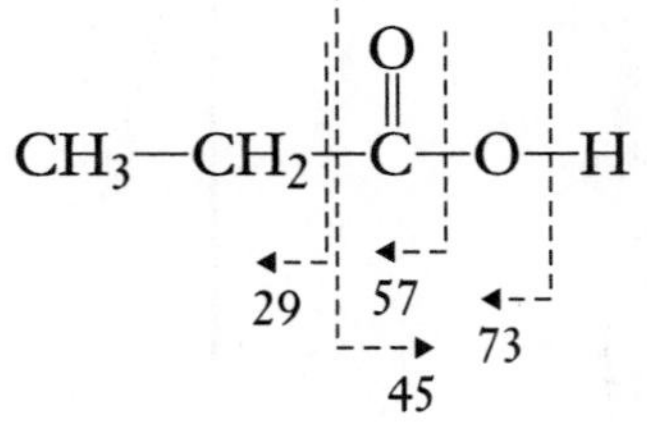

$$CH_3-CH_2-C(=O)-O-H$$

29 | 57
45 ---> 73

With short-chain carboxylic acids, the loss of OH and COOH through α-cleavage on either side of the C=O group may be observed. In the mass spectrum of propanoic acid, loss of OH gives rise to a peak at $m/e = 57$. Loss of COOH gives rise to a peak at $m/e = 29$. Loss of the alkyl group as a free radical, leaving the $COOH^+$ ion ($m/e = 45$), also occurs. The intense peak at $m/e = 28$ is due to additional fragmentation of the ethyl portion of the acid molecule.

Methyl butanoate; $C_5H_{10}O_2$, MW = 102 (Fig. 28.15)

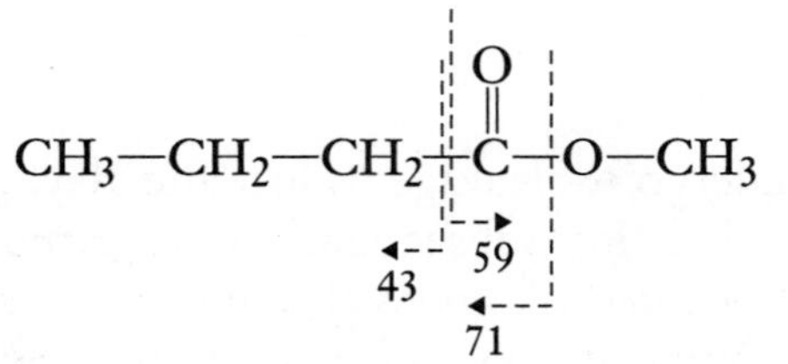

$$CH_3-CH_2-CH_2-C(=O)-O-CH_3$$

43 59
71

The most important of the α-cleavage reactions involves the loss of the alkoxy group from the ester to form the corresponding acylium ion, RCO^+. The acylium ion peak appears at $m/e = 71$ in the mass spectrum of methyl butanoate. A second important peak results from the loss of the alkyl group from the acyl portion of the ester molecule, leaving a fragment $CH_3-O-C=O^+$ that appears at $m/e = 59$. Loss of the carboxylate function group to leave the alkyl group as a cation gives rise to a peak at $m/e = 43$. The intense peak at $m/e = 74$ results from a rearrangement process (see Section 28.6).

1-Bromohexane; $C_6H_{13}Br$, MW = 165 (Fig. 28.16)

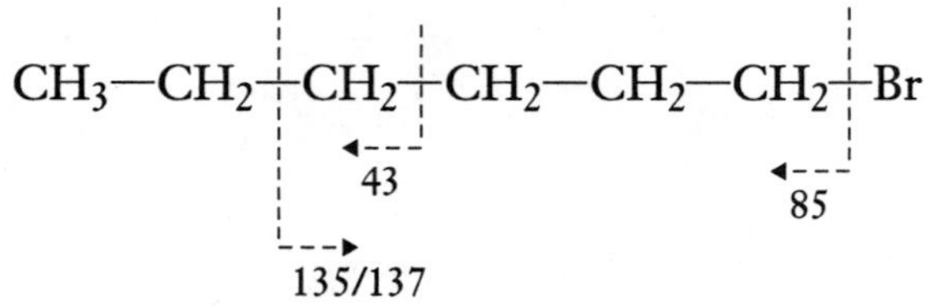

$$CH_3-CH_2-CH_2-CH_2-CH_2-CH_2-Br$$

43   85
135/137

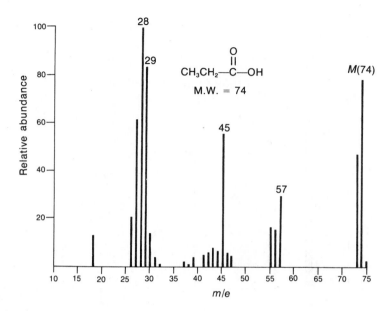

Figure 28.14 The mass spectrum of propanoic acid.

The most interesting characteristic of the mass spectrum of 1-bromohexane is the presence of the doublet in the molecular ion. These two peaks, of equal height, separated by two mass units, are strong evidence that bromine is present in the substance. Notice also that loss of the terminal ethyl group yields a fragment ion that still contains bromine (*m/e* = 135 and 137). The presence of the doublet demonstrates that this fragment contains bromine.

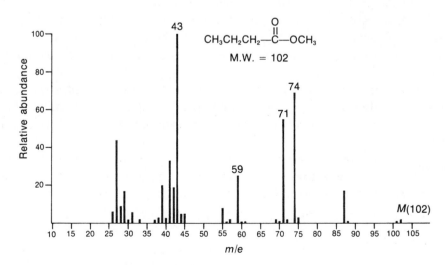

Figure 28.15 The mass spectrum of methyl butanoate.

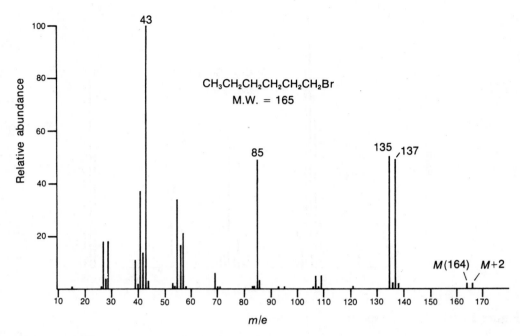

Figure 28.16 The mass spectrum of 1-bromohexane.

28.6 REARRANGEMENT REACTIONS

Because the fragment ions that are detected in a mass spectrum are cations, we can expect that these ions will exhibit behavior we are accustomed to associate with carbocations. It is well known that carbocations are prone to rearrangement reactions, converting a less-stable carbocation into a more stable one. These types of rearrangements are also observed in the mass spectrum. If the abundance of a cation is especially high, it is assumed that a rearrangement to yield a longer-lived cation must have occurred.

Other types of rearrangements are also known. An example of a rearrangement that is not normally observed in solution chemistry is the rearrangement of a benzyl cation to a tropylium ion. This rearrangement is seen in the mass spectrum of toluene (Fig. 28.9).

A particular type of rearrangement process that is unique to mass spectrometry is the **McLafferty rearrangement.** This type of rearrangement occurs when an alkyl chain of at least three carbons in length is attached to an energy-absorbing structure such as a phenyl or carbonyl group that can accept the transfer of a hydrogen ion. The mass spectrum of methyl butanoate (Fig. 28.15) contains a prominent peak at $m/e = 74$. This peak arises from a McLafferty rearrangement of the molecular ion.

$$m/e = 74$$

REFERENCES

Beynon, J. H. *Mass Spectrometry and Its Applications to Organic Chemistry.* Amsterdam: Elsevier, 1960.

Biemann, K. *Mass Spectrometry: Organic Chemical Applications.* New York: McGraw-Hill Book Company, 1962.

Budzikiewicz, H., Djerassi, C., and Williams, D. H. *Mass Spectrometry of Organic Compounds.* San Francisco: Holden-Day, 1967.

McLafferty, F. W., and Tureček, F. *Interpretation of Mass Spectra*, 4th ed. Mill Valley, CA: University Science Books, 1993.

Pavia, D. L., Lampman, G. M., and Kriz, G. S. *Introduction to Spectroscopy: A Guide for Students of Organic Chemistry*, 3rd ed. Philadelphia: Harcourt College Publishers, 2001.

Silverstein, R. M., and Webster, F. X. *Spectrometric Identification of Organic Compounds*, 6th ed. New York: John Wiley & Sons, 1998.

CHAPTER 29

Computational Chemistry

Since the beginnings of organic chemistry, somewhere in the middle of the nineteenth century, chemists have sought to visualize the three-dimensional characteristics of the all but invisible molecules that participate in chemical reactions. Concrete models that could be held in the hand were developed. Many kinds of model sets, such as framework, ball-and-stick, and space-filling models, were devised to allow people to visualize the spatial and directional relationships within molecules. These hand-held models were interactive, and they could be readily manipulated in space.

Today we can also use the computer to help us visualize these molecules. The computer images are also completely interactive, allowing us to rotate, scale, and change the type of model viewed at the press of a button or the click of a mouse. In addition, the computer can rapidly calculate many properties of the molecules that we view. This combination of visualization and calculation is often called **computational chemistry** or, more colloquially, **molecular modeling.**

There are two distinct methods of molecular modeling commonly used by organic chemists today. The first of these is **quantum mechanics,** and it involves the calculation of orbitals and their energies using solutions of the Schrödinger equation. The second method is not based on orbitals, but is founded on our knowledge of the way in which the bonds and angles in a molecule behave. Classical equations that describe the stretching of bonds and the bending of angles are used. This second approach is called **molecular mechanics.** The two types of calculation are used for different purposes and do not calculate the same types of molecular properties. In Part A of this chapter, molecular mechanics will be discussed; in Part B, the quantum mechanical methods, *ab initio* and semiempirical calculations, will be discussed.

Part A. Molecular Mechanics

29.1 MOLECULAR MECHANICS

Molecular mechanics (MM) was first developed in the early 1970s by two groups of chemical researchers: the Engler, Andose, and Schleyer group, and the Allinger group. In molecular mechanics, a mechanical **force field** is defined that is used to calculate an energy for the molecule under study. The energy calculated is often called the **strain energy** or **steric energy** of the molecule. The force field is composed of several components, such as bond-stretching energy, angle-bending energy, and bond-torsion energy. A typical force field expression[1] might be represented by the following composite expression:

$$E_{\text{strain}} = E_{\text{stretch}} + E_{\text{angle}} + E_{\text{torsion}} + E_{\text{oop}} + E_{\text{vdW}} + E_{\text{dipole}}$$

To calculate the final strain energy for a molecule, the computer systematically changes every bond length, bond angle, and torsional angle in the molecule, recalculating the strain energy each time, keeping each change that minimizes the total energy, and rejecting those that increase the energy. In other words, all the bond lengths and angles are changed until the energy of the molecule is *minimized*.

Each term contained in the composite expression (E_{strain}) is defined in Table 29.1. All these terms come from classical physics, not quantum mechanics. We will not discuss every term, but take E_{stretch} as an illustrative example. Classical mechanics says that a bond behaves like a spring. Each type of bond in a molecule can be assigned a normal bond length, x_0. If the bond is stretched or compressed, its potential energy will increase, and there will be a restoring force that attempts to restore the bond to its normal length. According to Hooke's Law, the restoring force is proportional to the size of the displacement

$$F = -k_i(x_f - x_0) \ \text{ or } \ F = -k_i(\Delta x)$$

where k_i is the **force constant** of the bond being studied (that is, the "stiffness" of the spring) and Δx is the change in bond length from the bond's normal length (x_0). The actual energy term that is minimized is given in Table 29.1. This equation indicates that all the bonds in the molecule contribute to the strain; it is a sum (Σ) starting with the first bond's contribution ($n = 1$) and proceeding through the contributions of all the other bonds (n_bonds).

These calculations are based on empirical data. To perform these calculations, the system must be **parameterized** with experimental data. To parameterize, a table of the normal bond lengths (x_0) and force constants (k_i) for every type of bond in the molecule has to be created. The program uses these experimental parameters to perform its calculations. The quality of the results from any molecular mechanics approach directly depends on how well the parameterization has been performed for each type of atom and bond that has to be considered. The MM procedure requires each of the factors in Table 29.1 to have its own parameter table.

Each of the first four terms in Table 29.1 is treated as a spring in the same manner as discussed for bond stretching. For instance, an angle also has a force constant k that resists a change in the size of the angle θ. In effect, in the first four terms the molecule

---

[1]Other force fields may be found that include more terms than this one, and that contain calculational methods more sophisticated than those shown here.

Table 29.1 Some of the Factors Contributing to a Molecular Force Field[a]

Type of Contribution	Illustration	Typical Equation
E_{stretch} (bond stretching)		$E_{\text{stretch}} = \displaystyle\sum_{i=1}^{n\_\text{bond}} (k_i/2)(x_i - x_0)^2$
E_{angle} (angle bending)		$E_{\text{angle}} = \displaystyle\sum_{j=1}^{n\_\text{angles}} (k_j/2)(\theta_j - \theta_0)^2$
E_{torsion} (bond torsion)		$E_{\text{torsion}} = \displaystyle\sum_{k=1}^{n\_\text{torsions}} (k_k/2)[1 + sp_k(\cos p_k\theta)]$
E_{oop} (out-of-plane bending)		$E_{\text{oop}} = \displaystyle\sum_{m=1}^{n\_\text{oops}} (k_m/2)d_m^{\,2}$
E_{vdW} (van der Waals repulsion)		$E_{\text{vdW}} = \displaystyle\sum_{i=1}^{n\_\text{atoms}} \sum_{j=1}^{n\_\text{atoms}} (E_i E_j)^{1/2}\left[\dfrac{1}{a_{ij}^{12}} - \dfrac{2}{a_{ij}^{6}}\right]$ $a_{ij} = r_{ij}/(R_i + R_j)$
E_{dipole} (electric dipole repulsion or attraction)		$E_{\text{vdW}} = K \displaystyle\sum_{i=1}^{n\_\text{atoms}} \sum_{j=i+1}^{n\_\text{atoms}} Q_i Q_j/r_{ij}^{2}$

[a] The factors selected here are similar to those in the "Tripos force field" used in the Alchemy III molecular modeling program.

is treated as a collection of interacting springs, and the energy of this collection of springs must be minimized. In contrast, the last two terms are based on electrostatic or "Coulomb" repulsions. Without going into detail for these terms, it should be understood that these terms must also be minimized.

29.2 MINIMIZATION AND CONFORMATION

The object of minimizing the strain energy is to find the lowest energy **conformation (equilibrium geometry)** of a molecule. Molecular mechanics does a very good job of finding conformations, because it varies bond distances, bond angles, torsional angles, and the positions of atoms in space. However, most minimizers have some limitations of which users must be aware. Many of the programs use a minimization procedure that will locate a local minimum in the energy, but will not necessarily find a global minimum. Figure 29.1 illustrates the problem.

In Figure 29.1, the molecule under consideration has two conformations that represent energy minima for the molecule. Many minimizers will not automatically find the lowest energy conformation, the **global minimum.** The global minimum will be found only when the structure of your starting molecule is already close to the global minimum's conformation. For instance, if the starting structure corresponds to the point labeled B on the curve, then the global minimum will be found. However, if your starting molecule is not close to the global minimum in structure, a **local minimum** (one nearby) may be found. In Figure 29.1, if your starting structure corresponds to the point labeled A, then a local minimum will be found, instead of the global minimum. Some of the more expensive programs always find the global minimum, because they use more sophisticated minimization procedures that depend on random (Monte Carlo) changes instead of sequential ones.

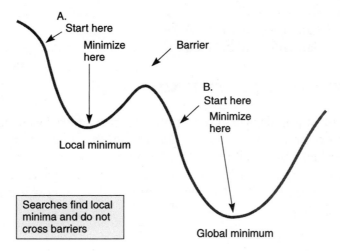

Figure 29.1 Global and local energy minima.

However, unless the program has specifically dealt with this problem, the user must be careful to avoid finding a false local minimum when the global minimum is expected. It may be necessary to use several different starting structures to discover the global minimum for a given molecule.

29.3 LIMITATIONS OF MOLECULAR MECHANICS

From our discussion thus far, it should be obvious that molecular mechanics was developed to find the lowest energy conformation of a given molecule, or to compare the energies of several conformations of the same molecule. Molecular mechanics calculates a "strain energy," not a thermodynamic energy such as a heat of formation. Procedures based on quantum mechanics and statistical mechanics are required to calculate thermodynamic energies. Therefore, it is very dangerous to compare the strain energies of two *different* molecules. For instance, molecular mechanics can make a good evaluation of the relative energies of *anti-* and *gauche*-butane conformations, but it cannot fruitfully compare butane and cyclobutane. Isomers can be compared only if they are very closely related. The *cis-* and *trans*-isomers of 1,2-dimethylcyclohexane, or those of 2-butene, can be compared. However, the isomers 1-butene and 2-butene cannot be compared; one is a monosubstituted alkene, whereas the other is disubstituted.

Molecular mechanics will perform the following tasks quite well:

1. It will give good estimates for the actual bond lengths and angles in a molecule.
2. It will find the best conformation for a molecule, but you must watch out for local minima!

Molecular mechanics will not calculate the following properties:

1. It will not calculate thermodynamic properties such as the heat of formation[2] of a molecule.
2. It will not calculate electron distributions, charges, or dipole moments.
3. It will not calculate molecular orbitals or their energies.
4. It will not calculate infrared, NMR, or ultraviolet spectra.

29.4 CURRENT IMPLEMENTATIONS

With time, the most popular version of molecular mechanics has become the version developed by Norman Allinger and his research group. The original program from this group was called MM1. The program has undergone constant revisions and improvements, and the current Allinger versions are now designated MM2 and MM3. However, many other versions of molecular mechanics are now available from both

---

[2]Some of the latest versions are now parameterized to give heats of formation.

private and commercial sources. Some popular commercial programs that now incorporate their own force fields and parameters include Alchemy III, Alchemy 2000, CAChe, Personal CAChe, HyperChem, Insight II, PC Model, MacroModel, Spartan, PC Spartan, MacSpartan, and Sybyl. You should also realize, however, that there are many modeling programs that do not have molecular mechanics or minimization. These programs will "clean up" a structure that you create by attempting to make every bond length and angle "perfect." With these programs every sp^3 carbon will have 109-degree angles, and every sp^2 carbon will have perfect 120-degree angles. Using one of these programs is equivalent to using a standard model set that has connectors and bond with perfect angles and lengths. If you intend to find a molecule's preferred conformation, be sure you use a program that has a force field and performs a true minimization procedure. Also remember that you may have to control the starting structure's geometry in order to find the correct result.

Part B. *Ab Initio* and Semiempirical Methods

29.5 QUANTUM MECHANICAL CALCULATIONS

In Part A, the application of **molecular mechanics** to solving chemical problems was discussed. Molecular mechanics is very good at providing estimates of the bond lengths and angles in a molecule and finding the best geometry or conformation of a molecule. However, the application of **quantum mechanics** is required to find good estimates of the thermodynamic, spectroscopic, and electronic properties of a molecule. In Part B, we will discuss the application of quantum mechanics to organic molecules.

Quantum mechanics computer programs can calculate heats of formation and the energies of transition states. The shapes of orbitals can be displayed in three dimensions. Important properties can be mapped onto the surface of a molecule. With these programs the chemist can visualize concepts and properties in a way that the mind cannot readily imagine. Often this visualization is the key to understanding or to solving a problem.

29.6 *Ab Initio* AND SEMIEMPIRICAL CALCULATIONS

To determine the electronic structure and energy of a molecule, quantum mechanics requires the formulation of a wavefunction Ψ (psi) that describes the distribution of all the electrons within the system. The nuclei are assumed to have relatively small motions and to be essentially fixed in their equilibrium positions (Born–Oppenheimer approximation). The average energy of the system is calculated by using the Schrödinger equations as

$$E = \int \Psi^* H \Psi \, d\tau \, / \int \Psi^* \Psi \, d\tau$$

where H, the Hamiltonian operator, is a multiterm function that evaluates all the potential energy contributions (electron–electron repulsions and nuclear–electron attractions) and the kinetic energy terms for each electron in the system.

Because we can never know the true wavefunction Ψ for the molecule, we must guess at the nature of this function. According to the **Variation Principle,** a cornerstone idea in quantum mechanics, we can continue to guess at this function forever, and never reach the true energy of the system, which will always be lower than our best guess. Because of the Variation Principle, we can formulate an approximate wavefunction and then consistently vary it until we minimize the energy of the system (as calculated using the Schrödinger equation). When we reach the variational minimum, the resulting wavefunction is often a good approximation of the system we are studying. Of course, you cannot just make any guess and get good results. It has taken theoretical chemists quite a few years to learn how to formulate both wavefunctions and Hamiltonian operators that yield results that agree quite closely with experiment. Today, however, most methods for performing these calculations have been well established, and computational chemists have devised easy-to-use computer programs, which can be used by any chemist to calculate molecular wavefunctions.

Molecular quantum-mechanical calculations can be divided into two classes: *ab initio* (Latin: "from the beginning" or "from first principles") and *semiempirical*.

1. ***Ab initio* calculations** use the fully correct Hamiltonian for the system and attempt a complete solution without using any experimental parameters.
2. **Semiempirical calculations** generally use a simplified Hamiltonian operator and incorporate experimental data or a set of parameters that can be adjusted to fit experimental data.

Ab initio calculations require a great deal of computer time and memory, because every term in the calculations is evaluated explicitly. Semiempirical calculations have more modest computer requirements, allowing the calculations to be completed in a shorter time, and making it possible to treat larger molecules. Chemists generally use semiempirical methods whenever possible, but it is useful to understand both methods when solving a problem.

29.7 SOLVING THE SCHRÖDINGER EQUATION

The Hamiltonian. The exact form of the Hamiltonian operator, which is a collection of potential energy (electrostatic attraction and repulsion terms) and kinetic energy terms, is standardized now and need not concern us here. However, all the programs require the **Cartesian coordinates** (locations in three-dimensional space) of all the atoms and a **connectivity matrix** that specifies which atoms are bonded and how (single, double, triple, H-bond, etc.). In modern programs, the user draws or constructs the molecule on the computer screen, and the program automatically constructs the atomic-coordinate and connectivity matrices.

The Wavefunction.   It is not necessary for the user to construct or guess at a trial wavefunction—the program will do this. However, it is important to understand how the wavefunctions are formulated, because the user frequently has a choice of methods. The complete molecular wavefunction is made up of a determinant of molecular orbitals:

$$\psi = \begin{vmatrix} \phi_1(1) & \phi_2(1) & \phi_3(1) & \text{........} & \phi_n(1) \\ \phi_1(2) & \phi_2(2) & \phi_3(2) & \text{........} & \phi_n(2) \\ \phi_1(n) & \phi_2(n) & \phi_3(n) & \text{......} & \phi_n(n) \end{vmatrix}$$

The molecular orbitals $\phi_i(n)$ must be made up from some type of mathematical function. They are usually made up by a **linear combination of atomic orbitals** χ_j (LCAO) from each of the atoms that make up the molecule.

$$\phi_i(n) = \Sigma_j c_{ji} \chi_j = c_1\chi_1 + c_2\chi_2 + c_3\chi_3 \cdots$$

This combination includes all the orbitals in the *core* and the *valence shell* of each atom in the molecule. The complete set of orbitals χ_j is called the **basis set** for the calculation. When an *ab initio* calculation is performed, most programs require the user to choose the basis set.

29.8 BASIS SET ORBITALS

It should be apparent that the most obvious basis set to use for an *ab initio* calculation is the set of hydrogen-like atomic orbitals, 1*s*, 2*s*, 2*p*, and so on, that we are all familiar with from atomic structure and bonding theory. Unfortunately, these "actual" orbitals present computational difficulties because they have radial nodes when they are associated with the higher shells of an atom. As a result, a more convenient set of functions was devised by Slater. These **Slater-type orbitals (STOs)** differ from the hydrogen-like orbitals in that they have no radial nodes, but they have the same angular terms and overall shape. More importantly, they give good results (those that agree with experiment) when used in semiempirical and *ab initio* calculations.

Slater-Type Orbitals.   The radial term of an STO is an exponential function with the form $R_{n1} = r^{(n-1)} e^{[-(-Z-s)r/n]}$, where Z is the nuclear charge of the atom, and s is a "screening constant" that reduces the nuclear charge Z that is "seen" by an electron. Slater formulated a set of rules to determine the values of s that are required to produce orbitals that agree in shape with the customary hydrogen-like orbitals.

Radial Expansion and Contraction.   A problem with simple STOs is that they do not have the ability to vary their radial size. Today it is common to use two or more simpler STOs so that expansion and contraction of the orbitals can occur during the calculation. For instance, if we take two functions such as $R(r) = r\, e^{(-\zeta r)}$ with different values of ζ (Fig. 29.2), the larger value of ζ gives an orbital more contracted around

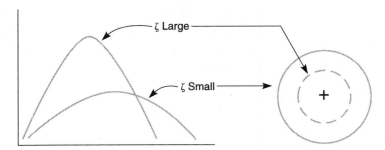

Figure 29.2 Variation of the radial size of an STO with the value of the exponent, ζ (zeta).

the nucleus (an inner STO), and the smaller value of ζ gives an orbital extended further out from the nucleus (an outer STO). By using these two functions in different combinations, any size STO can be generated.

Gaussian-Type Orbitals.   Eventually the original Slater-type orbitals were abandoned, and *simulated* STOs built from Gaussian functions were used. The most common basis set of this kind is the **STO-3G basis set,** which uses three Gaussian functions (3G) to simulate each one-electron orbital. A Gaussian function is of the type $R(r) = re^{(-\alpha r^2)}$.

In the STO-3G basis set, the coefficients of the Gaussian functions are selected so as to give the best fit to the corresponding Slater-type orbitals. In this formulation, for instance, a hydrogen electron is represented by a single STO (a $1s$ type orbital), which is simulated by a combination of three Gaussian functions. An electron on any Period 2 element (Li to Ne) will be represented by five STOs ($1s$, $2s$, $2p_x$, $2p_y$, $2p_z$) each simulated by three Gaussian functions. Each electron in a given molecule will have its own STO. (The molecule is literally built up by a series of one-electron orbitals. A spin function is also included so that no two of the one-electron orbitals are exactly the same.)

Split-Valence Basis Sets.   In a further step of evolution, it is now common to abandon trying to simulate the hydrogen-like orbitals with STOs and to simply use an optimized combination of the Gaussian functions themselves for the basis set. The 3-21G basis set has largely replaced the STO-3G basis set for all but the largest molecules. The 3-21G symbolism means that three Gaussian functions are used for the wavefunction of each core electron, but that the wavefunctions of the valence electrons are "split" two-to-one (21) between inner and outer Gaussian functions, allowing the valence shell to expand or contract in size as shown in Figure 29.3.

A larger basis set (and one that requires more calculation time) is 6-31G, which uses six Gaussian "primitives" and a three-to-one split in the valence shell orbitals.

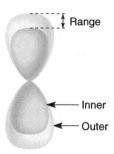

Figure 29.3 Split-valence orbitals.

Polarization Basis Sets. Both the 3-21G and 6-31G basis sets can be extended to 3-21G* and 6-31G*. The asterisk (*) indicates that these are **polarization sets,** where the next higher type of orbital is included (for instance a, p orbital can be polarized by adding a d orbital function). Polarization allows deformation of the orbital toward the bond on one side of the atom. This is illustrated in Figure 29.4.

The largest basis set in current use is 6-311G*. Because it is computationally intensive, it is used only for **single-point calculations** (a calculation on a fixed geometry—no minimization performed). Other basis sets include the 6-31G** (which includes six d orbitals per atom instead of the usual five) and the 6-31+G* or 6-31++G* sets, which include diffuse s functions (electrons at a larger distance from the nucleus) to better deal with anions.

29.9 SEMIEMPIRICAL METHODS

It would be quite impossible to give a short and complete overview of the various semi-empirical methods that have evolved over time. One must really get into the mathematical details of the method to understand what approximations have been made in each case, and what kinds of empirical data have been included. In many of these methods it is common to omit integrals that are expected (either from experience or for

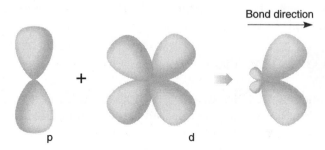

Figure 29.4 Polarization orbitals.

theoretical reasons) to have negligible values. Certain integrals are stored in a table and are not calculated each time the program is applied. For instance, the **frozen core approximation** is often used. This approximation assumes that the *completed shells* of the atom do not differ from one atom to another in the same period. All the core calculations are stored in a table, and they are simply looked up when needed. This makes the computation much easier to perform.

One of the more popular semiempirical methods in use today is AM-1. The parameters in this method work especially well for organic molecules. In fact, whenever possible, you should try to solve your problem using a semiempirical method such as AM-1 before you resort to an *ab initio* calculation. Also popular are MINDO/3 and MNDO, which are often found together in a computational package called MOPAC. If you are performing semiempirical calculations on inorganic molecules, you must make sure the method you use is optimized for transition metals. Two popular methods for inorganic chemists wishing to involve metals in their calculations are PM-3 and ZINDO.

29.10 Picking a Basis Set for *Ab Initio* Calculations

When you perform an *ab initio* calculation, it is not always easy to know which basis set to use. Normally you should not use more complexity than is needed to answer your question or solve the problem. In fact, it may be desirable to determine the approximate geometry of the molecule using *molecular mechanics*. Many programs will allow you to use the result of a molecular mechanics **geometry optimization** as a starting point for an *ab initio* calculation. If possible, you should do so to save computational time.

Most of the time 3-21G is a good starting point for an *ab initio* calculation, but if you have a very large molecule you may wish to use STO-3G, a simpler basis set. Avoid doing geometry optimizations with the larger basis sets. Often you can do the geometry optimization first with 3-21G (or a semiempirical method), and then polish up the result with a **single-point energy** calculation with a larger basis set, such as 6-31G. You should "move up the ladder": AM-1 to STO-3G to 3-21G to 6-31G, etc. If you do not see any change in the results as you move up to successively more complex basis sets, it is generally fruitless to continue. If you include elements beyond Period 2, use polarization sets (PM-3 for semiempirical). Some programs have special sets for cations and anions, or for radicals. If the result does not match experimental results, you may not have used the correct basis set.

29.11 Heats of Formation

In classical thermodynamics, the **heat of formation**, ΔH_f, is defined as the energy consumed (endothermic reaction) or released (exothermic reaction) when the molecule is formed from its elements at standard conditions of pressure and temperature. The elements are assumed to be in their standard states.

$$2 \text{ C (graphite)} + 3 \text{ H}_2 \text{ (g)} \rightarrow \text{C}_2\text{H}_6 \text{ (g)} + \Delta H_f \qquad (25°\text{C})$$

Both *ab initio* and semiempirical programs calculate the energy of a molecule as its "heat of formation." This heat of formation, however, is not identical to the thermodynamic function, and it is not always possible to make direct comparisons.

Heats of formation in semiempirical calculations are generally calculated in kcal/mole (1 kcal = 4.18 kJoules) and are similar but not identical to the thermodynamic function. The AM-1, PM-3, and MNDO methods are parameterized by fitting them to a set of experimentally determined enthalpies. They are calculated from the binding energy of the system. The **binding energy** is the energy released when molecules are formed from their separated electrons and nuclei. The semiempirical heat of formation is calculated by subtracting atomic heats of formation from the binding energy. For most organic molecules, AM-1 will calculate the heat of formation correctly to within a few kilocalories per mole.

In *ab initio* calculations the heat of formation is given in **hartrees** (1 hartree = 627.5 kcal/mole = 2625 kJoule/mole). In the *ab initio* calculation, the heat of formation is best defined as total energy. Like the binding energy, the **total energy** is the energy released when molecules are formed from their separated electrons and nuclei. This "heat of formation" always has a large negative value and does not relate well to the thermodynamic function.

Although these values are not related directly to the thermodynamic values, they can be used to compare the energies of isomers (molecules of the same formula), such as *cis*- and *trans*-2-butene, or of tautomers such as acetone in its enol and keto forms.

$$\Delta E = \Delta H_f(\text{isomer 2}) - \Delta H_f(\text{isomer 1})$$

It is also possible to compare the energies of balanced chemical equations by subtracting the energies of the products from the reactants.

$$\Delta E = [\Delta H_f(\text{product 1}) + \Delta H_f(\text{product 2})] - [\Delta H_f(\text{reactant 1}) + \Delta H_f(\text{reactant 2})]$$

29.12 GRAPHIC MODELS AND VISUALIZATION

Although the solution of the Schrödinger equation minimizes the *energy* of the system and gives a heat of formation, it also calculates the shapes and energies of all the molecular orbitals in the system. A big advantage of semiempirical and *ab initio* calculations, therefore, is the ability to determine the energies of the individual molecular orbitals and to plot their shapes in three dimensions. For chemists investigating chemical reactions, two molecular orbitals are of paramount interest: the HOMO and the LUMO.

The **HOMO,** the highest occupied molecular orbital, is the last orbital in the molecule to be filled with electrons. The **LUMO,** the lowest unoccupied molecular orbital, is the first empty orbital in the molecule. These two orbitals are often called **frontier orbitals** (Fig. 29.5).

The frontier orbitals are similar to the valence shell of the molecule. They are where most of the chemical reactions occur. For instance, if a reagent is going to react with a Lewis base, the electron pair of the base must be placed into an empty orbital of

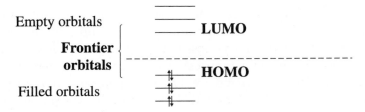

Figure 29.5 Frontier orbitals.

the acceptor molecule (Fig. 29.6A). The most available orbital is the LUMO. By examining the structure of the LUMO, one can determine the most likely spot where the addition will take place—usually at the atom where the LUMO has its biggest lobe. Conversely, if a Lewis acid attacks a molecule, it will bond to electrons that already exist in the molecule under attack (Fig. 29.6B). The most likely spot for this attack would be the atom where the HOMO has its biggest lobe (the electron density should be greatest at that site). Where it is not obvious which molecule is the electron pair donor, the HOMO that has the highest orbital energy will usually be the electron pair donor, placing electrons into the LUMO of the other molecule (Fig. 29.6C). The frontier orbitals, HOMO and LUMO, are where most chemical reactions occur.

29.13 SURFACES

Chemists use many kinds of hand-held models to visualize molecules. A framework model best represents the angles, lengths, and directions of bonds. A molecule's size and shape are probably best represented by a space-filling model. In quantum mechanics a model similar to the space-filling model can be generated by plotting a surface that represents all the points where the electron density of the molecule's wavefunction has a constant value. If this value is picked correctly, the resulting surface will resemble the surface of a space-filling model (Fig. 29.7A). This type of surface is called an **electron-density surface**. The electron-density surface is useful for visualizing the size and shape of the molecule, but it does not reveal the position of the nuclei, bond lengths, or angles,

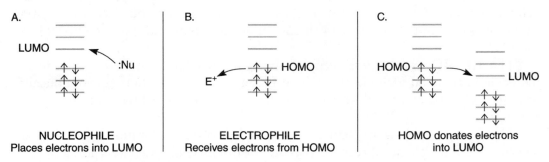

Figure 29.6 Reactions involving frontier orbitals.

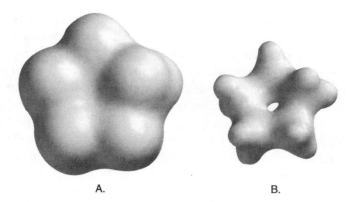

Figure 29.7 Cyclopentane. **(A)** Electron-density surface. **(B)** Bond-density surface.

because you cannot see inside of the surface. The electron-density value used to define this surface will be quite low because electron density falls off with increasing distance from the nucleus. If you choose a higher value of electron density when you plot this surface, a **bond-density surface** will be obtained (Fig. 29.7B). This surface will not give you an idea of the size or shape of the molecule but it will reveal where the bonds are located because the electron density will be higher where bonding is taking place.

29.14 MAPPING PROPERTIES ONTO A DENSITY SURFACE

It is also possible to map a calculated property onto an electron-density surface. Because all three Cartesian coordinates are used to define the points on the surface, the property must be mapped in color, with the colors of the spectrum, red–orange–yellow–green–blue, representing a range of values. In effect, this is a four-dimensional plot (*x*, *y*, *z*, + property mapped). One of the most common plots of this type is the **density–electrostatic potential,** or **density–elpot,** plot (Fig. 29.8A). The electrostatic

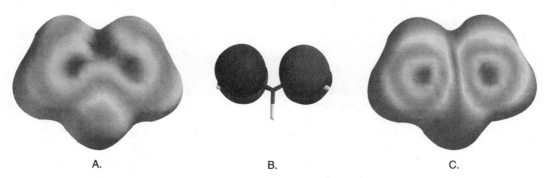

Figure 29.8 Allyl cation. **(A)** Density–electron potential surface. **(B)** LUMO surface. **(C)** Density–LUMO surface.

potential is determined by placing a unit positive charge at each point on the surface and measuring the interaction energy of this charge with the nuclei and electrons in the molecule. Depending on the magnitude of the interaction, that point on the surface is painted one of the colors of the spectrum. In the Spartan program, areas of high electron density are painted red or orange, and areas of lower electron density are plotted blue or green. When you view such a plot, the polarity of the molecule is immediately apparent.

The second common type of mapping plots values of one of the frontier orbitals (either the HOMO or the LUMO) in color on the density surface (Fig. 29.8C). The color values plotted correspond to the value of the orbital where it intersects the surface. For a density–LUMO plot, for instance, the "hot spot" would be where the LUMO (Fig. 29.8B) has its largest lobe. Because the LUMO is empty, this would be a bright blue area. In a density–HOMO plot, a bright red area would be the "hot spot."

REFERENCES

Molecular Mechanics

Casanova, J. "Computer-Based Molecular Modeling in the Curriculum." Computer Series 155. *Journal of Chemical Education, 70* (November 1993): 904.

Clark, T. *A Handbook of Computational Chemistry—A Practical Guide to Chemical Structure and Energy Calculations.* New York: John Wiley & Sons, 1985.

Lipkowitz, K. B. "Molecular Modeling in Organic Chemistry—Correlating Odors with Molecular Structure." *Journal of Chemical Education, 66* (April 1989): 275.

Tripos Associates. *Alchemy III—User's Guide.* St. Louis: Tripos Associates, 1992.

Ulrich, B., and Allinger, N. L. *Molecular Mechanics,* ACS Monograph 177. Washington, D.C.: American Chemical Society, 1982.

AB INITIO AND SEMIEMPIRICAL METHODS

Introductory

Hehre, W. J., Shusterman, A. J., and Nelson, J. E. *The Molecular Modeling Workbook for Organic Chemistry.* Wavefunction, Inc., 18401 Von Karman Ave., Suite 370, Irvine, CA 92612, 1998.

Hehre, W. J., Burke, L. D., Shusterman, A. J., and Pietro, W. J. *Experiments in Computational Organic Chemistry.* Wavefunction, Inc., 18401 Van Karman Avenue, Suite 370, Irvine, CA 92612, 1993.

Hypercube, Inc. *HyperChem Computational Chemistry.* HyperCube, Inc., 419 Phillip Street, Waterloo, Ontario, Canada N2L 3X2, 1996.

Shusterman, G. P., and Shusterman, A. J. "Teaching Chemistry with Electron Density Models." *Journal of Chemical Education, 74* (July 1997): 771.

Wavefunction, Inc. "PC-Spartan—Tutorial and User's Guide." Wavefunction, Inc., 18401 Von Karman Ave., Suite 370, Irvine, CA 92612, 1996.

Advanced

Clark, T. *Computational Chemistry.* New York: Wiley-Interscience, 1985.

Fleming, I. *Frontier Orbitals and Organic Chemical Reactions.* New York: John Wiley & Sons, 1976.

Fukui, K. *Accounts of Chemical Research, 4* (1971): 57.

Hehre, W. J., Random, L., Schleyer, P. v. R., and Pople, J. A. *Ab Initio Molecular Orbital Theory.* New York: Wiley-Interscience, 1986.

Woodward, R. B., and Hoffmann, R. *The Conservation of Orbital Symmetry.* Weinheim: Verlag Chemie, 1970.

Woodward, R. B., and Hoffmann, R. *Accounts of Chemical Research, 1* (1968): 17.

CHAPTER 30

Guide to the Chemical Literature

Often you may need to go beyond the information contained in the typical organic chemistry textbook and to use reference material in the library. At first glance, using library materials may seem formidable because of the numerous sources the library contains. If, however, you adopt a systematic approach, the task can prove rather useful. This description of various popular sources and an outline of logical steps to follow in the typical literature search should be helpful.

30.1 LOCATING PHYSICAL CONSTANTS: HANDBOOKS

To find information on routine physical constants, such as melting points, boiling points, indices of refraction, and densities, you should first consider a handbook. Examples of suitable handbooks are

Budavari, S., ed. *The Merck Index*, 12th ed. Whitehouse Station, NJ: Merck, 1996.
Dean, J. A., ed. *Lange's Handbook of Chemistry*, 14th ed. New York: McGraw-Hill, 1992.
Lide, D. R., ed. *CRC Handbook of Chemistry and Physics*, 80th ed. Boca Raton, FL: CRC Press, 1999.
Aldrich Handbook of Fine Chemicals. Milwaukee, WI: Sigma-Aldrich, 2000.

Each of these references is discussed in detail in Chapter 4. The *CRC Handbook* is the reference consulted most often because the book is so widely available. There are, however, distinct advantages to using the other handbooks. The *CRC Handbook* uses the *Chemical Abstracts* system of nomenclature that requires you to identify the parent name; 3-methyl-1-butanol is listed as 1-butanol, 3-methyl.

The Merck Index has fewer compounds listed, but for the ones listed, there is far more information provided. If the compound is a medicinal or natural product, this is the reference of choice. This handbook contains literature references on the isolation and synthesis of a compound, along with certain properties of medicinal interest, such as toxicity. *Lange's Handbook* and the *Aldrich Handbook* list compounds in alphabetical order; 3-methyl-1-butanol is listed as 3-methyl-1-butanol.

A more complete handbook that is usually housed in the library is

Buckingham, J., ed. *Dictionary of Organic Compounds*. New York: Chapman & Hall/Methuen, 1982–1992.

This is a revised version of an earlier four-volume handbook, edited by I. M. Heilbron and H. M. Bunbury. In its present form, it consists of seven volumes with 10 supplements.

30.2 GENERAL SYNTHETIC METHODS

Many standard introductory textbooks in organic chemistry provide tables that summarize most of the common reactions, including side reactions, for a given class of compounds. These books also describe alternative methods of preparing compounds.

Brown, W. H., and Foote, C. *Organic Chemistry*, 2nd ed. Philadelphia: Saunders College Publishing, 1998.

Carey, F. A. *Organic Chemistry*, 4th ed. New York: McGraw-Hill, 2000.

Ege, S. *Organic Chemistry*, 4th ed. Boston: Houghton-Mifflin, 1999.

Fessenden, R. J., and Fessenden, J. S. *Organic Chemistry*, 6th ed. Pacific Grove, CA: Brooks/Cole, 1998.

Fox, M. A., and Whitesell, J. K. *Organic Chemistry*, 2nd ed. Boston: Jones & Bartlett, 1997.

Jones, M., Jr. *Organic Chemistry*, New York: W. W. Norton, 2000.

Loudon, G. M. *Organic Chemistry*, 3rd ed. Menlo Park, CA: Benjamin/Cummings, 1995.

McMurry, J. *Organic Chemistry*, 5th ed. Pacific Grove, CA: Brooks/Cole, 2000.

Morrison, R. T., and Boyd, R. N. *Organic Chemistry*, 7th ed. Englewood Cliffs, NJ: Prentice-Hall, 1999.

Smith, M. B., and March, J. *Advanced Organic Chemistry*, 5th ed. New York: John Wiley & Sons, 2000.

Solomons, T. W. G., and Fryhle, C. *Organic Chemistry*, 7th ed. New York: John Wiley & Sons, 2000.

Streitwieser, A., Heathcock, C. H., and Kosower, E. M. *Introduction to Organic Chemistry*, 4th ed. New York: Prentice Hall, 1992.

Vollhardt, K. P. C., and Schore, N. E. *Organic Chemistry*, 3rd ed. New York: W. H. Freeman, 1999.

Wade, L. G., Jr. *Organic Chemistry*, 4th ed. Englewood Cliffs, NJ: Prentice-Hall, 1999.

30.3 SEARCHING THE CHEMICAL LITERATURE

If the information you are seeking is not available in any of the handbooks mentioned in Section 30.1 or if you are searching for more detailed information than they can provide, then a proper literature search is in order. Although an examination of standard textbooks can provide some help, you often must use all the resources of the library, including journals, reference collections, and abstracts. The following sections of this chapter outline how the various types of sources should be used and what sort of information can be obtained from them.

The methods for searching the literature discussed in this chapter use mainly printed materials. Modern search methods also make use of computerized databases and are discussed in Section 30.11. These are vast collections of data and bibliographic materials that can be scanned very rapidly from remote computer terminals. Although computerized searching is widely available, its use may not be readily accessible to undergraduate students. The following references provide excellent introductions to the literature of organic chemistry:

Carr, C. "Teaching and Using Chemical Information." *Journal of Chemical Education*, *70* (September 1993): 719.

Maizell, R. E. *How to Find Chemical Information*, 3rd ed. New York: John Wiley & Sons, 1998.

Smith, M. B., and March, J. *Advanced Organic Chemistry*, 5th ed. New York: John Wiley & Sons, 2000.

Somerville, A. N. "Information Sources for Organic Chemistry, 1: Searching by Name Reaction and Reaction Type." *Journal of Chemical Education*, *68* (July 1991): 553.

Somerville, A. N. "Information Sources for Organic Chemistry, 2: Searching by Functional Group." *Journal of Chemical Education*, *68* (October 1991): 842.

Somerville, A. N. "Information Sources for Organic Chemistry, 3: Searching by Reagent." *Journal of Chemical Education, 69* (May 1992): 379.

Wiggins, G. *Chemical Information Sources*. New York: McGraw-Hill, 1991. Integrates printed materials and computer sources of information.

30.4 COLLECTIONS OF SPECTRA

Collections of infrared, nuclear magnetic resonance, and mass spectra can be found in the following catalogues of spectra:

Cornu, A., and Massot, R. *Compilation of Mass Spectral Data*, 2nd ed. London: Heyden and Sons, Ltd., 1975.

High-Resolution NMR Spectra Catalog. Palo Alto, CA: Varian Associates. Vol. 1, 1962; Vol. 2, 1963.

Johnson, L. F., and Jankowski, W. C. *Carbon-13 NMR Spectra*. New York: John Wiley & Sons, 1972.

Pouchert, C. J. *Aldrich Library of Infrared Spectra*, 3rd ed. Milwaukee: Aldrich Chemical Co., 1981.

Pouchert, C. J. *Aldrich Library of NMR Spectra*, 2nd ed. Milwaukee: Aldrich Chemical Co., 1983.

Pouchert, C. J. *Aldrich Library of FT-IR Spectra*, 2nd ed. Milwaukee: Aldrich Chemical Co., 1997.

Pouchert, C. J., and Behnke, J. *Aldrich Library of ^{13}C and ^{1}H FT NMR Spectra*. Milwaukee: Aldrich Chemical Co., 1993.

Sadtler Standard Spectra. Philadelphia: Sadtler Research Laboratories. Continuing collection.

Stenhagen, E., Abrahamsson, S., and McLafferty, F. W. *Registry of Mass Spectral Data*. New York: Wiley-Interscience, 1974. Four-volume set.

The American Petroleum Institute has also published collections of infrared, nuclear magnetic resonance, and mass spectra.

30.5 ADVANCED TEXTBOOKS

Much information about synthetic methods, reaction mechanisms, and reactions of organic compounds is available in any of the many current advanced textbooks in organic chemistry. Examples of such books are

Carey, F. A., and Sundberg, R. J. *Advanced Organic Chemistry. Part A. Structure and Mechanisms: Part B. Reactions and Synthesis*, 4th ed. New York: Plenum Press, 2000.

Carruthers, W. *Some Modern Methods of Organic Synthesis*, 3rd ed. Cambridge, UK: Cambridge University Press, 1986.

Corey, E. J., and Cheng, Xue-Min. *The Logic of Chemical Synthesis*. New York: John Wiley & Sons, 1989.

Fieser, L. F., and Fieser, M. *Advanced Organic Chemistry*. New York: Reinhold, 1961.

Finar, I. L. *Organic Chemistry*, 6th ed. London: Longman Group, Ltd., 1986.

House, H. O. *Modern Synthetic Reactions*, 2nd ed. Menlo Park, CA: W. H. Benjamin, 1972.

Noller, C. R. *Chemistry of Organic Compounds*, 3rd ed. Philadelphia: W. B. Saunders, 1965.

Smith, M. B. *Organic Synthesis*. New York: McGraw-Hill, 1994.

Smith, M. B., and March, J. *Advanced Organic Chemistry*, 5th ed. New York: John Wiley & Sons, 2000.

Stowell, J. C. *Intermediate Organic Chemistry*, 2nd ed. New York: John Wiley & Sons, 1993.

Warren, S. *Organic Synthesis: The Disconnection Approach*. New York: John Wiley & Sons, 1982.

These books often contain references to original papers in the literature for students wanting to follow the subject further. Consequently you obtain not only a review of the subject from such a textbook but also a key reference that is helpful toward a more extensive literature search. The textbook by Smith and March is particularly useful for this purpose.

30.6 SPECIFIC SYNTHETIC METHODS

Anyone interested in locating information about a particular method of synthesizing a compound should first consult one of the many general textbooks on the subject. Useful ones are

Anand, N., Bindra, J. S., and Ranganathan, S. *Art in Organic Synthesis*, 2nd ed. New York: John Wiley & Sons, 1988.

Barton, D., and Ollis, W. D., eds. *Comprehensive Organic Chemistry*. Oxford: Pergamon Press, 1979. Six-volume set.

Buehler, C. A., and Pearson, D. E. *Survey of Organic Syntheses*. New York: Wiley-Interscience, 1970 and 1977. Two-volume set.

Carey, F. A., and Sundberg, R. J. *Advanced Organic Chemistry. Part B. Reactions and Synthesis*, 4th ed. New York: Plenum Press, 2000.

Compendium of Organic Synthetic Methods. New York: Wiley-Interscience, 1971–1995. This is a continuing series, now in eight volumes.

Fieser, L. F., and Fieser, M. *Reagents for Organic Synthesis*. New York: Wiley-Interscience, 1967–1999. This is a continuing series, now in 19 volumes.

Greene, T. W., and Wuts, P. G. M. *Protective Groups in Organic Synthesis*, 3rd ed. New York: John Wiley & Sons, 1999.

House, H. O. *Modern Synthetic Reactions*, 2nd ed. Menlo Park, CA: W. H. Benjamin, 1972.

Larock, R. C. *Comprehensive Organic Transformations*, 2nd ed. New York: Wiley-VCH, 1999.

Mundy, B. P., and Ellerd, M. G. *Name Reactions and Reagents in Organic Synthesis*. New York: John Wiley & Sons, 1988.

Patai, S., ed. *The Chemistry of the Functional Groups*. London: Interscience, 1964–present. This series consists of many volumes, each one specializing in a particular functional group.

Smith, M. B., and March, J. *Advanced Organic Chemistry*, 5th ed. New York: John Wiley & Sons, 2000.

Trost, B. M., and Fleming, I. *Comprehensive Organic Synthesis*. Amsterdam: Pergamon/Elsevier Science, 1992. This series consist of nine volumes plus supplements.

Vogel, A. I. Revised by members of the School of Chemistry, Thames Polytechnic. *Vogel's Textbook of Practical Organic Chemistry, Including Qualitative Organic Analysis*, 5th ed. London: Longman Group, Ltd., 1989.

Wagner, R. B., and Zook, H. D. *Synthetic Organic Chemistry*. New York: John Wiley & Sons, 1956.

More specific information, including actual reaction conditions, exists in collections specializing in organic synthetic methods. The most important of these is

Organic Syntheses. New York: John Wiley & Sons, 1921–present. Published annually.

Organic Syntheses, Collective Volumes. New York: John Wiley & Sons, 1941–1993.

Vol. 1, 1941, Annual Volumes 1–9
Vol. 2, 1943, Annual Volumes 10–19
Vol. 3, 1955, Annual Volumes 20–29
Vol. 4, 1963, Annual Volumes 30–39
Vol. 5, 1973, Annual Volumes 40–49
Vol. 6, 1988, Annual Volumes 50–59
Vol. 7, 1990, Annual Volumes 60–64
Vol. 8, 1993, Annual Volumes 65–69
Vol. 9, 1998, Annual Volumes 70–74

It is much more convenient to use the collective volumes where the earlier annual volumes of *Organic Syntheses* are combined in groups of nine or ten in the first six collective volumes (Volumes 1–6), and then in groups of five for the next three volumes

(Volumes 7, 8, and 9). Useful indices are included at the end of each of the collective volumes that classify methods according to the type of reaction, type of compound prepared, formula of compound prepared, preparation or purification of solvents and reagents, and use of various types of specialized apparatus.

The main advantage of using one of the *Organic Syntheses* procedures is that they have been tested to make sure that they work as written. Often, an organic chemist will adapt one of these tested procedures to the preparation of another compound. One of the features of the advanced organic textbook by Smith and March is that it includes references to specific preparative methods contained in *Organic Syntheses*.

More advanced material on organic chemical reactions and synthetic methods may be found in any one of a number of annual publications that review the original literature and summarize it. Examples include

Advances in Organic Chemistry: Methods and Results. New York: John Wiley & Sons, 1960–present.
Annual Reports of the Chemical Society, Section B. London: Chemical Society, 1905–present. Specifically, the section on *Synthetic Methods*.
Annual Reports in Organic Synthesis. Orlando, FL: Academic Press, 1985–1995.
Progress in Organic Chemistry. New York: John Wiley & Sons, 1952–1973.
Organic Reactions. New York: John Wiley & Sons, 1942–present.

Each of these publications contains a great many citations to the appropriate articles in the original literature.

30.7 ADVANCED LABORATORY TECHNIQUES

The student who is interested in reading about more advanced techniques than those described in this textbook, or in more complete descriptions of techniques, should consult one of the advanced textbooks specializing in organic laboratory techniques. Besides focusing on apparatus construction and the performance of complex reactions, these books also provide advice on purifying reagents and solvents. Useful sources of information on organic laboratory techniques include

Bates, R. B., and Schaefer, J. P. *Research Techniques in Organic Chemistry*. Englewood Cliffs, NJ: Prentice-Hall, 1971.
Krubsack, A. J. *Experimental Organic Chemistry*. Boston: Allyn and Bacon, 1973.
Leonard, J., Lygo, B., and Procter, G. *Advanced Practical Organic Chemistry*, 2nd ed. London: Chapman and Hall, 1995.
Monson, R. S. *Advanced Organic Synthesis: Methods and Techniques*. New York: Academic Press, 1971.
Techniques of Chemistry. New York: John Wiley & Sons, 1970–present. Currently 21 volumes. The successor to *Technique of Organic Chemistry*, this series covers experimental methods of chemistry, such as purification of solvents, spectral methods, and kinetic methods.
Weissberger, A., et al., eds. *Technique of Organic Chemistry*, 3rd ed. New York: Wiley-Interscience, 1959–1969. This work is in 14 volumes.
Wiberg, K. B. *Laboratory Technique in Organic Chemistry*, New York: McGraw-Hill, 1960.

Numerous works and some general textbooks specialize in particular techniques. The preceding list is only representative of the most common books in this category. The following books deal specifically with microscale and semimicroscale techniques.

Cheronis, N. D. "Micro and Semimicro Methods." In A. Weissberger, ed., *Technique of Organic Chemistry*, Vol. 6. New York: Wiley-Interscience, 1954.

Cheronis, N. D., and Ma, T. S. *Organic Functional Group Analysis by Micro and Semimicro Methods.* New York: Wiley-Interscience, 1964.

Ma, T. S., and Horak, V. *Microscale Manipulations in Chemistry.* New York: Wiley-Interscience, 1976.

30.8 REACTION MECHANISMS

As with the case of locating information on synthetic methods, you can obtain a great deal of information about reaction mechanisms by consulting one of the common textbooks on physical organic chemistry. The textbooks listed here provide a general description of mechanisms, but they do not contain specific literature citations. Very general textbooks include

Miller, A., and Solomon, P. *Writing Reaction Mechanisms in Organic Chemistry*, 2nd ed. San Diego: Academic Press, 1999.

Sykes, P. *A Primer to Mechanisms in Organic Chemistry.* Menlo Park, CA: Benjamin/Cummings, 1995.

More advanced textbooks include

Carey, F. A., and Sundberg, R. J. *Advanced Organic Chemistry. Part A. Structure and Mechanisms*, 4th ed. New York: Plenum Press, 2000.

Hammett, L. P. *Physical Organic Chemistry: Reaction Rates, Equilibria, and Mechanisms*, 2nd ed. New York: McGraw-Hill, 1970.

Hine, J. *Physical Organic Chemistry*, 2nd ed. New York: McGraw-Hill, 1962.

Ingold, C. K. *Structure and Mechanism in Organic Chemistry*, 2nd ed. Ithaca, NY: Cornell University Press, 1969.

Isaacs, N. S. *Physical Organic Chemistry*, 2nd ed. New York: John Wiley & Sons, 1995.

Jones, R. A. Y. *Physical and Mechanistic Organic Chemistry*, 2nd ed. Cambridge, U.K.: Cambridge University Press, 1984.

Lowry, T. H., and Richardson, K. S. *Mechanism and Theory in Organic Chemistry*, 3rd ed. New York: Harper & Row, 1987.

Moore, J. W., and Pearson, R. G. *Kinetics and Mechanism*, 3rd ed. New York: John Wiley & Sons, 1981.

Smith, M. B., and March, J. *Advanced Organic Chemistry*, 5th ed. New York: John Wiley & Sons, 2000.

These books include extensive bibliographies that permit the reader to delve more deeply into the subject.

Most libraries also subscribe to annual series of publications that specialize in articles dealing with reaction mechanisms. Among these are

Advances in Physical Organic Chemistry. London: Academic Press, 1963–preseFnt.

Annual Reports of the Chemical Society. Section B. London: Chemical Society, 1905–present. Specifically, the section on *Reaction Mechanisms*.

Organic Reaction Mechanisms. Chichester, U.K.: John Wiley & Sons, 1965–present.

Progress in Physical Organic Chemistry. New York: Interscience, 1963–present.

These publications provide the reader with citations from the original literature that can be very useful in an extensive literature search.

30.9 ORGANIC QUALITATIVE ANALYSIS

Many laboratory manuals provide basic procedures for identifying organic compounds through a series of chemical tests and reactions. Occasionally you might require a more complete description of analytical methods or a more complete set of tables of derivatives. Textbooks specializing in organic qualitative analysis should fill this need. Examples of sources for such information include

Cheronis, N. D., and Entriken, J. B. *Identification of Organic Compounds: A Student's Text Using Semimicro Techniques.* New York: Interscience, 1963.

Pasto, D. J., and Johnson, C. R. *Laboratory Text for Organic Chemistry: A Source Book of Chemical and Physical Techniques.* Englewood Cliffs, NJ: Prentice-Hall, 1979.

Rappoport, Z. ed. *Handbook of Tables for Organic Compound Identification*, 3rd ed. Cleveland: Chemical Rubber Co., 1967.

Shriner, R. L., Hermann, C. K. F., Merrill, T. C., Curtin, D. Y., and Fuson, R. C. *The Systematic Identification of Organic Compounds*, 7th ed. New York: John Wiley & Sons, 1998.

Vogel, A. I. *Elementary Practical Organic Chemistry. Part 2. Qualitative Organic Analysis*, 2nd ed. New York: John Wiley & Sons, 1966.

Vogel, A. I. Revised by members of the School of Chemistry, Thames Polytechnic. *Vogel's Textbook of Practical Organic Chemistry, Including Qualitative Organic Analysis*, 5th ed. London: Longman Group, Ltd., 1989.

30.10 *BEILSTEIN* AND *CHEMICAL ABSTRACTS*

One of the most useful sources of information about the physical properties, synthesis, and reactions of organic compounds is *Beilsteins Handbuch der Organischen Chemie*. This is a monumental work, initially edited by Friedrich Konrad Beilstein, and updated through several revisions by the Beilstein Institute in Frankfurt am Main, Germany. The original edition (the *Hauptwerk*, abbreviated H) was published in 1918 and covers completely the literature to 1909. Five supplementary series (*Ergänzungswerken*) have been published since that time. The first supplement (*Erstes Ergänzungswerk*, abbreviated E I) covers the literature from 1910–1919; the second supplement (*Zweites Ergänzungswerk*, E II) covers 1920–1929; the third supplement (*Drittes Ergänzungswerk*, E III) covers 1930–1949; the fourth supplement (*Viertes Ergänzungswerk*, E IV) covers 1950–1959; and the fifth supplement (in English) covers 1960–1979. Volumes 17–27 of supplementary series III and IV, covering heterocyclic compounds, are combined in a joint issue, E III/IV. Supplementary series III, IV, and V are not complete, so the coverage of *Handbuch der Organischen Chemie* can be considered complete to 1929, with partial coverage to 1979.

Beilsteins Handbuch der Organischen Chemie, usually referred to simply as *Beilstein*, also contains two types of cumulative indices. The first of these is a name index (*Sachregister*), and the second is a formula index (*Formelregister*). These indices are particularly useful for a person wishing to locate a compound in *Beilstein*.

The principal difficulty in using *Beilstein* is that it is written in German through the fourth supplement. The fifth supplement is in English. Although some reading knowledge of German is useful, you can obtain information from the work by learning a few key phrases. For example, *Bildung* is "formation" or "structure." *Darst* or *Darstellung* is "preparation," K_P or *Siedepunkt* is "boiling point," and *F* or *Schmelzpunkt* is "melting point." Furthermore, the names of some compounds in German are not cognates of the English names. Some examples are *Apfelsäure* for "malic acid" (*säure* means "acid"), *Harnstoff* for "urea," *Jod* for "iodine," and *Zimtsäure* for "cinnamic acid." If you have access to a German–English dictionary for chemists, many of these difficulties can be overcome. The best such dictionary is

Patterson, A. M. *German–English Dictionary for Chemists*, 4th ed. New York: John Wiley & Sons, 1991.

Beilstein is organized according to a very sophisticated and complicated system. However, most students do not wish to become experts on *Beilstein* to this extent. A simpler, though slightly less reliable, method is to look for the compound in the formula index that accompanies the second supplement. Under the molecular formula, you will find the names of compounds that have that formula. After that name will be a series of numbers that indicate the pages and volume in which that compound is listed. Suppose, as an example, that you are searching for information on *p*-nitroaniline. This compound has the molecular formula $C_6H_6N_2O_2$. Searching for this formula in the formula index to the second supplement, you find

4-Nitro-anilin **12** 711, **I** 349, **II** 383

This information tells us that *p*-nitroaniline is listed in the main edition *Hauptwerk)* in Volume 12, page 711. Locating this particular volume, which is devoted to isocyclic monoamines, we turn to page 711 and find the beginning of the section on *p*-nitroaniline. At the left side of the top of this page, we find "Syst. No. 1671." This is the system number given to compounds in this part of Volume 12. The system number is useful, as it can help you find entries for this compound in subsequent supplements. The organization of *Beilstein* is such that all entries on *p*-nitroaniline in each of the supplements will be found in Volume 12. The entry in the formula index also indicates that material on this compound may be found in the first supplement on page 349 and in the second supplement on page 383. On page 349 of Volume 12 of the first supplement, there is a heading, "**XII, 710–712,**" and on the left is "Syst. No. 1671." Material on *p*-nitroaniline is found in each supplement on a page that is headed with the volume and page of the *Hauptwerk* in which the same compound is found. On page 383 of Volume 12 of the second supplement, the heading in the center of the top of the page is "**H12, 710–712.**" On the left you find "Syst. No. 1671." Again, because *p*-nitroaniline appeared in Volume 12, page 711, of the main edition, you can locate it by searching through Volume 12 of any supplement until you find a page with the heading corresponding to Volume 12, page 711. Because the third and fourth supplements are not complete, there is no comprehensive formula index for these supplements. However, you can still find material on *p*-nitroaniline by using the system number and the volume and page in the main work. In the third supplement, because the amount of information available has grown so much since the early days of Beilstein's work, Volume 12 has now expanded so that it is found in several bound parts. However, you select the part that includes system number 1671. In this part of Volume 12, you look through the pages until you find a page headed "Syst. No. 1671/H711." The information on *p*-nitroaniline is found on this page (page 1580). If Volume 12 of the fourth supplement was available, you would go on in the same way to locate more recent data on *p*-nitroaniline. This example is meant to illustrate how you can locate information on particular compounds without having to learn the *Beilstein* system of classification. You might do well to test your ability at finding compounds in *Beilstein* as we have described here.

Guidebooks to using *Beilstein*, which include a description of the *Beilstein* system, are recommended for anyone who wants to work extensively with *Beilstein*. Among such sources are

Heller, S. R. *The Beilstein System: Strategies for Effective Searching.* New York: Oxford University Press, 1997.

Huntress, E. H. *A Brief Introduction to the Use of Beilsteins Handbuch der Organischen Chemie*, 2nd ed. New York: John Wiley & Sons, 1938.

How to Use Beilstein. Beilstein Institute, Frankfurt am Main. Berlin: Springer-Verlag.

Weissbach, O. *The Beilstein Guide: A Manual for the Use of Beilsteins Handbuch der Organischen Chemie.* New York: Springer-Verlag, 1976.

Beilstein reference numbers are listed in such handbooks as *CRC Handbook of Chemistry and Physics* and *Lange's Handbook of Chemistry*. Additionally, *Beilstein* numbers are included in the *Aldrich Catalog Handbook of Fine Chemicals*, issued by the Aldrich Chemical Company. If the compound you are seeking is listed in one of these handbooks, you will find that using *Beilstein* is simplified.

Another very useful publication for finding references for research on a particular topic is *Chemical Abstracts*, published by the Chemical Abstracts Service of the American Chemical Society. *Chemical Abstracts* contains abstracts of articles appearing in more than 10,000 journals from virtually every country conducting scientific research. These abstracts list the authors, the journal in which the article appeared, the title of the paper, and a short summary of the contents of the article. Abstracts of articles that appeared originally in a foreign language are provided in English, with a notation indicating the original language.

To use *Chemical Abstracts*, you must know how to use the various indices that accompany it. At the end of each volume there appears a set of indices, including a formula index, a general subject index, a chemical substances index, an author index, and a patent index. The listings in each index refer the reader to the appropriate abstract according to the number assigned to it. There are also collective indices that combine all the indexed material appearing in a 5-year period (10-year period before 1956). In the collective indices, the listings include the volume number as well as the abstract number.

For material after 1929, *Chemical Abstracts* provides the most complete coverage of the literature. For material before 1929, use *Beilstein* before consulting *Chemical Abstracts*. *Chemical Abstracts* has the advantage that it is written entirely in English. Nevertheless, most students perform a literature search to find a relatively simple compound. Finding the desired entry for a simple compound is much easier in *Beilstein* than in *Chemical Abstracts*. For simple compounds, the indices in *Chemical*

Abstracts are likely to contain very many entries. To locate the desired information, you must comb through this multitude of listings—potentially a very time-consuming task.

The opening pages of each index in *Chemical Abstracts* contain a brief set of instructions on using that index. If you want a more complete guide to *Chemical Abstracts*, consult a textbook designed to familiarize you with these abstracts and indices. Two such books are

CAS Printed Access Tools: A Workbook. *Washington, DC: Chemical Abstracts Service, American Chemical Society, 1977.*

How to Search Printed CA. *Washington, DC: Chemical Abstracts Service, American Chemical Society, 1989.*

Chemical Abstracts Service maintains a computerized database that permits users to search through *Chemical Abstracts* very rapidly and thoroughly. This service, which is called *CA Online*, is described in Section 30.11. *Beilstein* is also available for online searching by computer.

30.11 COMPUTER ONLINE SEARCHING

You can search a number of chemistry databases online by using a computer and modem or a direct internet connection. Many academic and industrial libraries can access these databases through their computers. One organization that maintains a large number of databases is the Scientific and Technical Information Network (STN International). The fee charged to the library for this service depends on the total time used in making the search, the type of information being asked for, the time of day when the search is being conducted, and the type of database being searched.

The Chemical Abstracts Service database *(CA Online)* is one of many databases available on STN. It is particularly useful to chemists. Unfortunately, this database extends back only to about 1967, although some earlier references are available. Searches earlier than 1967 must be made with printed abstracts (Section 30.10). An online search is much faster than searching in the printed abstracts. In addition, you can tailor the search in a number of ways by using keywords and the Chemical Abstracts Substance Registry Number (CAS number) as part of the search routine. The CAS number is a specific number assigned to every compound listed in the *Chemical Abstracts* database. The CAS number is used as a key in an online search in order to locate information about the compound. For the more common organic compounds, you can easily obtain CAS numbers from the catalogues of most of the companies that supply chemicals. Another advantage of performing an online search is that the *Chemical Abstracts* files are updated much more quickly than the printed versions of abstracts. This means that your search is more likely to reveal the most current information available.

Other useful databases available from STN include *Beilstein* and *CASREACTS*. As described in Section 30.10, *Beilstein* is very useful to organic chemists. Currently, there are over 3.5 million compounds listed in the database. You can use the CAS Registry Numbers to help in a search that has the potential of going back to 1830.

CASREACTS is a chemical reactions database derived from over 100 journals covered by *Chemical Abstracts*, starting in 1985. With this database, you can specify a starting material and a product using the CAS Registry Numbers. Further information on *CA Online*, *Beilstein*, *CASREACTS*, and other databases can be obtained from the following references:

Smith, M. B., and March, J. *Advanced Organic Chemistry*, 5th ed. New York: John Wiley & Sons, 2000.

Somerville, A. N. "Information Sources for Organic Chemistry, 2: Searching by Functional Group." *Journal of Chemical Education*, *68* (October 1991): 842.

Somerville, A. N. "Subject Searching of Chemical Abstracts Online." *Journal of Chemical Education*, *70* (March 1993): 200.

Wiggins, G. *Chemical Information Sources*. New York: McGraw-Hill, 1990. Integrates printed materials and computer sources of information.

30.12 SCIENTIFIC JOURNALS

Ultimately, someone wanting information about a particular area of research will be required to read articles from the scientific journals. These journals are of two basic types: review journals and primary scientific journals. Journals that specialize in review articles summarize all the work that bears on the particular topic. These articles may focus on the contributions of one particular researcher but often consider the contributions of many researchers to the subject. These articles also contain extensive bibliographies, which refer you to the original research articles. Among the important journals devoted, at least partly, to review articles are

Accounts of Chemical Research
Angewandte Chemie (International Edition, in English)
Chemical Reviews
Chemical Society Reviews (formerly known as *Quarterly Reviews*)
Nature
Science

The details of the research of interest appear in the primary scientific journals. Although there are thousands of journals published in the world, a few important journals specializing in articles dealing with organic chemistry might be mentioned here. These are

Canadian Journal of Chemistry
European Journal of Organic Chemistry (formerly known as *Chemische Berichte*)
Journal of Organic Chemistry
Journal of the American Chemical Society
Journal of the Chemical Society, Chemical Communications
Journal of the Chemical Society, Perkin Transactions (Parts I and II)
Journal of Organometallic Chemistry
Organic Letters
Organometallics
Synlett
Synthesis
Tetrahedron
Tetrahedron Letters

30.13 TOPICS OF CURRENT INTEREST

The following journals and magazines are good sources for topics of educational and current interest. They specialize in news articles and focus on current events in chemistry or in science in general. Articles in these journals (magazines) can be useful in keeping you abreast of developments in science that are not part of your normal specialized scientific reading.

American Scientist
Chemical and Engineering News
Chemistry and Industry
Chemistry in Britain
Chemtech
Discover
Journal of Chemical Education
Nature
New Scientist
Omni
Science
Science Digest
Scientific American
SciQuest (formerly *Chemistry*)

Other sources for topics of current interest include the following:

Encyclopedia of Chemical Technology, also called *Kirk-Othmer Encyclopedia of Chemical Technology*, 4th ed., 1992. (25 volumes plus index and supplements.)
McGraw-Hill Encyclopedia of Science and Technology, 1997 (20 volumes and supplements).

30.14 HOW TO CONDUCT A LITERATURE SEARCH

The easiest method to follow in searching the literature is to begin with secondary sources and then go to the primary sources. In other words, you would try to locate material in a textbook, *Beilstein*, or *Chemical Abstracts*. From the results of that search, you would then consult one of the primary scientific journals.

A literature search that ultimately requires you to read one or more papers in the scientific journals is best conducted if you can identify a particular paper central to the study. Often you can obtain this reference from a textbook or a review article on the subject. If this is not available, a search through *Beilstein* is required. A search through one of the handbooks that provides *Beilstein* reference numbers (see Section 30.10) may be helpful. Searching through *Chemical Abstracts* would be considered the next logical step. From these sources, you should be able to identify citations from the original literature on the subject.

Additional citations may be found in the references cited in the journal article. In this way, the background leading to the research can be examined. It is also possible to conduct a search forward in time from the date of the journal article through the *Science Citation Index*. This publication provides the service of listing articles and the papers in which these articles were cited. Although the *Science Citation Index* consists of several types of indices, the *Citation Index* is most useful for the purposes described here. A person who knows of a particular key reference on a subject can examine the *Science Citation Index* to obtain a list of papers that have used that seminal reference in support of the work described. The *Citation Index* lists papers by their senior author, journal, volume, page, and date, followed by citations of papers that have referred to that article, author, journal, volume, page, and date of each. The *Citation Index* is published in annual volumes, with quarterly supplements issued during the current year. Each volume contains a complete list of the citations of the key articles made during that year. A disadvantage is that *Science Citation Index* has been available only since 1961. An additional disadvantage is that you may miss journal articles on the subject of interest if they failed to cite that particular key reference in their bibliographies—a reasonably likely possibility.

You can, of course, conduct a literature search by a "brute force" method, by beginning the search with *Beilstein* or even with the indices in *Chemical Abstracts*. However, the task can be made much easier by performing a computer search (Section 30.11) or by starting with a book or an article of general and broad coverage, which can provide a few citations for starting points in the search.

The following guides to using the chemical literature are provided for the reader who is interested in going further into this subject.

Bottle, R. T., and Rowland, J. F. B., eds. *Information Sources in Chemistry*, 4th ed. New York: Bowker-Saur, 1992.

Mellon, M. G. *Chemical Publications*, 5th ed. New York: McGraw-Hill, 1982.

Maizell, R. E. *How to Find Chemical Information: A Guide for Practicing Chemists, Educators, and Students*, 3rd ed. New York: John Wiley & Sons, 1998.

Wiggins, G. *Chemical Information Sources*. New York: McGraw-Hill, 1991. Integrates printed materials and computer sources of information.

PROBLEMS

1. Find the following compounds in the formula index for the *Second Supplement of Beilstein* (Section 30.10). (i) List the page numbers from the Main Work and the Supplements (First and Second). (ii) Using these page numbers, look up the System Number (Syst. No.) and the Main Work number (Hauptwerk number, H) for each compound in the Main Work, and the First and Second supplements. In some cases, a compound may not be found in all three places. (iii) Now use the System Number and Main Work number to find each of these compounds in the Third and Fourth Supplements. List the page numbers where these compounds are found.

    (a) 2,5-hexanedione (acetonylacetone)

    (b) 3-nitroacetophenone

    (c) 4-*tert*-butylcyclohexanone

    (d) 4-phenylbutanoic acid (4-phenylbutyric acid, γ-phenylbuttersäure)

2. Using the *Science Citation Index* (Section 30.14), list five research papers by complete title and journal citation for each of the following chemists who have been awarded the Nobel Prize. Use the *Five-Year Cumulation Source Index* for the years 1980–1984 as your source.
   (a) H. C. Brown
   (b) R. B. Woodward
   (c) D. J. Cram
   (d) G. Olah

3. The reference book by Smith and March is listed in Section 30.2. Using Appendix B in this book, give two methods for preparing the following functional groups. You will need to provide equations.
   (a) carboxylic acids
   (b) aldehydes
   (c) esters (carboxylic esters)

4. *Organic Synthesis* is described in Section 30.6. There are currently nine collective volumes in the series, each with its own index. Find the compounds listed below and provide the equations for preparing each compound.
   (a) 2-methylcyclopentane-1,3-dione
   (b) *cis*-Δ^4-tetrahydrophthalic anhydride (listed as tetrahydrophthalic anhydride)

5. Provide four ways that may be used to oxidize an alcohol to an aldehyde. Give complete literature references for each method as well as equations. Use the *Compendium of Organic Synthetic Methods* or *Survey of Organic Syntheses* by Buehler and Pearson (Section 30.6).

Index

elution sequence for com-
pounds 293
elution techniques 303
flash chromatography 310
packing the column
295–302
packing the column,
macroscale 297
packing the column, mi-
croscale 299
packing the column, semi-
microscale 298
principles of separation 291
recovering the separated
compounds 307
sample application 302
separation of a mixture 289
solvent flow rate 295
solvent reservoirs 304
solvents 292
Computational chemistry 492,
497
electrostatic potential maps
505
heats of formation 502
Condenser
air 38
cold–finger 273
macroscale 37
microscale 38
water 67, 69, 72, 73, 75, 207
water, macroscale 37
water, microscale 38
Conical vial
thin–walled 38
Conical vials 38
measuring with 63
methods of sealing 35
use in extraction 171–177
Cooling methods 77
cold baths 77
Correlation charts
NMR spectroscopy 421
C–13 NMR spectroscopy
452, 453
Correlation table
infrared spectroscopy 388
Coupling constant 427
Craig tube 38, 113, 150
centrifugation 114–115
*CRC Handbook of Chemistry
and Physics* 42
Crystallization 139–162
common solvents 155

Craig tube 150
inducing crystal formation
158
macroscale 143–149
microscale 150–153
mixed solvents 160
solvent selection 153
solvent selection by testing
156
summary of steps 149
Crystallization tube
Craig tube 38
CW–NMR spectrometer 410
Cyclohexanol
carbon–13 NMR spectrum
458
Cyclohexanone
carbon–13 NMR spectrum
459
Cyclohexene
carbon–13 NMR spectrum
459
Cyclopentane
mass spectrum 481
Cylinder
graduated 52

D

Dean–Stark water separator
238
Decane
infrared spectrum 392
Decolorization 157
by column chromatography
307
using a column 158
Decomposition point 123
Density
determination 198
Density electrostatic potential
map 505
Desiccator 160
Deuteriochloroform 412
hazards 18
Deuterium oxide 414
Dextrorotatory 360
Diatomaceous earth 109
1,2–Dichlorobenzene
C–13 NMR spectrum 462
infrared spectrum 396
1,3–Dichlorobenzene
C–13 NMR spectrum 462
1,4–Dichlorobenzene
C–13 NMR spectrum 462

Dichloromethane
See Methylene chloride
*Dictionary of Organic Com-
pounds* 507
Diethyl ether 135
hazards 19
1,2–Dimethoxyethane
hazards 18
2,2–Dimethylbutane
carbon–13 NMR spectrum
457
Dioxane
hazards 18
Dispensing pumps 53
Disposable pipets
See Pasteur pipets
Distillate 209
Distillation 201–268, 277–285
See also Fractional, Simple,
Steam, and Vacuum Dis-
tillations
bulb–to–bulb 260
simple 201
steam 277
vacuum 243
Distillation head 207
Claisen 214
Claisen, macroscale 37
Claisen, microscale 38
Hickman 38, 209
macroscale 37
microscale 38
Distribution coefficient 164
Dopamine
mass spectrum 471
Downfield 416
Dry film method 375
Dry ice 77
Drying agents 178
table 180
Drying tube 88
macroscale 37
microscale 38
DSS 415

E

Ebulliator 245
Electron–density surface 504
Elpot map 505
Eluates 289
Eluents 289
Elutants 289
Emulsions 181
Enantiomeric excess 361

Infrared Absorption Bands

	Type of Vibration		Frequency (cm^{-1})	Intensity
C–H	Alkanes	(stretch)	3000–2850	s
	–CH$_3$	(bend)	1450 and 1375	m
	–CH$_2$–	(bend)	1465	m
	Alkenes	(stretch)	3100–3000	m
		(out-of-plane bend)	1000–650	s
	Aromatics	(stretch)	3150–3050	s
		(out-of-plane bend)	900–690	s
	Alkyne	(stretch)	ca. 3300	s
	Aldehyde		2900–2800	w
			2800–2700	w
O–H	Alcohol, phenols			
	Free		3650–3600	m
	H-bonded		3400–3200	m
	Carboxylic acids		3400–2400	m
N–H	Primary and secondary amines and amides			
	(stretch)		3500–3100	m
	(bend)		1640–1550	m–s
C≡C	Alkyne		2250–2100	m–w
C≡N	Nitriles		2260–2240	m
C=C	Alkene		1680–1600	m–w
	Aromatic		1600 and 1475	m–w
N=O	Nitro (R–NO$_2$)		1550 and 1350	s
C=O	Aldehyde		1740–1720	s
	Ketone		1725–1705	s
	Carboxylic acid		1725–1700	s
	Ester		1750–1730	s
	Amide		1680–1630	s
	Anhydride		1810 and 1760	s
	Acid chloride		1800	s
C–O	Alcohols, ethers, esters, carboxylic acids, anhydrides		1300–1000	s
C–N	Amines		1350–1000	m–s
C–X	Fluoride		1400–1000	s
	Chloride		785–540	s
	Bromide, iodide		<667	s